Solutions Manual For THE ESSENTIAL LOGIC OF ORGANIC CHEMISTRY, (aka, How to Cure the Benzene Blues)

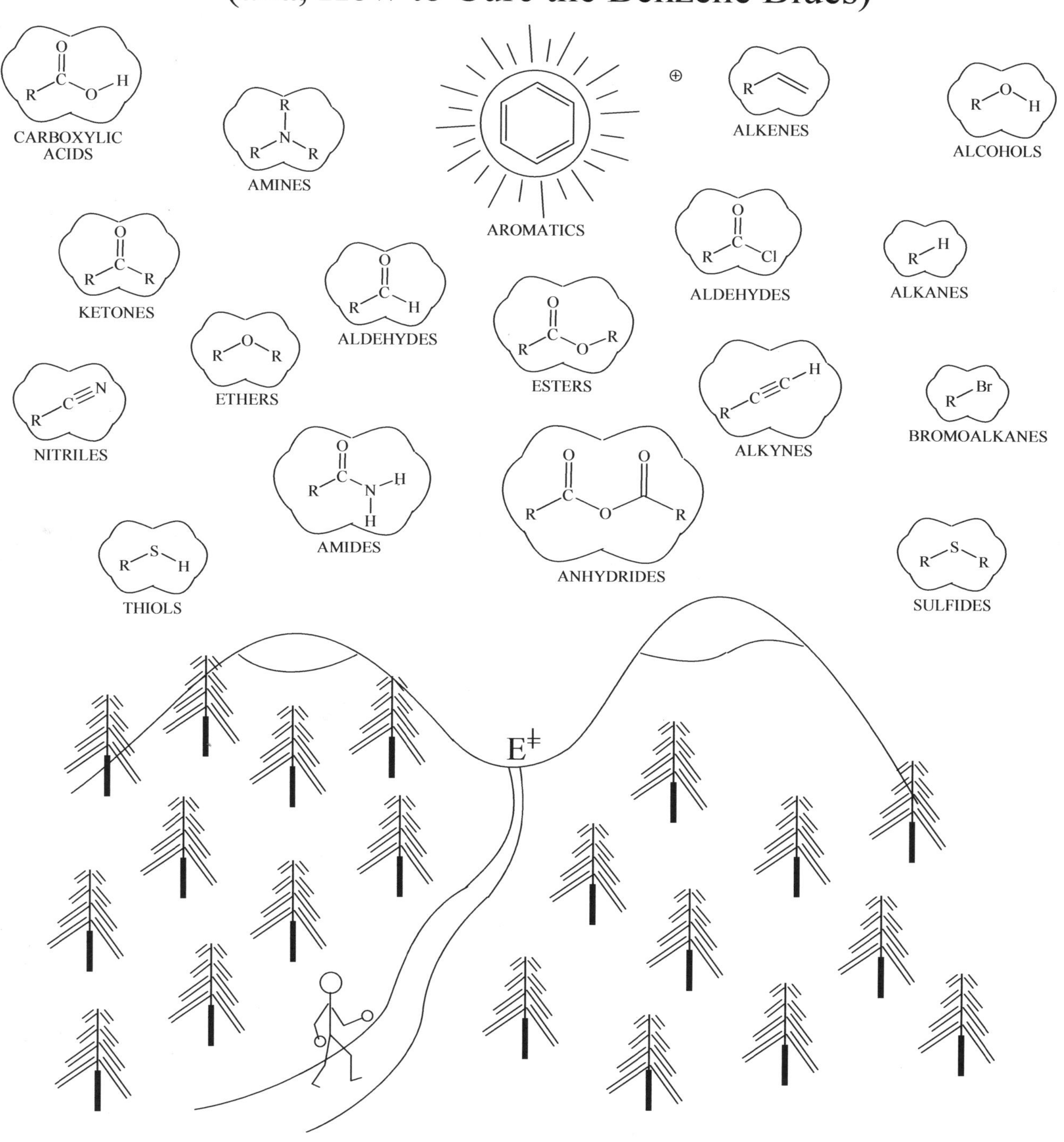

PHIL BEAUCHAMP

The Essential Logic of Organic Chemistry

CreateSpace

First Edition (December, 2016)

ISBN-13: 978-1541124660
ISBN-10: 1541124669

Nonfiction > Reference > Organic Chemistry

Chapter 1

The keys are done quickly. I do make mistakes. If you think you found an error, please bring it to my attention so I can correct it for others. Thank you.

Problem 1 (p 7) – Write and atomic configuration for H, B, C, N, O, F, Ne, S, Cl, Br, I.

hydrogen: $1s^1$
helium: $1s^2$
lithium: $1s^2, 2s^1$
beryllium: $1s^2, 2s^2$
boron: $1s^2, 2s^2, 2p^1$
carbon: $1s^2, 2s^2, 2p^2$
nitrogen: $1s^2, 2s^2, 2p^3$
oxygen: $1s^2, 2s^2, 2p^4$
fluorine: $1s^2, 2s^2, 2p^5$
neon: $1s^2, 2s^2, 2p^6$

sodium: $1s^2, 2s^2, 2p^6, 3s^1$
magnesium: $1s^2, 2s^2, 2p^6, 3s^2$
sulfur: $1s^2, 2s^2, 2p^6, 3s^2, 3p^4$
chlorine: $1s^2, 2s^2, 2p^6, 3s^2, 3p^5$
bromine: $1s^2, 2s^2, 2p^6, 3s^2, 3p^6, 3d^{10}, 4s^2, 4p^5$
Iodine: $1s^2, 2s^2, 2p^6, 3s^2, 3p^6, 3d^{10}, 4s^2, 4p^6, 4d^{10}, 5s^2, 5p^5$

Problem 2 (p 8) – What is the total nuclear charge and effective nuclear charge for each of the atoms below? How does this affect the electron attracting ability of an atom?

	H							He
Z_{total} =	+1							+2
core electrons =	0							0
$Z_{effective}$ =	+1							+2
	Li	**Be**	**B**	**C**	**N**	**O**	**F**	**Ne**
Z_{total} =	+3	+4	+5	+6	+7	+8	+9	+10
core electrons =	2	2	2	2	2	2	2	2
$Z_{effective}$ =	+1	+2	+3	+4	+5	+6	+7	+8
	Na	**Mg**	**Al**	**Si**	**P**	**S**	**Cl**	**Ar**
Z_{total} =	+11	+12	+13	+14	+15	+16	+17	+18
core electrons =	10	10	10	10	10	10	10	10
$Z_{effective}$ =	+1	+2	+3	+4	+5	+6	+7	+8
	K	**Ca**	**Ga**	**Ge**	**As**	**Se**	**Br**	**Kr**
Z_{total} =	+19	+20	+31	+32	+33	+34	+35	+36
core electrons =	18	18	28	28	28	28	28	28
$Z_{effective}$ =	+1	+2	+3	+4	+5	+6	+7	+8

Atoms with higher Z_{eff} tend to hold onto their valence elctrons tighter. Atoms with the same Z_{eff}, but lower valence shells also hold tighter because the valence electrons are closer to the nucleus having the same Z_{eff}.

Problem 3 (p 10) - Which atom in each pair below probably requires more energy to steal away an electron (also called ionization potential IP)? Why? Are any of the comparisons ambiguous? Why? Check your answers with the data in the I.P. table on page 7.

a. C vs N b. N vs O c. O vs F d. F vs Ne

Since all comparisons are in a row, the answers use essentially the same reasoning.

C	N

Both C and N are in the same row and use the same n=2 valence shell. N has a Z_{eff} = +5 and holds onto its electrons tighter than C with a Z_{eff} = +4.

N	O

Both N and O are in the same row and use the same n=2 valence shell. O has a Z_{eff} = +6 and holds onto its electrons tighter than N with a Z_{eff} = +5.

O	F

Both O and F are in the same row and use the same n=2 valence shell. F has a Z_{eff} = +7 and holds onto its electrons tighter than O with a Z_{eff} = +6.

F	Ne

Both F and Ne are in the same row and use the same n=2 valence shell. Ne has a Z_{eff} = +8 and holds onto its electrons tighter than F with a Z_{eff} = +7.

Problem 4 (p 10) - Explain the atomic trends in ionization potential in a row (Na vs Si vs Cl) and in a column (F vs Cl vs Br).

These three elements (Na vs Si vs Cl) are all in the same row. The Z_{eff} increases across a row (in the same shell) so the rightmost element should hold onto its electron the tightest and have the highest ionization potential, which it does.

Na = 118 > Si = 189 > Cl = 300 all kcal/mole

These three elements (F vs Cl vs Br) are all in the same column. The Z_{eff} is +7 for all of them. Fluorine's valence electrons are in the n=2 shell (closest to the +7 Z_{eff}), chlorine's in the n=3 and bromine's in n=4 (farthest from the +7 Z_{eff}). Being closer to the Z_{eff} should produce a stronger attraction, thus fluorine has the highest ionization potential

F = 402 > Cl = 300 > Br = 273 all kcal/mole

Problem 5 (p 10) - Explain the atomic trends in atomic radii in a row (Li vs C vs F) and in a column (C vs Si vs Ge).

Table 2

Neutral atomic radii in picometers (pm) = 10^{-12} m [100 pm = 1 angstrom]

H = 53								He = 31
Li = 167	Be = 112		B = 87	C = 67	N = 56	O = 48	F = 42	Ne = 38
Na = 190	Mg = 145		Al = 118	Si = 111	P = 98	S = 88	Cl = 79	Ar = 71
K = 243	Ca = 194	3d elements	Ga = 136	Ge = 125	As = 114	Se = 103	Br = 94	Kr = 88
Rb = 265	Sr = 219	4d elements	In = 156	Sn = 145	Sb = 133	Te = 123	I = 115	Xe = 108

Across a row Z_{eff} increases, in the same shell and the valence electrons are held tighter and have smaller radii (Li = 167, C = 67, F = 42). In a column, each atom has the same Z_{eff}, but there are additional layers, so larger radii are found (C = 67, Si = 111, Ge = 125).

Cation and anion radii in picometers (pm) = 10^{-12} m [100 pm = 1 angstrom]

Li^{+1} = 90	Be^{+2} = 59		B^{+3} = 41	C =	N^{-3} = 132	O^{-2} = 126	F^{-1} = 119
Na^{+1} = 116	Mg^{+2} = 86		Al^{+3} = 68	Si =	P =	S^{-2} = 170	Cl^{-1} = 167
K^{+1} = 152	Ca^{+2} = 114	3d elements	Ga^{+3} = 76	Ge =	As =	Se^{-2} = 184	Br^{-1} = 182
Rb^{+1} = 166	Sr^{+2} = 132	4d elements	In^{+3} = 94	Sn =	Sb =	Te^{-2} = 207	I^{-1} = 206

Problem 6 (p 10) - In the tables above:

a. Explain the cation distances compared to the atomic distances.

Each time an electron is lost from an element the positive charge goes up and the element holds on tighter to its remaining electrons. Also, electron/electron repulsion goes down as each electron is lost. This should contract the electron cloud more and more.

b. Explain the anion distances compared to the atomic distances.

Each time an electron is added to an element the negative charge goes up and there is more and more electron/electron repulsion. This should expand the electron cloud more and more.

c. Explain the cation distances in a row: Li = 90pm, Be = 59pm, B = 41pm, (pm = 10^{-12} m).

Similar to a. Each time an electron is lost from an element the positive charge goes up and the element holds on tighter to its remaining electrons. Also, electron/electron repulsion goes down as each electron is lost. This should contract the electron cloud more and more.

d. Explain the anion distances in a row (N = 132pm, O = 126pm, F = 119pm).

On the right side, in a row, there is an excess of electrons, which repel each other. There is greater excess on the nitrogen, Z_{eff} = +5, than the oxygen, Z_{eff} = +6, than the fluorine, Z_{eff} = +7. The nitrogen electron cloud expands the most and has the largest radius.

Problem 7 (p 12) Classify each bond type below as pure covalent, polar covalent or ionic according to our simplistic guidelines above. If a bond is ionic, rewrite it showing correct charges.

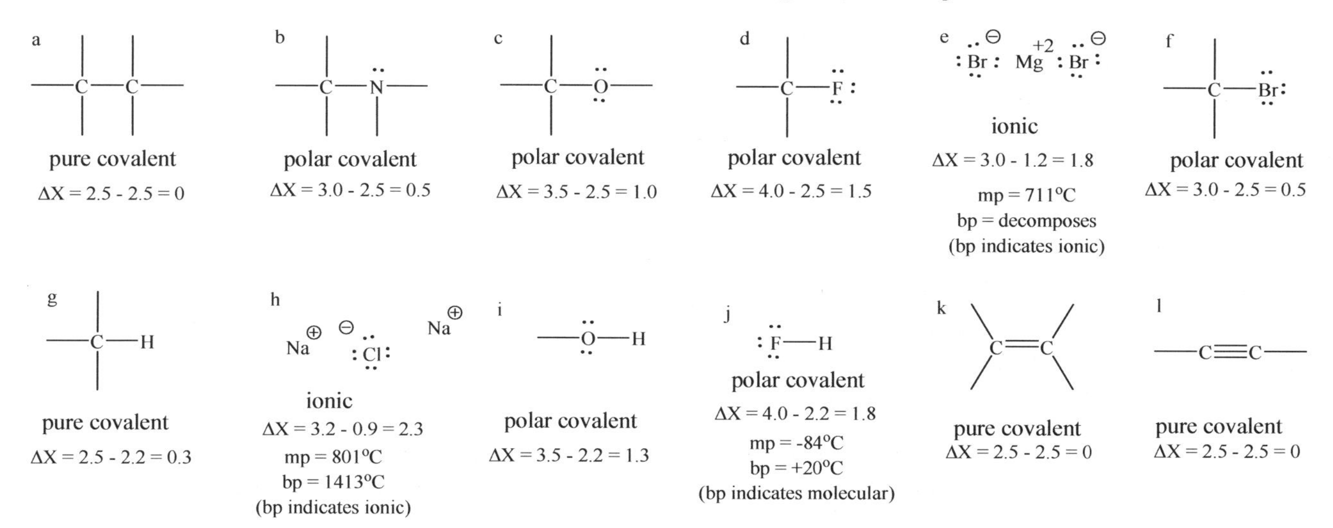

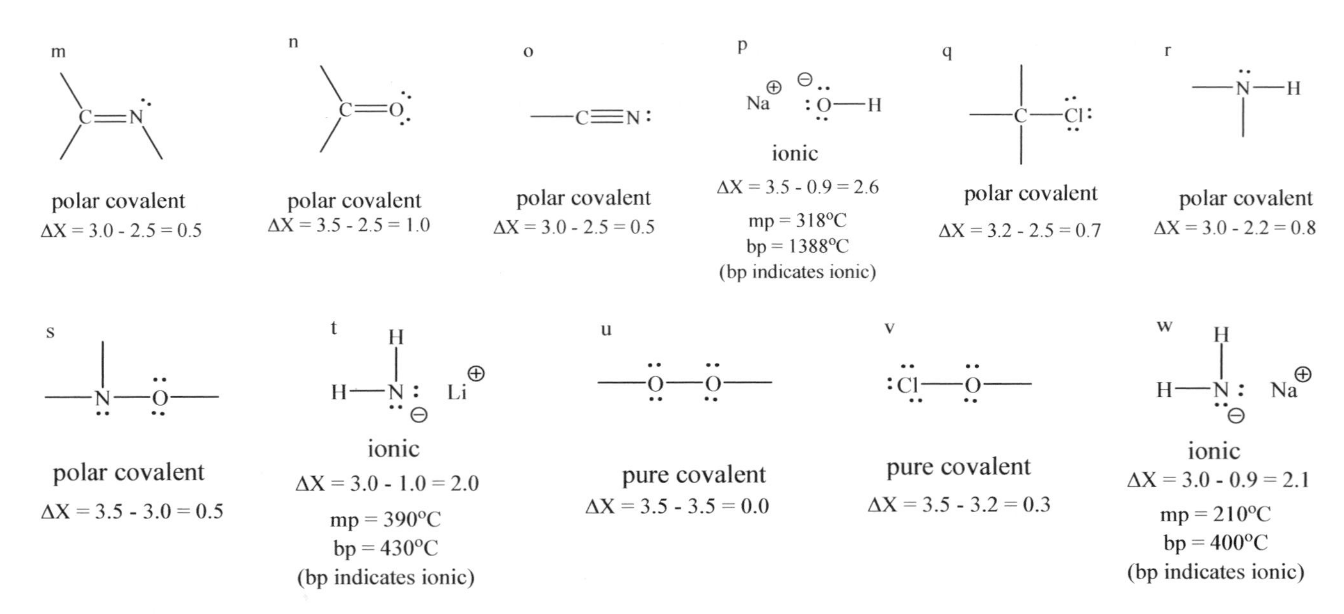

Problem 8 (p 17) – Draw a 3D representation or hydrogen cyanide, HCN. Show lines for the sigma bond skeleton and the lone pair of electrons in its proper location. Show two dots for the lone pair. Also show pi bonds represented in a manner similar to above. What is different about this structure compared with ethyne above (# bonds, lone pairs, polarity, all of them)? Would such a drawing work if oxygen is switched in for nitrogen? What would be different?

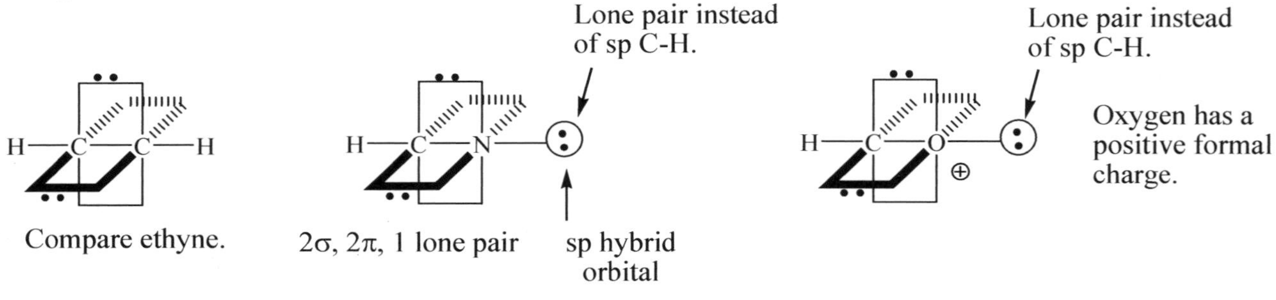

Problem 9 (p 18) – Draw a 3D representation of methanal (common name = formaldehyde), $H_2C{=}O$. Show lines for the sigma bond skeleton and the lone pairs of electrons with two dots for each lone pair. Also show pi bonds represented in a manner similar to above. What is different about this structure compared with ethene above (# bonds, lone pairs, polarity, all of these)? Draw a 3D representation of $H_2C{=}NH$. Can fluorine be switched in for oxygen? What would change?

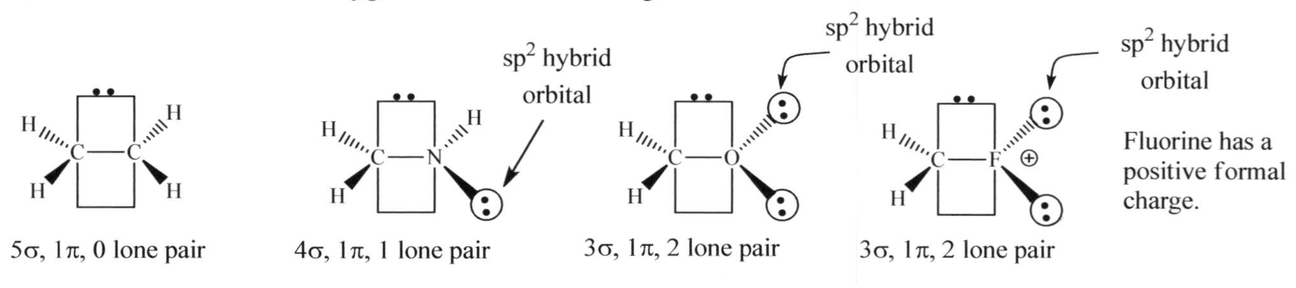

Problem 10 (p 20) – Draw a 3D representation or hydrogen methanol, H_3COH or CH_3OH, methanamine, H_3CNH_2 or CH_3NH_2 and fluoromethane CH_3F or H_3CF. Show lines for the sigma bond skeleton and a line with two dots for lone pairs, in a manner similar to above. What is different about this structure compared with ethene above (# bonds, lone pairs, polarity, all of these)?

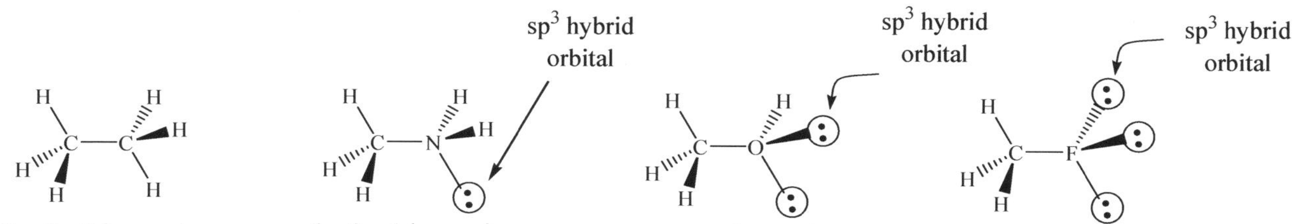

sp^3 hybrid orbital — sp^3 hybrid orbital — sp^3 hybrid orbital

7σ, 0π, 0 lone pair — 6σ, 0π, 1 lone pair — 5σ, 0π, 2 lone pair — 4σ, 0π, 3 lone pair

Problem 11 (p 23) - What is the hybridization of all carbon atoms in the structure below? What are the bond angles, shapes, number of sigma bonds, number of pi bonds and number's of attached hydrogen atoms? Bond line formulas are shorthand, symbolic representations of organic structures. Each bend represents a carbon, each end of a line represents a carbon and each dot represents a carbon. All carbon/carbon bonds are shown. The number of hydrogen atoms on a carbon is determined by the difference between four and the number of bonds shown.

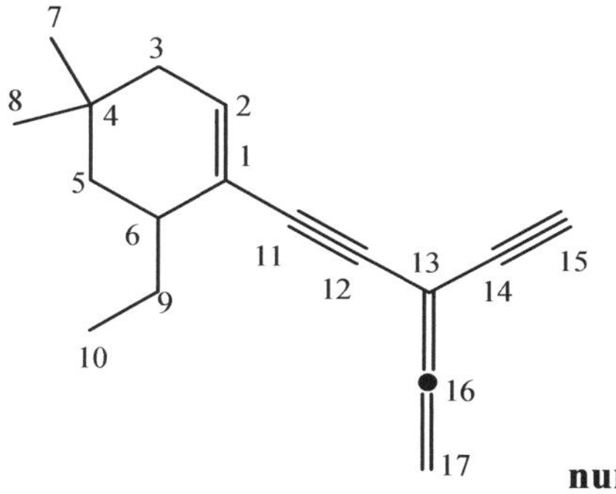

Carbon #	hybridization	bond angles	shape	number of hydrogen atoms	σ bonds	π bonds	lone pairs
1	sp^2	120	trigonal planar	0	3	1	0
2	sp^2	120	tetrahedral	1	3	1	0
3	sp^3	109	tetrahedral	2	4	0	0
4	sp^3	109	tetrahedral	0	4	0	0
5	sp^3	109	tetrahedral	2	4	0	0
6	sp^3	109	tetrahedral	1	4	0	0
7	sp^3	109	tetrahedral	3	4	0	0
8	sp^3	109	tetrahedral	3	4	0	0
9	sp^3	109	tetrahedral	2	4	0	0
10	sp^3	109	tetrahedral	3	4	0	0
11	sp	180	linear	0	2	2	0
12	sp	180	linear	0	2	2	0
13	sp^2	120	trigonal planar	0	3	1	0
14	sp	180	linear	0	2	2	0
15	sp	180	linear	1	2	2	0
16	sp	180	linear	0	2	2	0
17	sp^2	120	trigonal planar	2	3	1	0

Problem 12 (p 24) - What types of orbitals do the lone pair electrons occupy in each example above, according to our hybridization model? Hint: What is the hybridization of the atom? It has to use the same kind of hybrid orbital to hold the lone pair electrons…unless it is part of a resonant system (discussed later).

sp^3 sp^2 sp

No resonance is indicated so hybrid atom = hybrid orbitals.

Problem 13 (p 24) – Can a nitrogen atom bond in the second sp pattern above for carbon (four bonds)? Can an oxygen atom bond in the first sp pattern above for carbon (three bonds)? How about the second sp pattern of carbon (four bonds)? Can a fluorine atom bond in the sp^2 pattern for carbon (two bonds)? Are any of these reasonable possibilities? Have you ever seen a structure with four bonds to a nitrogen atom (NH_4^+) or three bonds to an oxygen atom (H_3O^+) or two bonds to a fluorine atom (H_2F^+)? Would there be any necessary changes for the atoms in such arrangements (notice the positive charge)? We'll discuss the answers to these questions soon.

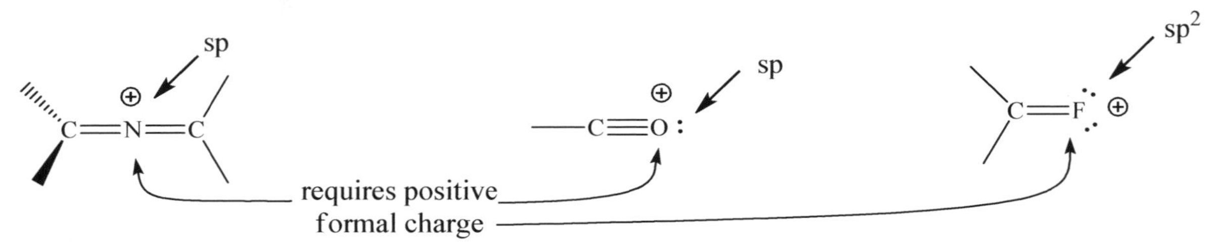

Problem 14 (p 25) - What is the hybridization of all carbon atoms in the structure below? What are the bond angles, shapes, number of sigma bonds, number of pi bonds, number of lone pairs and number's of attached hydrogen atoms? Bond line formulas are shorthand, symbolic representations of organic structures. Each bend represents a carbon, each end of a line represents a carbon and each dot represents a carbon. All carbon/carbon bonds are shown. The number of hydrogen atoms on a carbon is determined by the difference between four and the number of bonds shown.

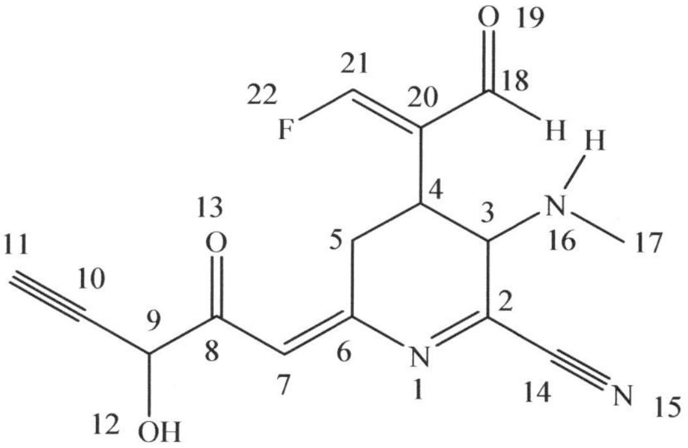

Carbon #	hybridization	bond angles	shape	number of hydrogen atoms	σ bond	π bonds	lone pairs	hybridization of lone pairs
1	sp^2	120	trigonal planar	0	2	1	1	sp^2
2	sp^2	120	trigonal planar	0	3	1	0	sp^2
3	sp^3	109	tetrahedral	1	4	0	0	sp^3
4	sp^3	109	tetrahedral	1	4	0	0	sp^3
5	sp^3	109	tetrahedral	2	4	0	0	sp^3
6	sp^2	120	trigonal planar	0	3	1	0	sp^2
7	sp^2	120	trigonal planar	1	3	1	0	sp^2
8	sp^2	120	trigonal planar	0	3	1	0	sp^2
9	sp^3	109	tetrahedral	1	4	0	0	sp^3
10	sp	180	linear	0	2	2	0	sp
11	sp	180	linear	1	2	2	0	sp
12	sp^3	109	tetrahedral	1	2	0	2	sp^3
13	sp^2	120	trigonal planar	0	1	1	2	sp^2
14	sp	180	linear	0	2	2	0	sp
15	sp	180	linear	0	1	2	1	sp
16	sp^3	109	tetrahedral	1	3	0	1	sp^3
17	sp^3	109	tetrahedral	3	4	0	0	sp^3
18	sp^2	120	trigonal planar	0	3	1	0	sp^2
19	sp^2	120	trigonal planar	0	1	1	2	sp^2
20	sp^2	120	trigonal planar	0	3	1	0	sp^2
21	sp^2	120	trigonal planar	1	3	1	0	sp^2
22	sp^3	109	tetrahedral	0	1	0	3	sp^3

Problem 15 (p 29-31) – Write in the formal charge wherever present in the atoms below. Resonance structures are shown within parentheses and have double headed arrows between them. All of the resonance contributors must be considered together to evaluate the true nature of a particular structure. You should include curved arrows to show necessary electron movement to form each subsequent resonance structure (no arrows on the last structure).

a. CH_3NO_2

1 x C = 4
3 x H = 3
1 x N = 5
2 x O = 12
total e- = 24

b. CH_2N_2

1 x C = 4
2 x H = 2
2 x N = 10
total e- =16

c. $NH_3CH_2CO_2$

2 x C = 8
5 x H = 5
1 x N = 5
2 x O = 12
total e- = 30

d. CH_3NO_2

1 x C = 4
3 x H = 3
1 x N = 5
2 x O = 12
total e- = 24

e. CO_3^{-2}

1 x C = 4
3 x O = 18
-2 = 2
total e- = 24

f. $C(NH_2)_3^{\oplus}$

1 x C = 4
3 x N = 15
6 x H = 6
+1 = -1
total e- = 24

resonance is possible for all of these structures, but only one additional structure is drawn for each structure.

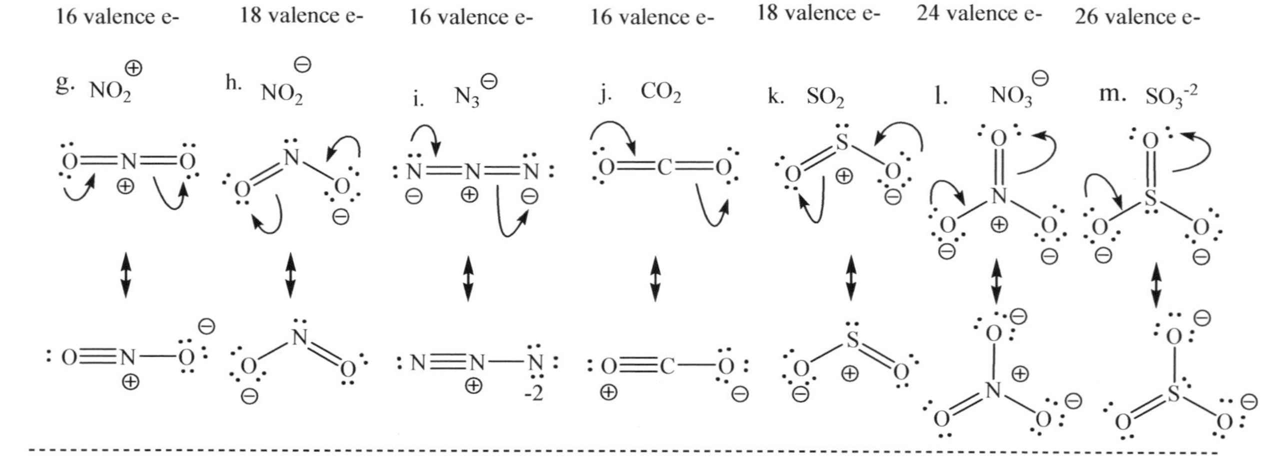

resonance is possible for all of these structures, but only one additional structure is drawn for each structure.

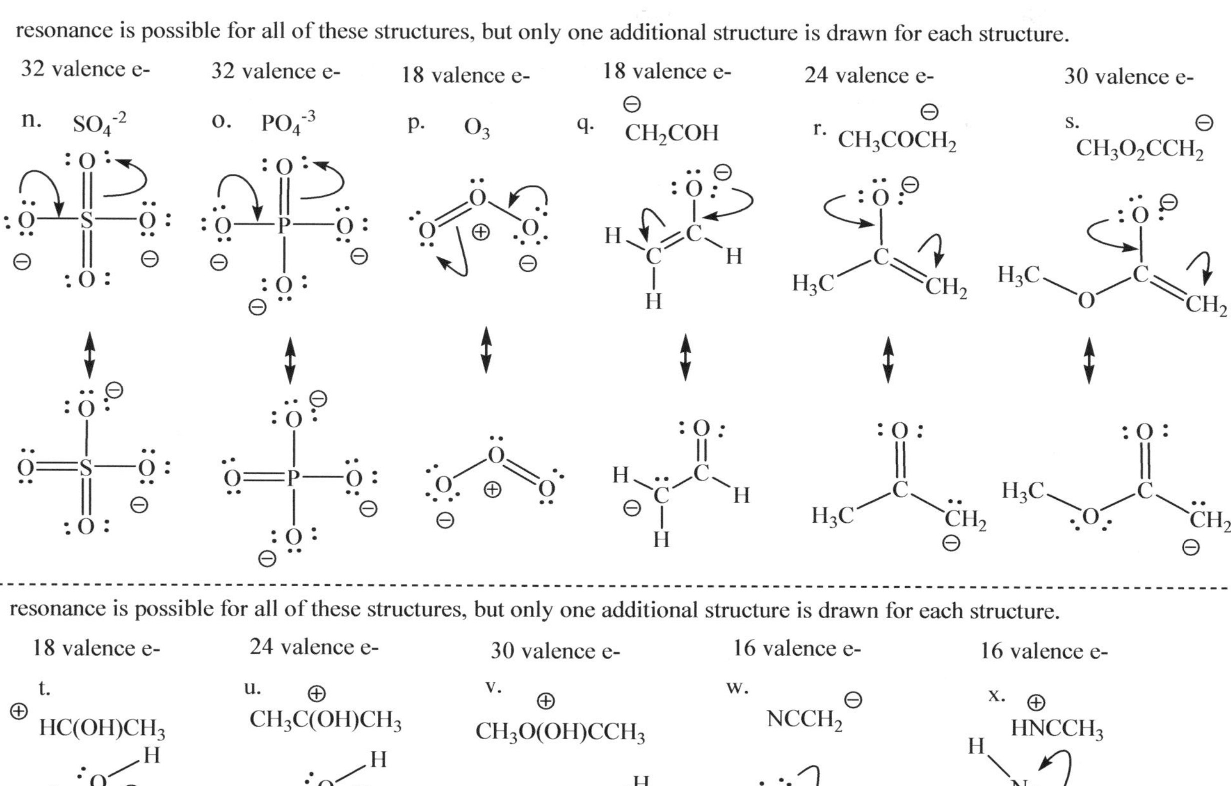

resonance is possible for all of these structures, but only one additional structure is drawn for each structure.

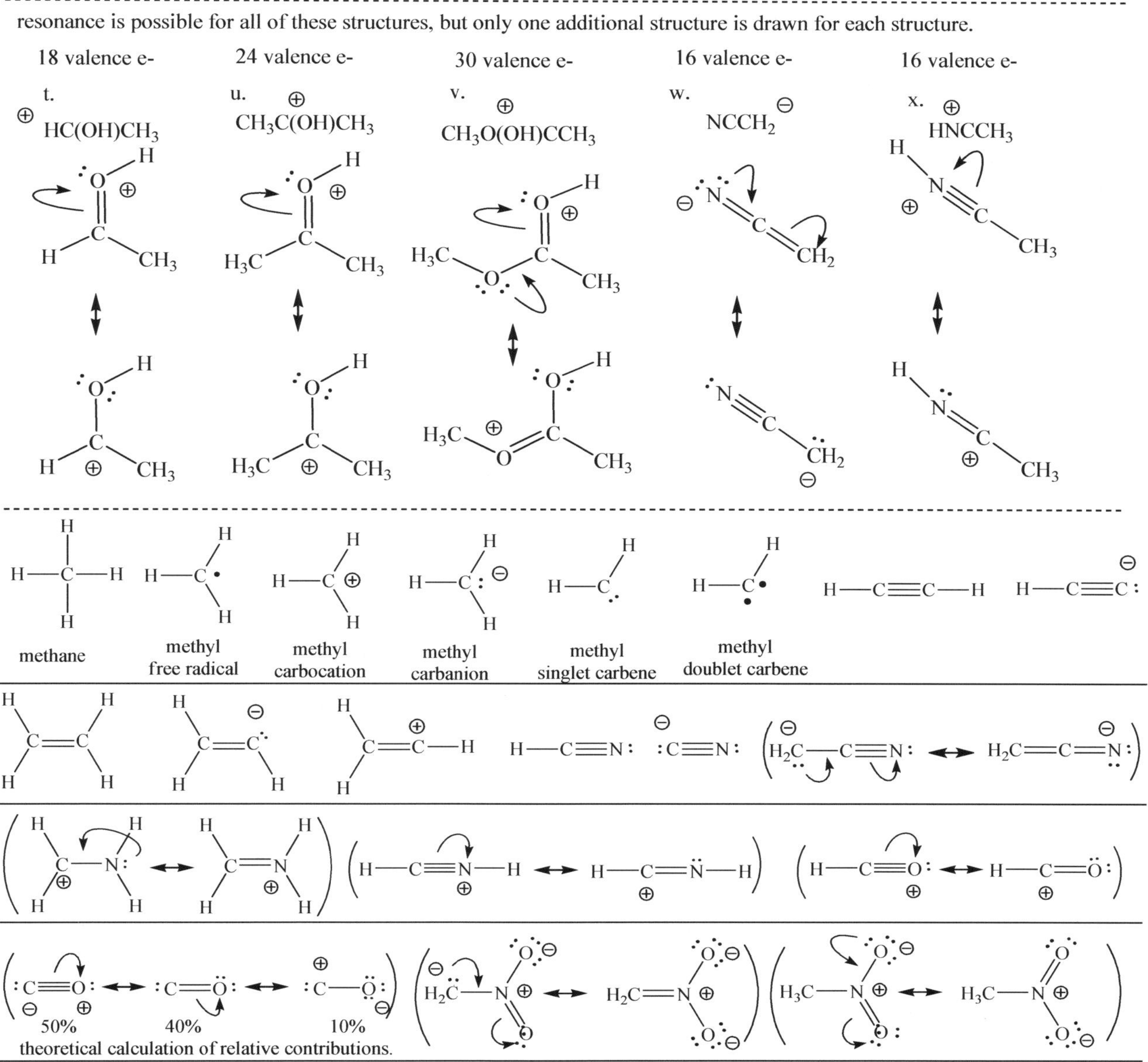

50% 40% 10%
theoretical calculation of relative contributions.

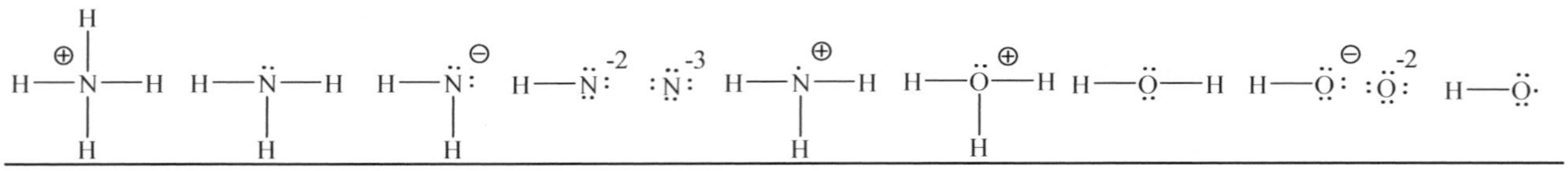

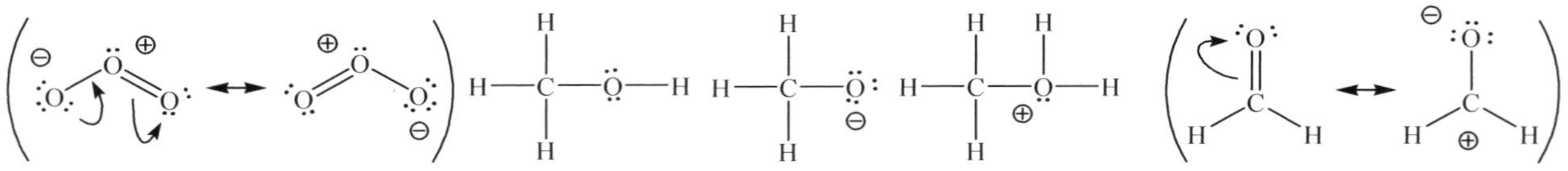

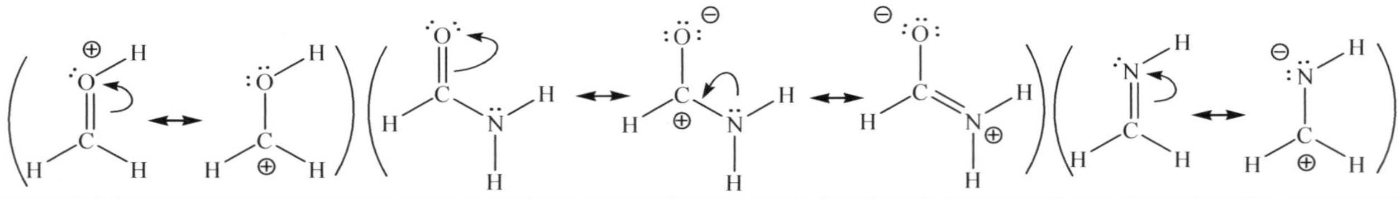

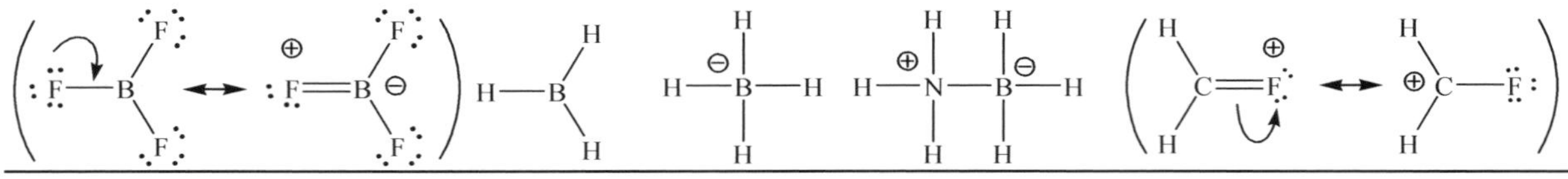

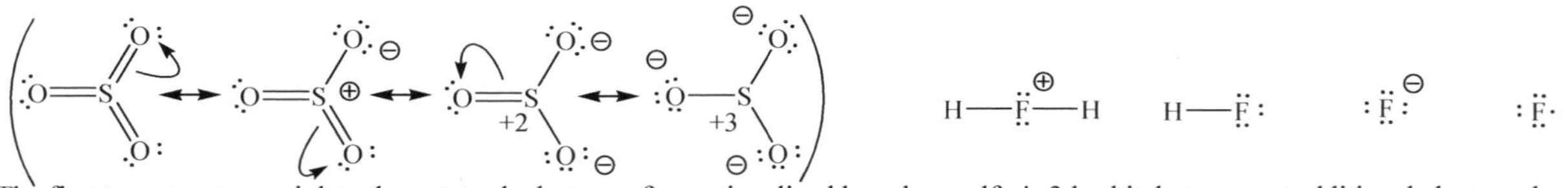

The first two structures violate the octet rule, but are often rationalized based on sulfur's 3d orbitals to accept additional electron density.

Problem 16 (p 35) - Draw two dimensional Lewis structures for the following structures. Include two dots for any lone pair electrons. What is the hybridization, bond angles and descriptive shape for each nonhydrogen atom below?

a alkane

b 1° bromoalkane

c ketone

d alkene 1° amide

e alkene

f 2° amine

g aldehyde

h nitrile

i 1° amine

j 3° amine

k carboxylic acid

l carboxylic acid

m 1° alcohol

n alkene alkyne

o ester

p acid bromide

q aromatic

r aromatic

s acid aldehyde

t alkyne

Problem 17 (p 36) - Draw two dimensional Lewis structures for the following structures. Include two dots for any lone pair electrons. Some of these have formal charges. What is the hybridization, bond angles and descriptive shape for each nonhydrogen atom below?

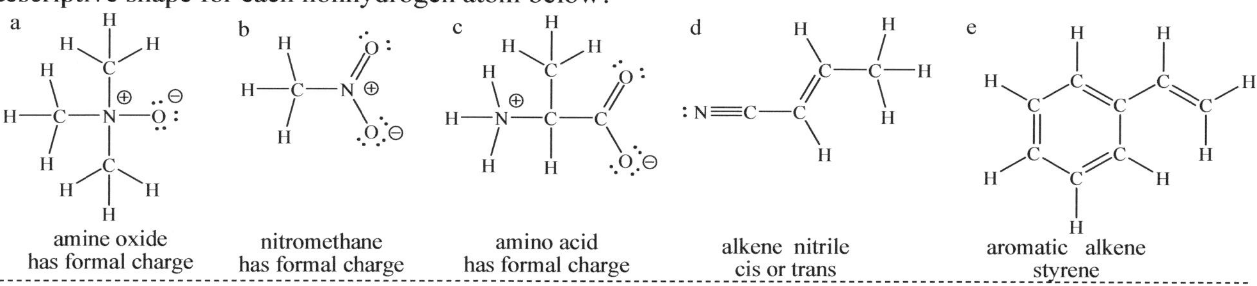

a amine oxide has formal charge

b nitromethane has formal charge

c amino acid has formal charge

d alkene nitrile cis or trans

e aromatic alkene styrene

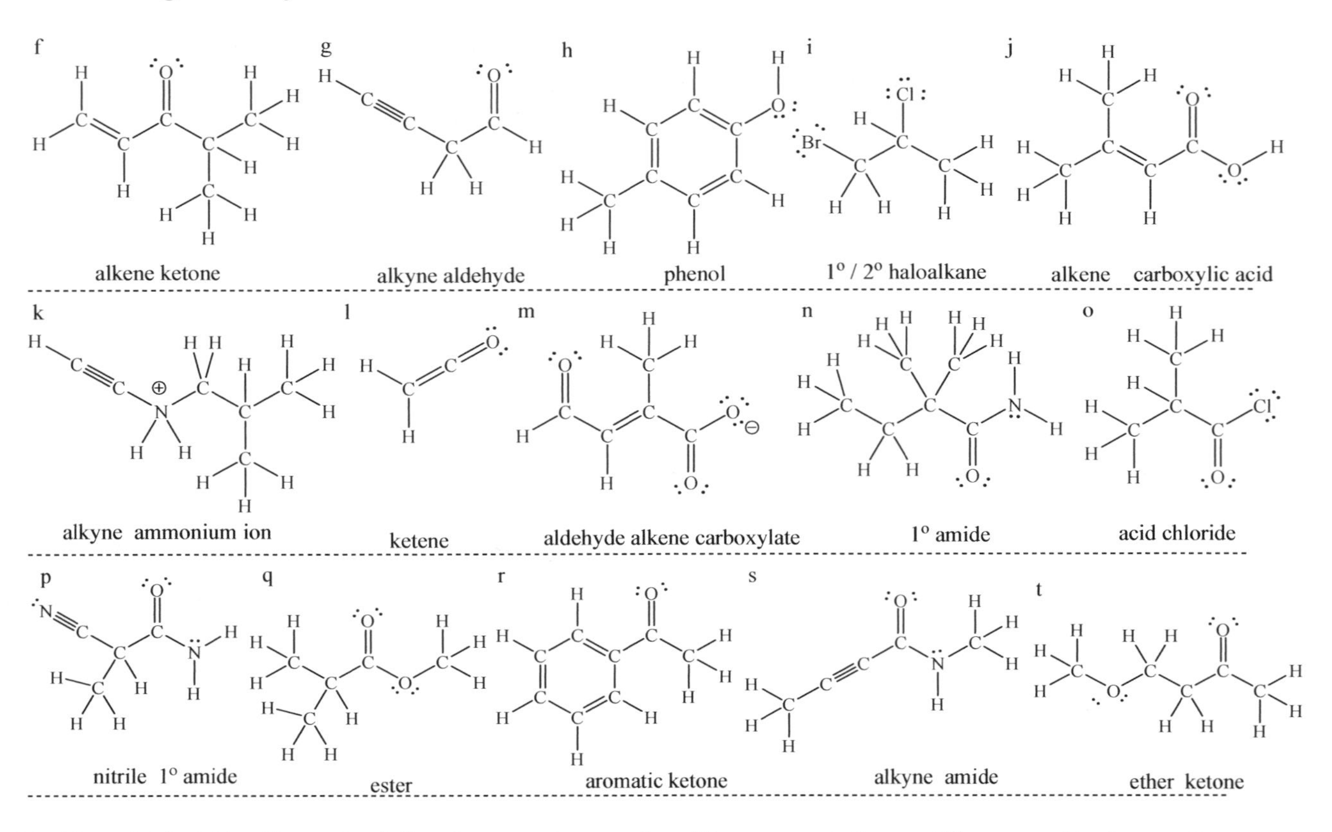

Problem 18 (p 38) – Draw the following structures using 3D representations. (If you need help, use the keys.)

a R—C≡C—C≡C—R

b R—C≡C—C≡N:

c

d

e

Problem 19 (p 43) – First, convert the condensed line formulas of the following hydrocarbons into 2D Lewis structures. Next, draw 3D structures for each of the 2D structures. You should show the bonds in front of the page as wedges and bonds in back of the page with dashed lines and bonds in the plane of the page as simple lines. Show the 2p orbitals for pi bonds along with their electrons. If you cannot figure out how the atoms are connected, there are some clues on the next couple of pages.

a.

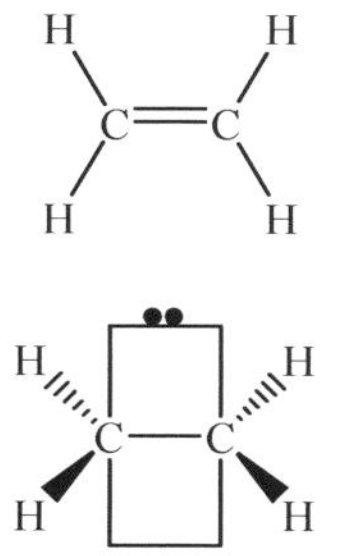

b.

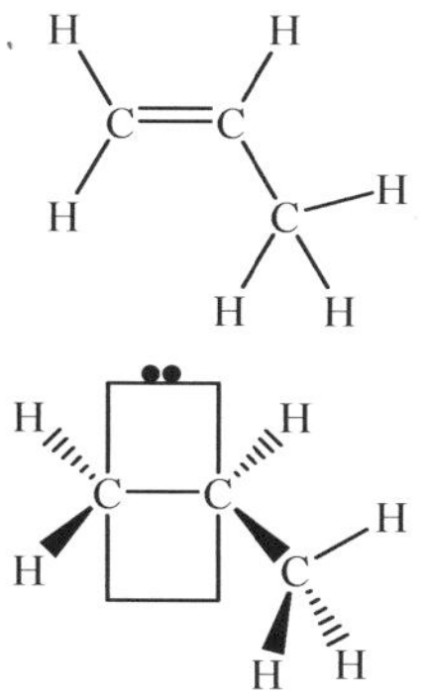

c.

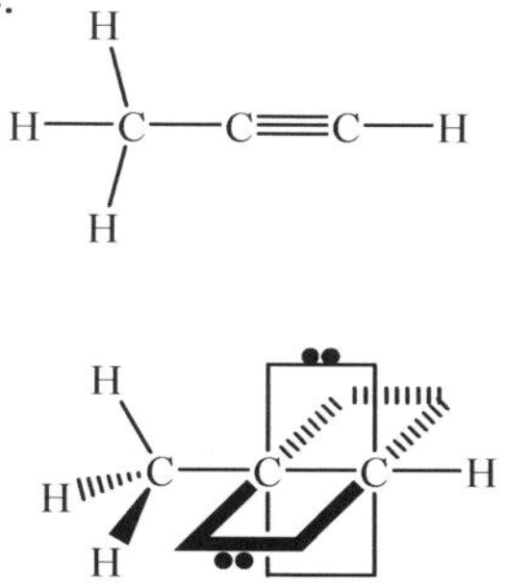

d.

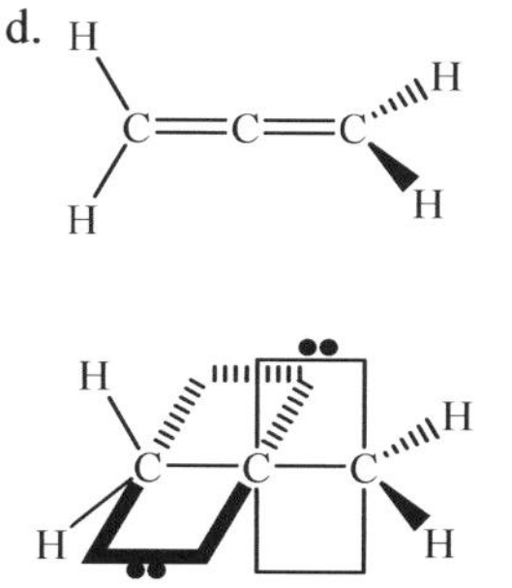

e

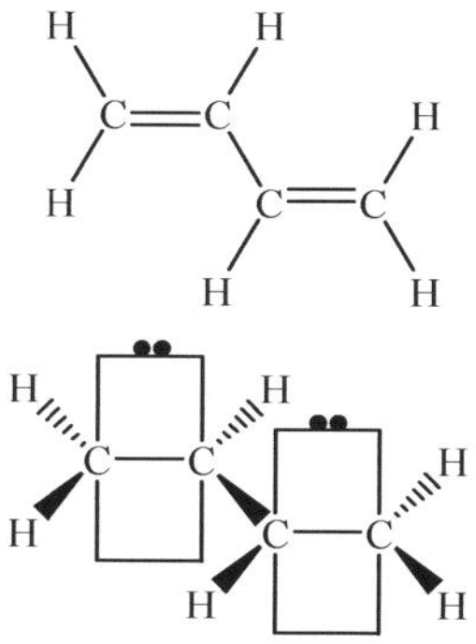

f

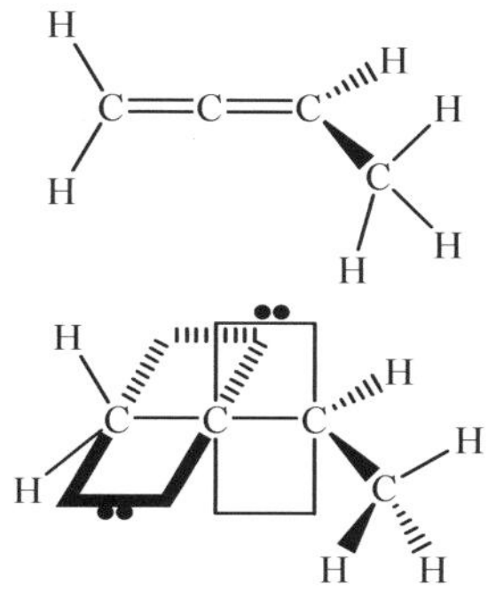

g

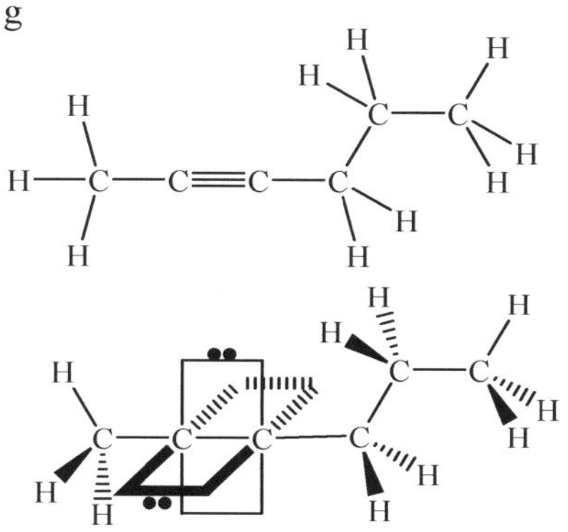

h

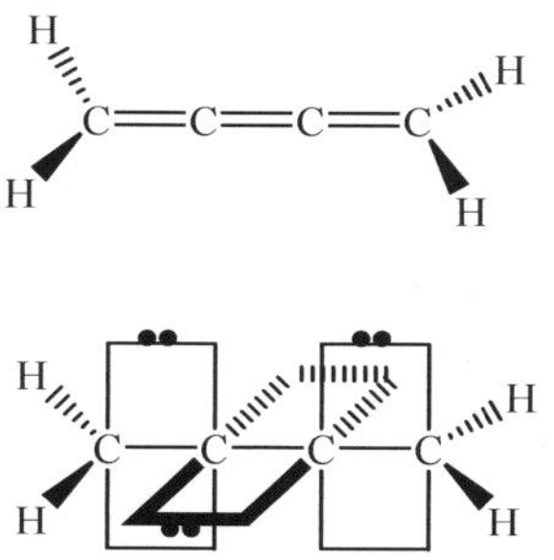

i

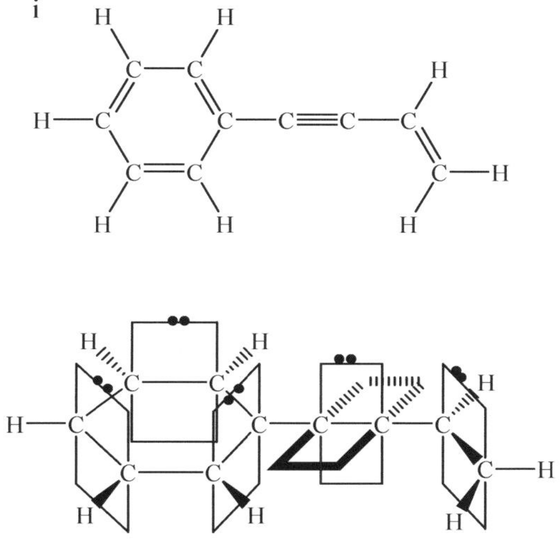

j.

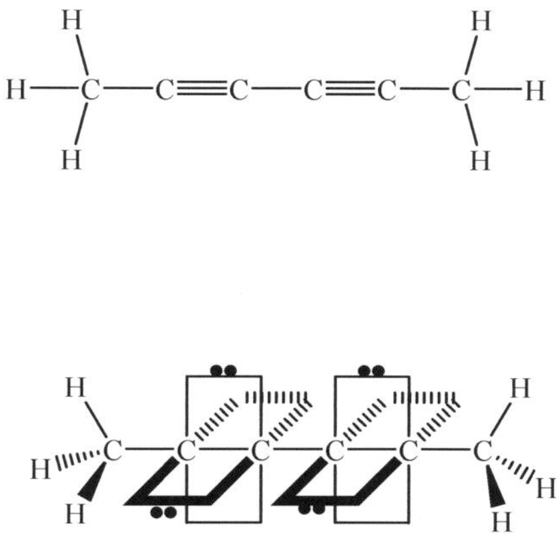

k.

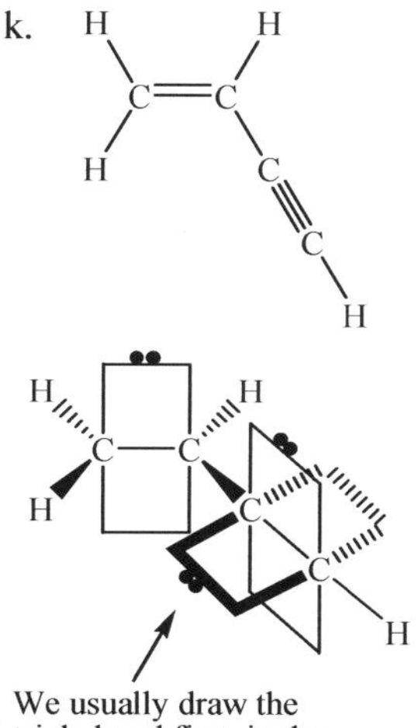

We usually draw the triple bond first, in the plane of the paper.

Problem 20 (p 44) – This problem is very similar to the above problem, except heteroatoms (N, O and F) are substituted in for some of the carbon atoms and some structures have formal charge. If resonance structures are present, decide an atom's hybridization based on the resonance structure where it has its maximum number of bonds. If you cannot figure out how the atoms are connected, there are some clues on the next couple of pages.

a

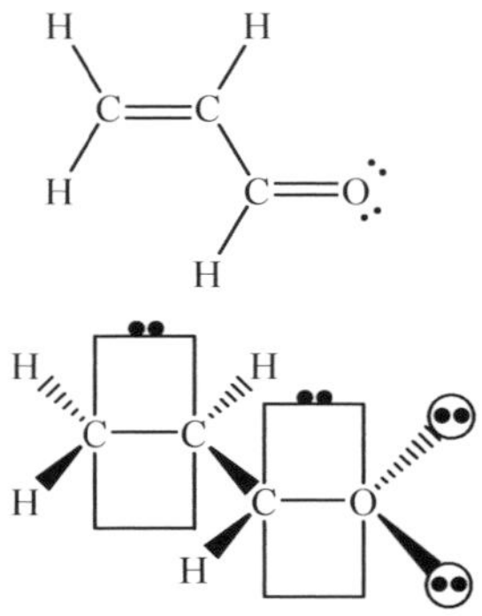

b

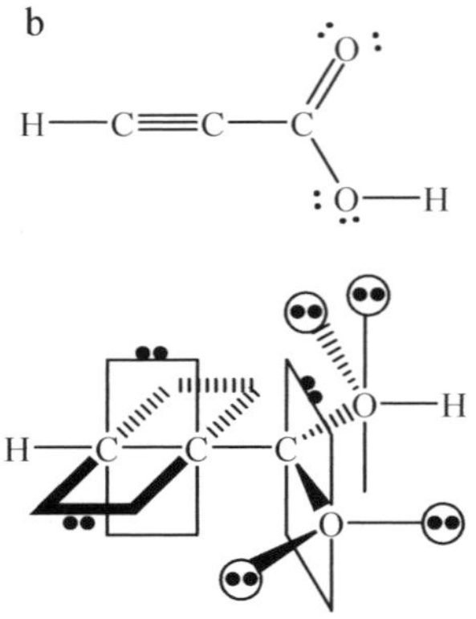

c

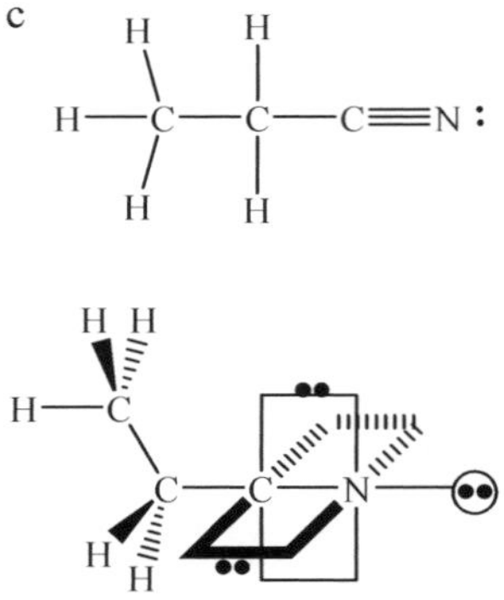

d

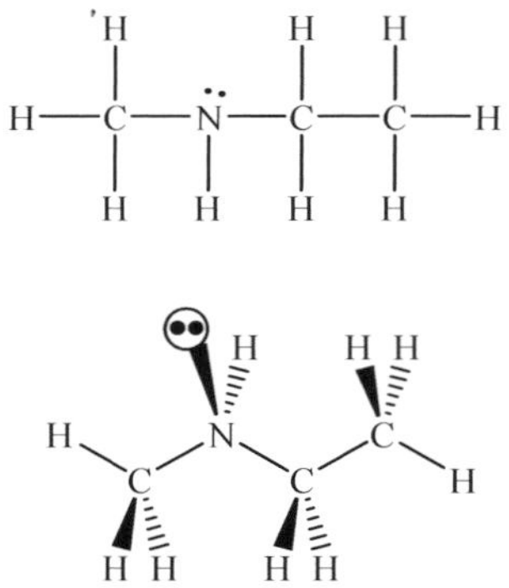

e

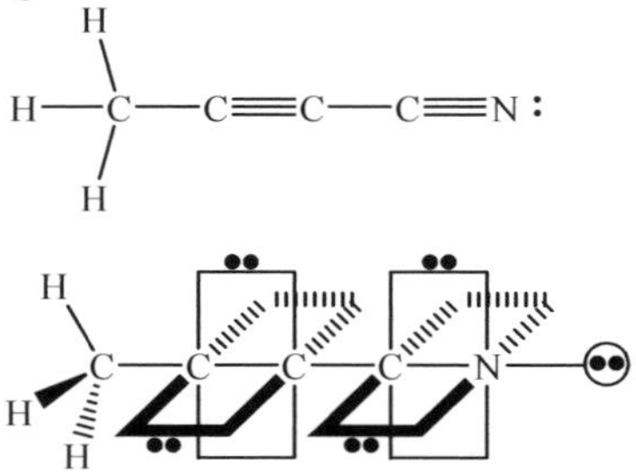

f

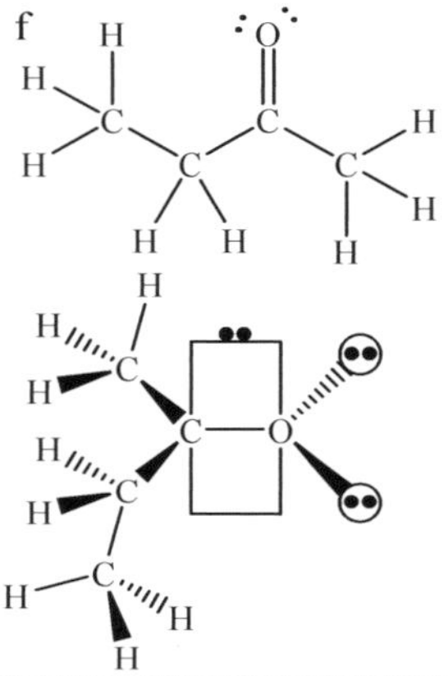

g

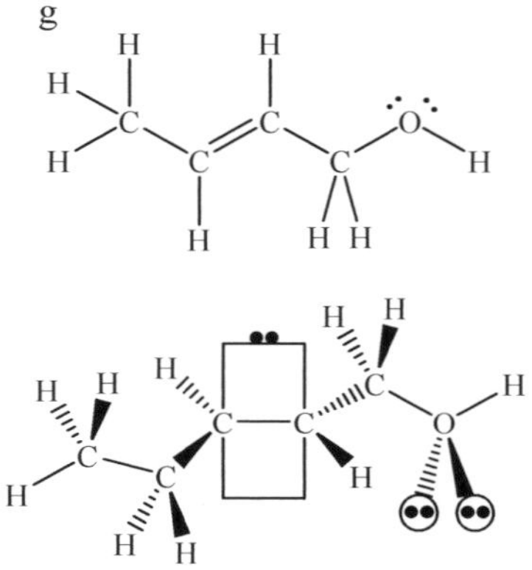

h

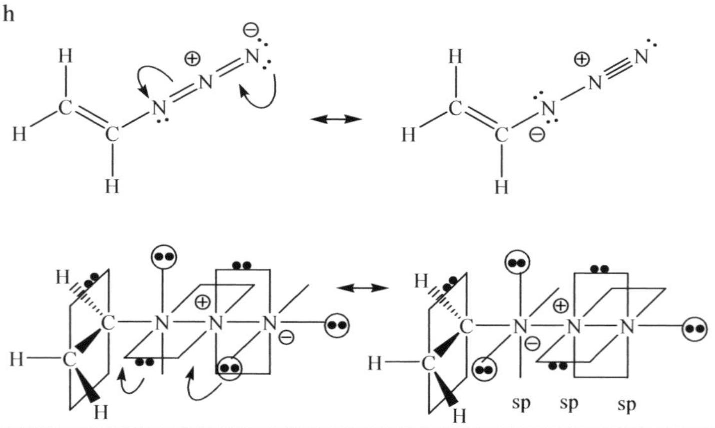

sp sp sp

i

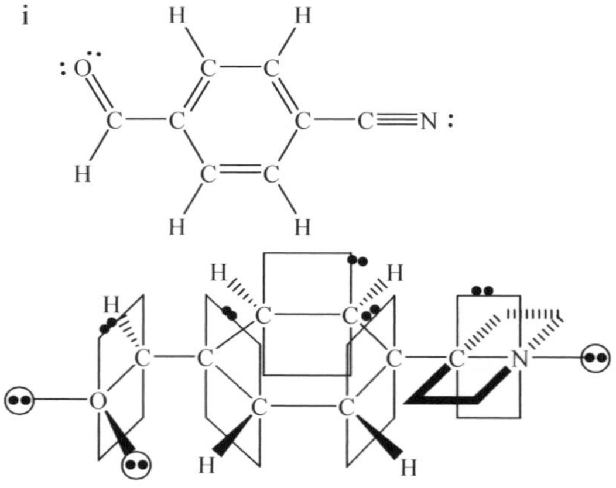

j. CH_3CHN_2 has formal charge

k. $CH_2CHC_6H_4CCCO_2CH_3$ six atom ring

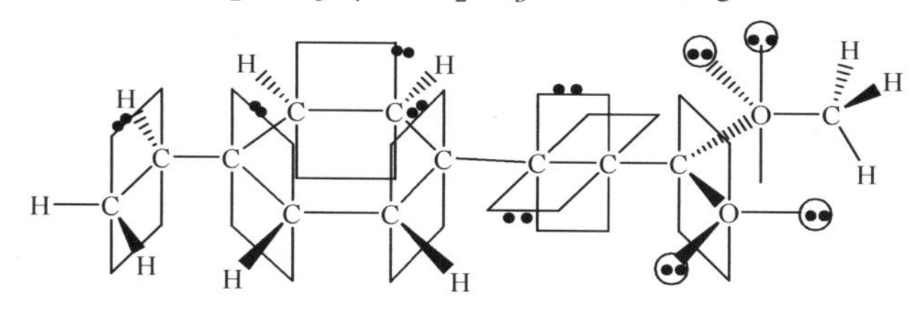

l. $ClCH_2CH_2CONHCH_3$

m. $O_2NCHCHCCCH_3$ has formal charge

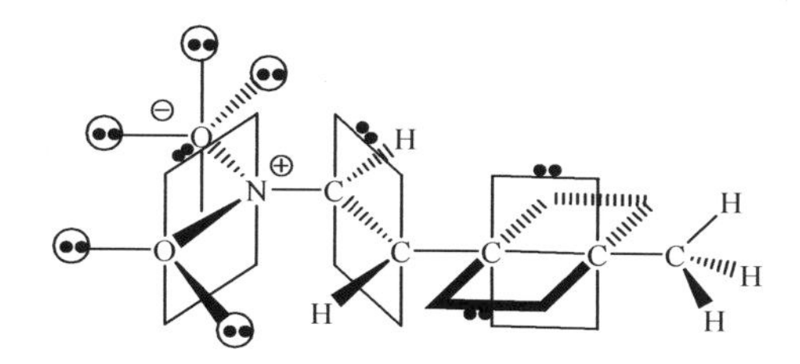

n. $CH_3C_5NH_4$ six atom heterocyclic ring

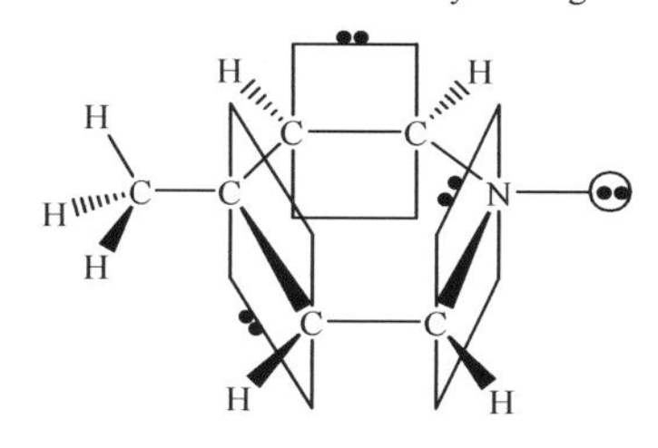

o. CH_3CHCO

Chapter 2

Problem 1 (p 51) - Predict the formula for the combination of the following pairs of ions. What kinds of melting points would be expected for these salts?

	$F^{\ominus}$	$NO_2^{\ominus}$	$NO_3^{\ominus}$	O^{-2}	$HCO_3^{\ominus}$	PO_4^{-3}	CO_3^{-2}
$K^{\oplus}$	KF	KNO_2	KNO_3	K_2O	$KHCO_3$	K_3PO_4	K_2CO_3
Ba^{+2}	BaF_2	$Ba(NO_2)_2$	$Ba(NO_3)_2$	BaO	$Ba(HCO_3)_2$	$Ba_3(PO_4)_2$	$BaCO_3$
Zn^{+2}	ZnF_2	$Zn(NO_2)_2$	$Zn(NO_3)_2$	ZnO	$Zn(HCO_3)_2$	$Zn_3(PO_4)_2$	$ZnCO_3$
Al^{+3}	AlF_3	$Al(NO_2)_3$	$Al(NO_3)_3$	Al_2O_3	$Al(HCO_3)_3$	$Al(PO_4)$	$Al_2(CO_3)_3$

Problem 2 (p 52) – Supply dipole arrows to any polar bonds above (according to our arbitrary rules). Make sure they point in the right direction.

single bonds - many types and possibilities

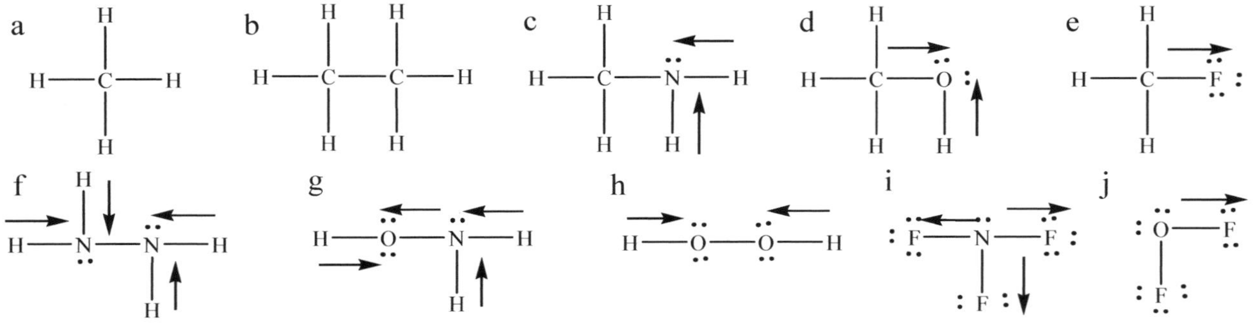

double bonds - many types and possibilities

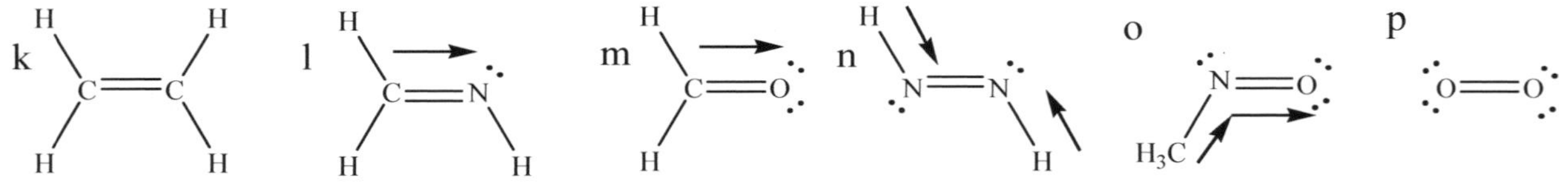

two double bonds - many types and possibilities

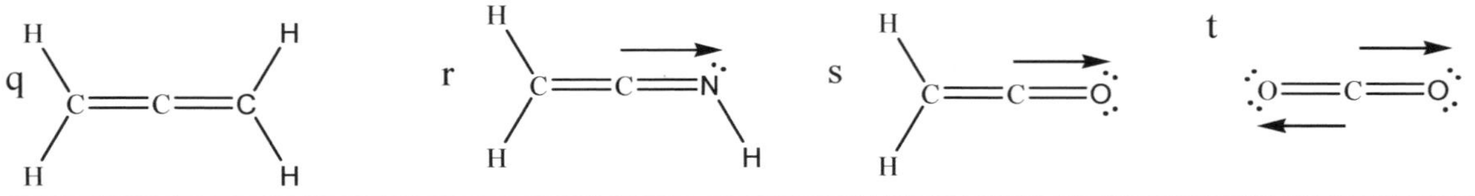

triple bonds - many types and possibilities

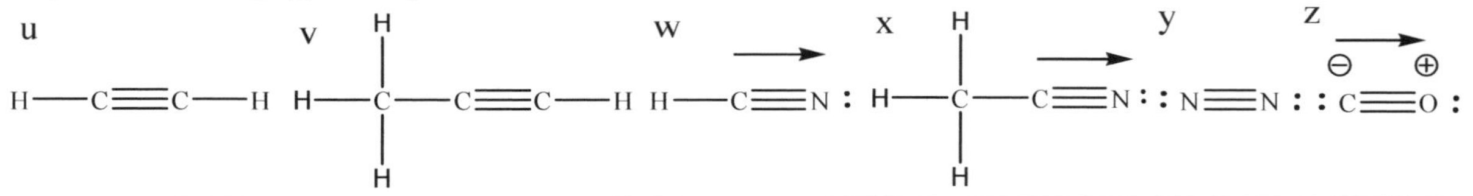

rings - many types and possibilities

aa (cycloheptane ring: H_2C, CH_2 units) bb (ring with :NH) cc (ring with :O:) dd (ring with C=C and H substituents)

Problem 3 (p 57) - The following pairs of molecules have the same formula; they are isomers. Yet, they have different melting points. Match the melting points with the correct structure and provide an explanation for the difference. Hint: single bonds can rotate and pi bonds tend to be rigid, fixed shapes.

a. $H_3C—C\equiv C—CH_3$
mp = -32°C
closer lattice packing

$H_3C—H_2C—C\equiv C—H$
mp = -126°C

b. mp = -105°C
better packing

mp = -185°C

c. C_6H_{14}
mp = -95°C
many irregular shapes

C_6H_{12}
mp = +6°C
one shape
better packing

d. mp = +79°C
one shape, close packing

mp = -88°C
flexible chain,
many shapes

Problem 4 (p 61) – Provide an explanation for the different boiling points in each column.

4B	5B	6B	7B	8B
CH_4 mp = -182°C bp = -164°C μ = 0.0 D ΔX = 0.3	NH_3 mp = -78°C bp = -33°C μ = 1.42 D ΔX = 0.8	H_2O mp = 0°C bp = +100°C μ = 1.80 D ΔX = 1.2	HF mp = -84°C bp = +20°C μ = 1.86 D ΔX = 1.8	He mp = -272°C bp = -269°C μ = 0.0 D ΔX = NA
SiH_4 mp = -185°C bp = -112°C μ = 0.0 D ΔX = 0.3	PH_3 mp = -132°C bp = -88°C μ = 0.0 D ΔX = 0.0	H_2S mp = -82°C bp = -60°C μ = 1.0 D ΔX = 0.4	HCl mp = -114°C bp = -85°C μ = 1.0 D ΔX = 1.0	Ne mp = -249°C bp = -246°C μ = 0.0 D ΔX = NA
GeH_4 mp = -165°C bp = -88°C μ = 0.0 D ΔX = 0.2	AsH_3 mp = -111°C bp = -62°C μ = 0.0 D ΔX = 0.0	H_2Se mp = -66°C bp = -41°C μ = ? D ΔX = 0.4	HBr mp = -87°C bp = -67°C μ = 0.8 D ΔX = 0.8	Ar mp = -189°C bp = -185°C μ = 0.0 D ΔX = NA
SnH_4 mp = -146°C bp = -52°C μ = 0.0 D ΔX = 0.2	SbH_3 mp = -88°C bp = -17°C μ = 0.0 D ΔX = 0.2	H_2Te mp = -49°C bp = -2°C μ = ? D ΔX = 0.1	HI mp = -51°C bp = -35°C μ = 0.4 D ΔX = 0.5	Kr mp = -157°C bp = -153°C μ = 0.0 D ΔX = NA

Answer: Each dotted line represents the boiling points in the hydrides in a column in the periodic table, plus the Noble gases. In the Noble gases there is a continuing trend towards higher boiling point as the main atom gets larger with more and more polarizable electron clouds. Where bonds to hydrogen become polar, hydrogen bonding becomes important, leading to stronger attractions for neighbor molecules and higher boiling points deviate from the trends observed due to dispersion forces. Deviations are clearly seen with NH_3, H_2O and HF. Melting points are not plotted.

Problem 5 (p 61) – Provide an explanation for the different boiling points in each series. In part b, an H-F bond is expected to be more polar than an H-O bond, so why does HF boil lower than H_2O? Think of an "H bond" donor site as a hand that can grab and a lone pair as a handle that can be held onto. If you only had one hand, it really doesn't make any difference how many handles are available to grab hold of and if you only have one handle, it doesn't make any difference how many hands you have. But, if you have two hands and two handles a much stronger grip is possible with twice the handholds.

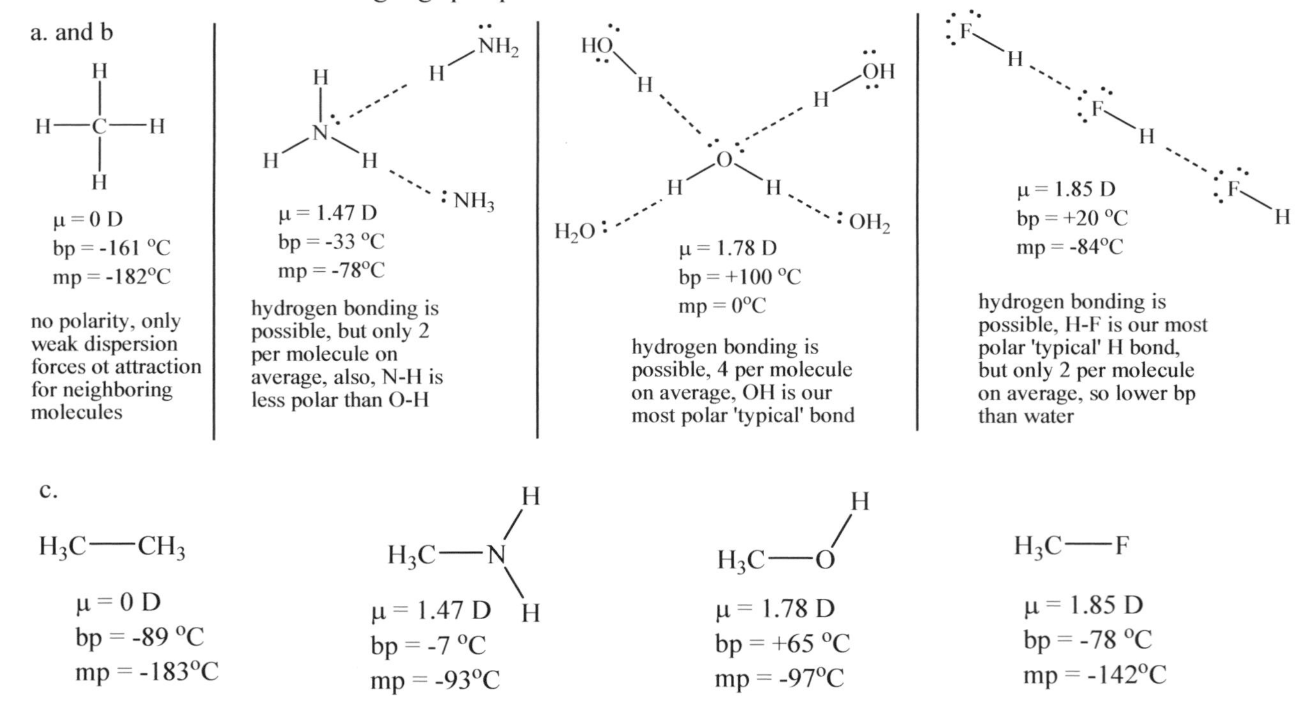

H bonding is the most important factor affecting bp's in these molecules. CH_3OH makes stronger H bonds than CH_3NH_2 and the other two do not have H bonds. While CH_3F has the most polar bond, fluorine is so electronegative that it does not share its lone pairs much and is not very polarizable and there is no polarized proton to hydrogen bond with. Ethane has no polarity and only has weak dispersion forces to form attractions with neighbor molecules, so has the lowest bp.

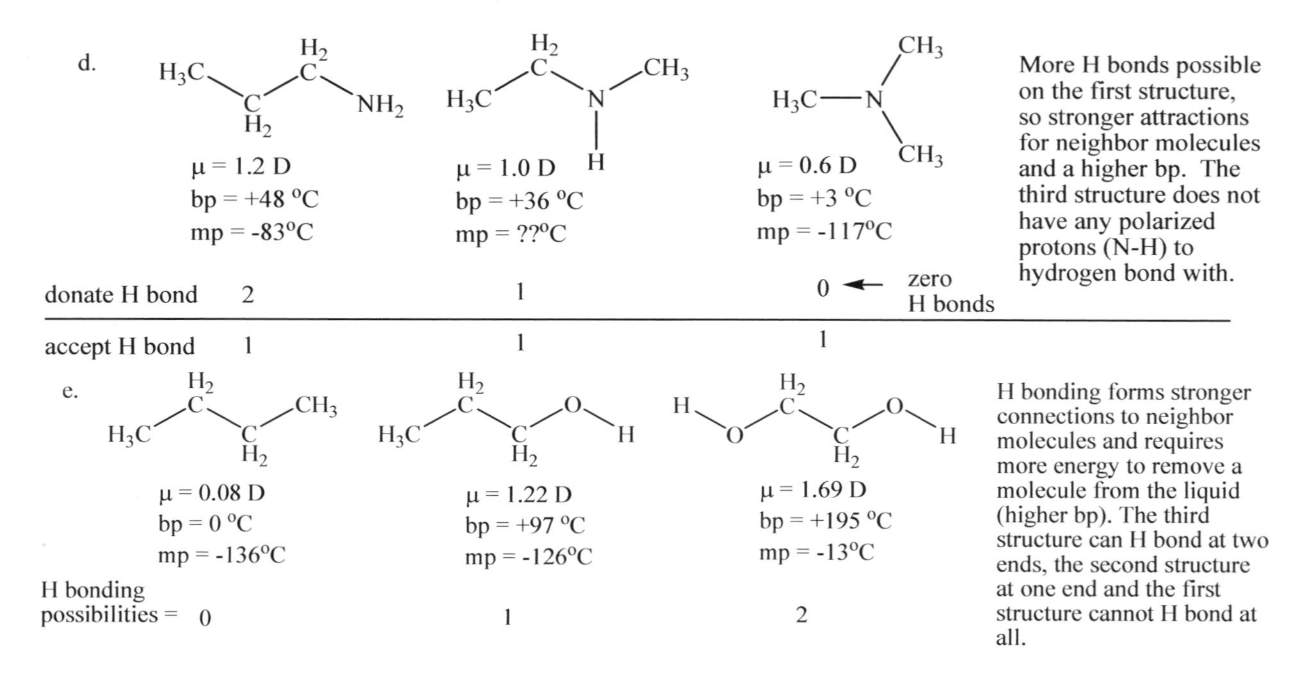

f.

Compound	μ	bp	mp
$HCHO$ (H–C(=O)–H)	μ = 2.33 D	bp = -19 °C	mp = -92 °C
CH_3OH	μ = 1.78 D	bp = +65 °C	mp = -97°C
CH_3CHO	μ = 2.7 D	bp = +20 °C	mp = -123 °C
CH_3CH_2OH	μ = 1.69 D	bp = +78 °C	mp = -114 °C
$(CH_3)_2CO$	μ = 2.91 D	bp = +56 °C	mp = -95 °C
$(CH_3)_2CHOH$	μ = 1.66 D	bp = +82 °C	mp = -89 °C

Δ = 84°C Δ = 58°C Δ = 26°C

In each part we are comparing two similar structures, one with a polar C=O bond with a polar C-O bond and H bonding possibilities. The alcohols, with H bonding, have stronger attractions for the neighbor molecules, thus higher bp's. The differences are larger on the smaller moleules, where the H bonding represents a larger fraction of the total interactions.

Problem 6 (p 63) – Provide an explanation for the different boiling points in each series.

a.

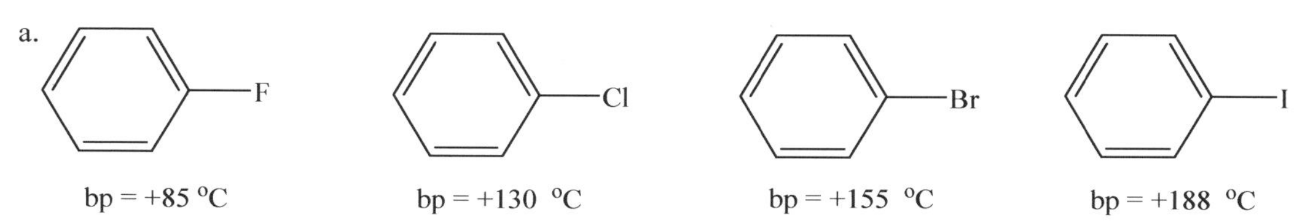

bp = +85 °C bp = +130 °C bp = +155 °C bp = +188 °C

Similar molecules. The highest boiling points go with the largest dispersion forces (greatest polarizability, I > Br > Cl > F)

---

b.

bp = -0.5 °C bp = +5 °C bp = +78 °C

Structure 3 has H bonding and polarity (strongest attractions), structure 2 has polar bonds and structure 1 only has dispersion forces (weakest attractions)

---

c.

bp = -47 °C bp = +20 °C bp = +101 °C bp = +210 °C O═C═O bp = -78 °C (sublimes)

Structure 1 only has weak dispersion forces, structure 2 had polar bonds (stronger), structure 3 has H bonds too (stronger), structure 4 has the strongest polarity and H bonds with both N-H bonds (highest). Finally CO_2 has polar bonds, but its symmetry cancels out its polarity. The oxygen atoms hold so tightly to their electrons that they are less polarizable than structure 1 and sublimes at a lower temperature.

---

d.

bp = -42 °C bp = -23 °C bp = +82 °C

Structure 1 (propane) only has weak dispersion forces, and has a bent shape that prevents closer contact with neighbor molecules.Structure 2 (propyne) only has weak dispersion forces, but has a linear shape that allows closer contact with neighbor molecules, so stronger attractions. Structure 3 has the shape of propyne, and relatively strong polarity which makes stronger attractions and a higher bp.

---

e.

$H_3C—CH_3$ bp = -89 °C

$H_3C—CH_2—CH_2—CH_2—CH_3$ bp = +36 °C

$H_3C—CH(CH_3)—CH_2—CH_3$ bp = +30 °C

$C(CH_3)_4$ bp = +10 °C

More carbons have greater contact surface area so stronger dispersion forces (C2 < C5). Straight chains can approach closer than when branches are present (branches tend to push neighbor molecules away = weaker dispersion forces). More branches weaken those attractions even more (D < C < A).

e.

CH_4	CH_3Cl	CH_2Cl_2	$CHCl_3$	CCl_4
bp = -164 °C	bp = -24 °C	bp = +40 °C	bp = +61 °C	bp = +77 °C

Chlorine is a larger atom (than C) and has polarizable lone pairs. More chlorines = more dispersion forces = higher bp. B, C and D have polarity too. A and E are not polar, but E has four chlorine atoms and larger dispersion forces dominate here (highest bp).

---

f.

$H_3C\text{-}CH_2\text{-}CH_3$	$H_3C\text{-}CH_2\text{-}NH_2$	$H_3C\text{-}CH_2\text{-}OH$
bp = -42 °C	bp = +17 °C	bp = +78 °C

Bp's follow H bonding possibilities. Structure 3 has the strongest H bonds because O-H is more polar. Structure 2 has some H bonding possibilitiies so a higher bp than structure 1 which only has dispersion forces. All 3 molecues are similar in size, so comparisons are fair.

---

g.

hexane, C_6H_{14}	cyclohexane, C_6H_{12}	benzene, C_6H_6
mp = -91 °C	mp = +6 °C	mp = +5 °C
bp = +69 °C	bp = +81 °C	bp = +80 °C

All 3 structures only have weak dispersion forces and are approximately the same size. The rings are more regular in shape and probably pack closer in the solid, thus the much higher mp's (similar to one another) and are also closer in the liquid with no branches wildly swinging around to bump neighbor molecules farther away, thus the higher bp's too (again, rings similar to one another).

Problem 7 (p 64) – Match the given boiling points with the structures below and give a short reason for your answers. (-7°C, +31 °C, +80 °C, +141°C, 1420°C)

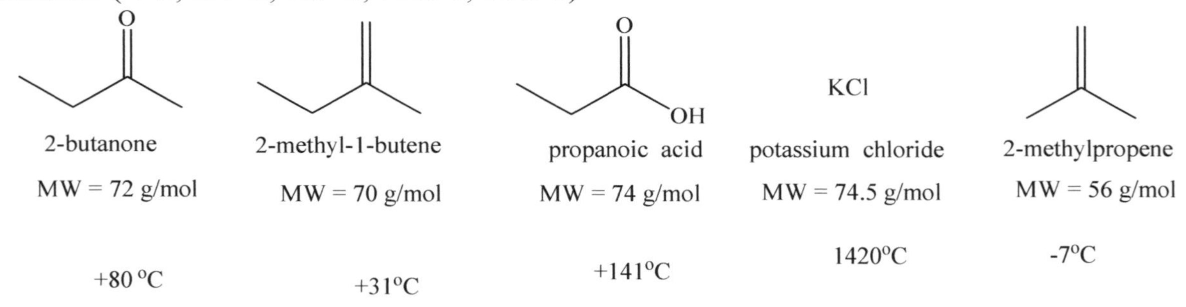

Structure 4 has ionic bonds, by far the strongest forces to overcome and has an extremely high bp. Structure 3 has strong H bonding possiblities and the second high bp. Structure 1 has polar C=O bond so higher than structures 2 and 5, which only have weak dispersion forces. Structure 2 is larger and has a larger surface area and more dispersion forces than structure 5, so has a higher bp.

Problem 8 (p 67) – Point out the polar hydrogen in methanol. What is it about dimethyl sulfoxide (DMSO) that makes it polar? Draw a simplistic picture showing how methanol interacts with a cation and an anion. Also use DMSO (below) and draw a simplistic picture showing the interaction with cations and anions. Explain the difference from the methanol picture.

polar H

Methanol is good at solvating positive and negative charge.

⊕ = cations are smaller because they have lost electrons and hold onto the remaining electrons tighter.

⊖ = anions are larger because they have extra electron density that is repulsive and expands the electron clouds

strong bond dipole

DMSO is very good at solvating positive charge but it is poor at solvaing n egative charge.

Problem 9 (p 68) –
a. Hexane (density = 0.65 g/ml) and water (density = 1.0 g/ml) do not mix. Which layer is on top?
b. Carbon tetrachloride (density = 1.59 g/ml) and water (density = 1.0 g/ml) do not mix. Which layer is on top?

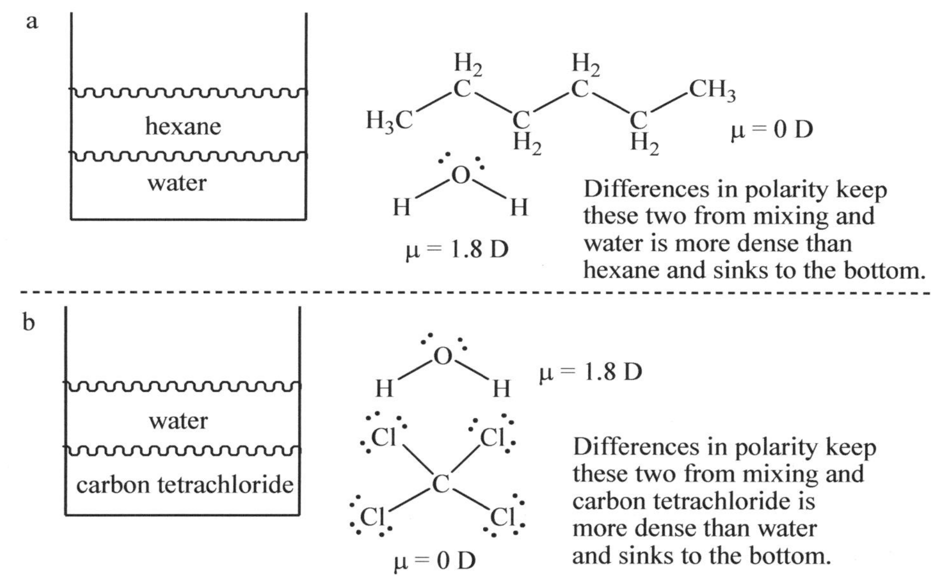
a
hexane
water
μ = 0 D
μ = 1.8 D
Differences in polarity keep these two from mixing and water is more dense than hexane and sinks to the bottom.
b
water
carbon tetrachloride
μ = 1.8 D
μ = 0 D
Differences in polarity keep these two from mixing and carbon tetrachloride is more dense than water and sinks to the bottom.

Problem 10 (p 68) - The melting point of NaCl is very high (≈ 800°C) and the boiling point is even higher (> 1400°C). Does this imply strong, moderate or weak forces of attraction between the ions? Considering your answer, is it surprising that NaCl dissolves so easily in water? Why does this occur? Consider another chloride salt, AgCl. How does your analysis work here? What changed?

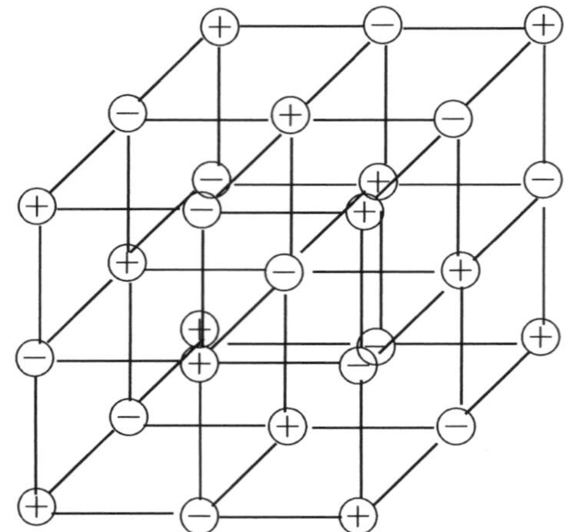

Lattic structure - depends on the size and charge of the ions.

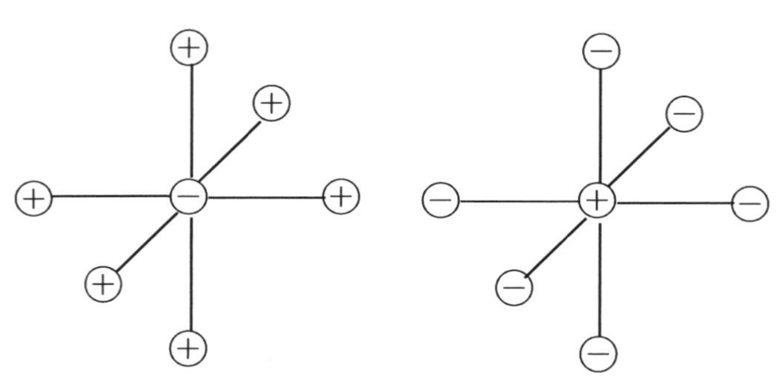

Each ion is surrounded on many sides by oppositely charged ions. To introduce the disorder of a liquid (melt) or a gas (boil) requires a very large input of energy (mp indicates the amount of energy required to break down the ordered lattice structure and boiling point indicates the amount of energy required to remove an ion pair from the influence of all neighbors. Ionic bonds (ionic attractions on all sides) can only be bronke at great expense in energy.

However, many water molucules working together can break down this structure by solvating both the cations and anions (solvation energy). When solvation energy > lattice energy, the salt will dissolve (NaCl). When solvation energy < lattice energy the salt will not dissolve (AgCl).

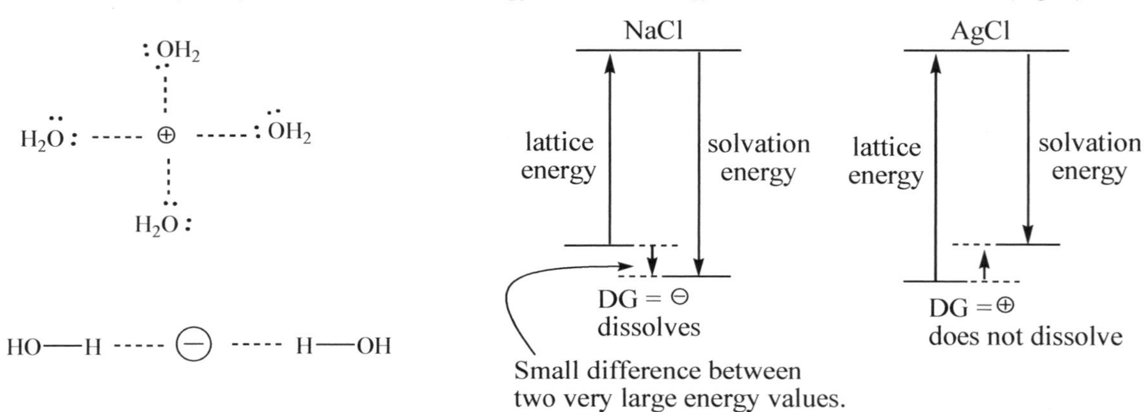

Problem 11 (p 68) – a. Which solvent do you suspect would dissolve NaCl better, DMSO or hexane? Explain your choice? b. Which solvent do you suspect would dissolve NaCl better, methanol or benzene? Explain your choice?

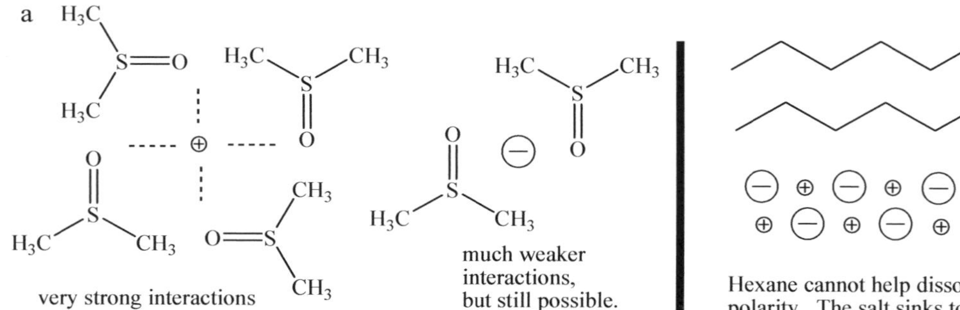

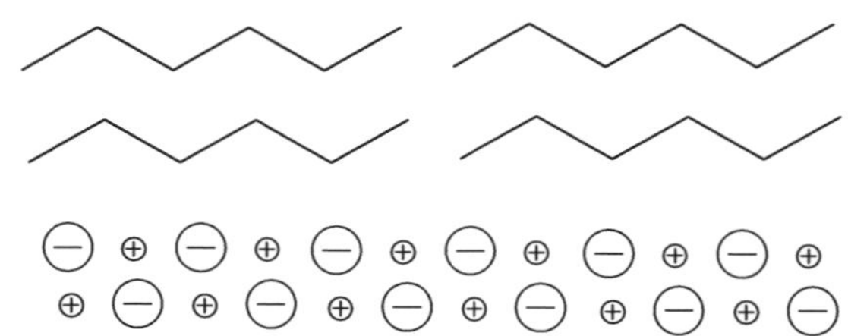

Hexane cannot help dissolve ions at all because it has no polarity. The salt sinks to the bottom (if more dense).

b

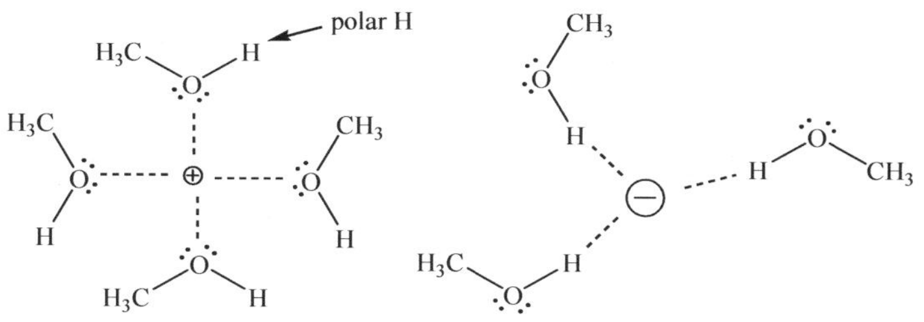

Methanol is good at solvating positive and negative charge.

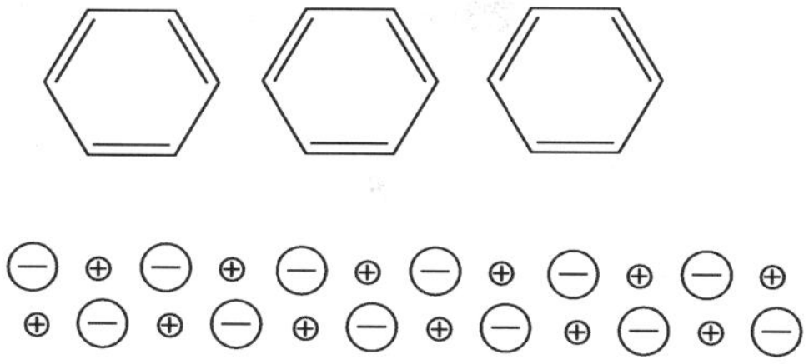

Benzene is not good at dissolving ions at all because it has no polarity, although its pi electrons can weakly interact with cations. The salt sinks to the bottom (if more dense).

Problem 12 (p 68) – The terms "hydrophilic" and "hydrophobic" are frequently used to describe structures that mix well or poorly with water, respectively. Biological molecules are often classified in a similar vein as water soluble (hydrophilic) or fat soluble (hydrophobic). The following list of well known biomolecules are often classified as fat soluble or water soluble. Examine each structure and place it in one of these two categories. Explain you reasoning.

a vitamin A

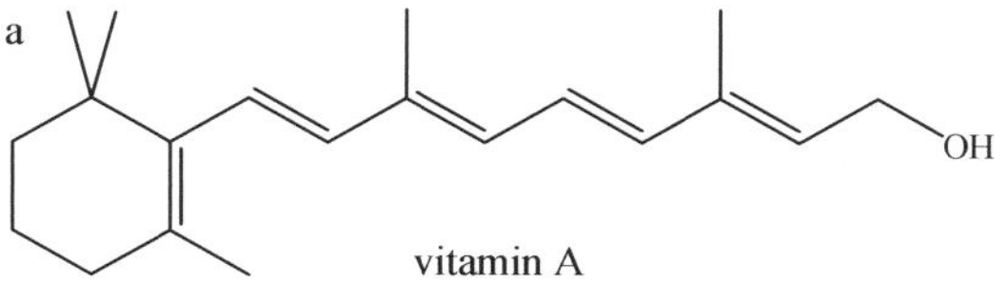

Hydrophobic - mostly nonpolar

b vitamin B2 (riboflavin)

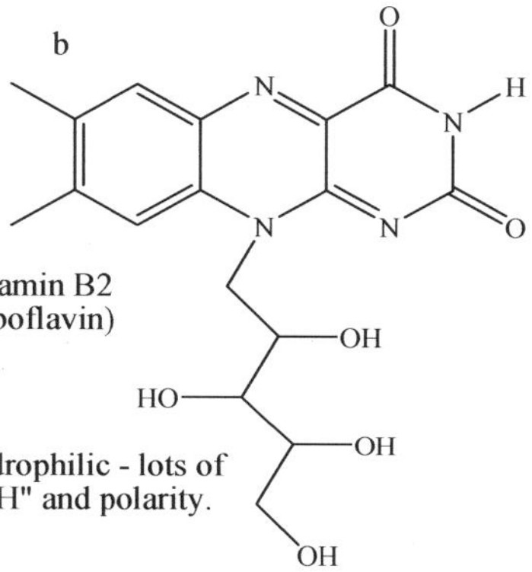

Hydrophilic - lots of "OH" and polarity.

c vitamin B6 (pyridoxine)

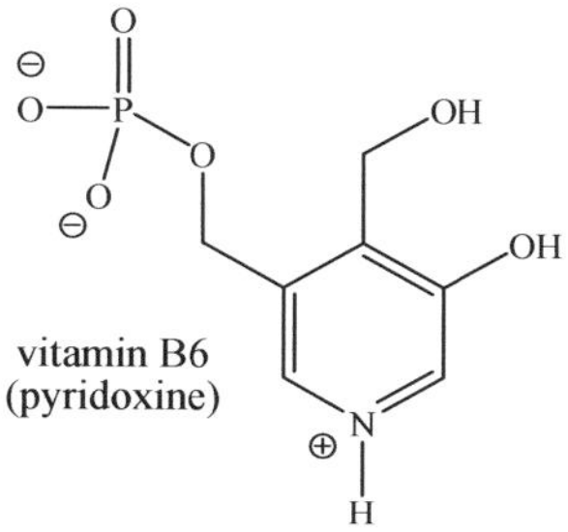

Hydrophilic - lots of "OH" and polarity and charge.

d ATP

Hydrophilic - lots of "OH" and polarity and charge.

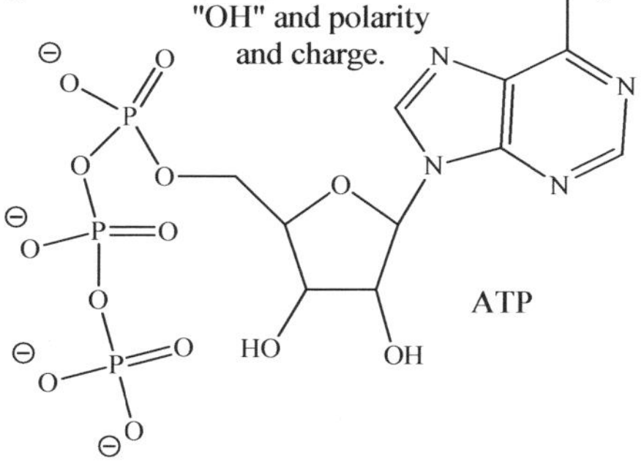

e vitamin C (ascorbic acid)

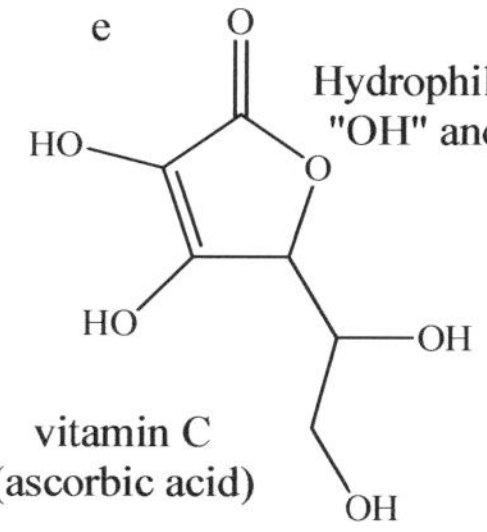

Hydrophilic - lots of "OH" and polarity.

f vitamin D2

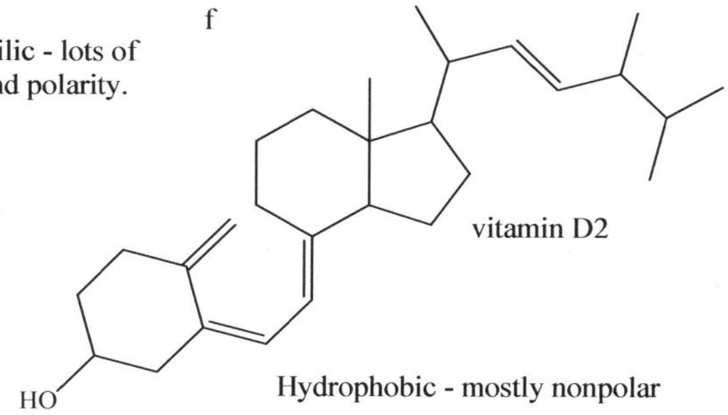

Hydrophobic - mostly nonpolar

g vitamin E (α-tocopherol)

Hydrophobic - mostly nonpolar

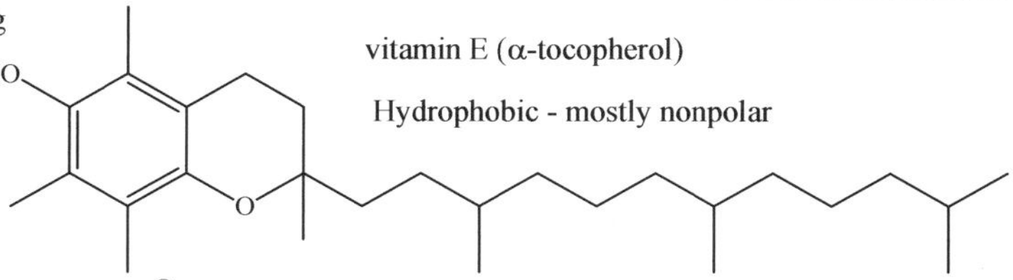

h 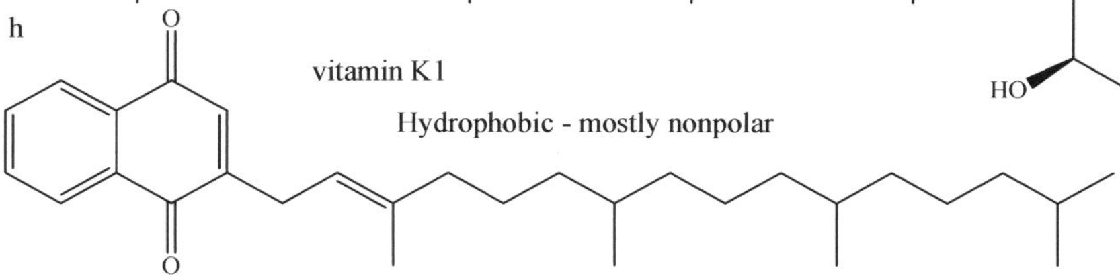 vitamin K1

Hydrophobic - mostly nonpolar

i cholesterol

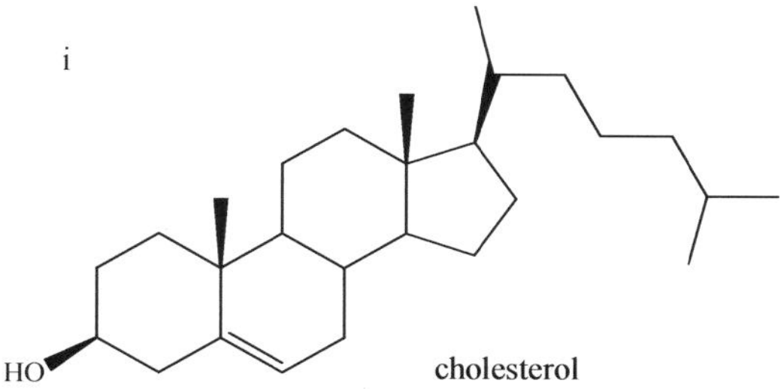

Hydrophobic - mostly nonpolar

Problem 13 (p 69) – a. Carbohydrates are very water soluble and fats do not mix well with water. Below, glucose is shown below as a typical hydrophilic carbohydrate, and a triglyceride is used as a typical hydrophobic fat. Point out why each is classified in the manner indicated.

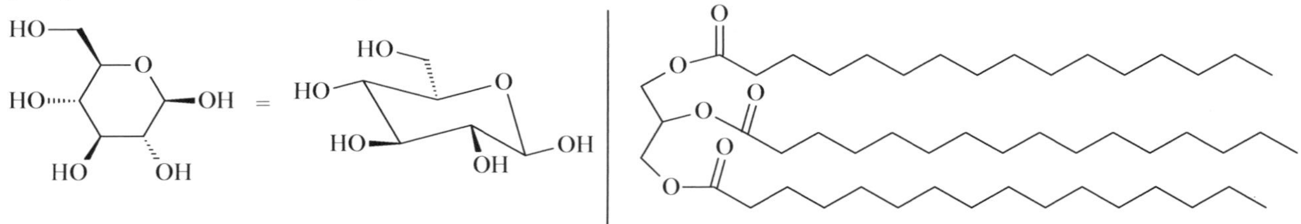

glucose (carbohydrate)

Hydrophilic - lots of "OH" and polarity.

typical saturated triglyceride (fat)

Hydrophobic = lipophilic - mostly nonpolar

b. All of the "OH" groups in glucose can be methylated. What do you think this will do to the solubility of glucose? Why? One of these structures is soluble in carbon tetrachloride the other one is not. Which one is it and why?

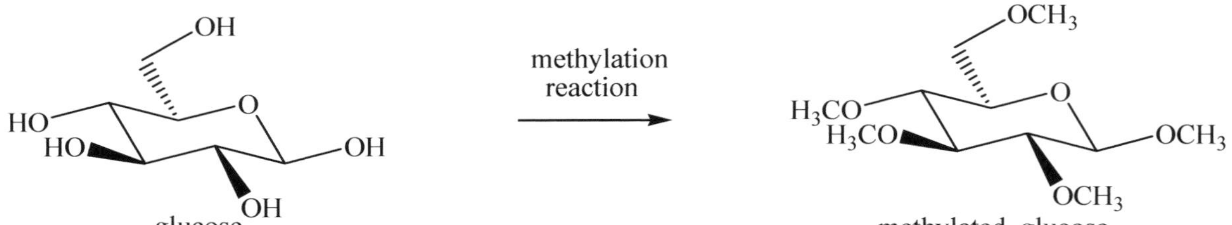

glucose → (methylation reaction) → methylated glucose

The methylation reaction converts the alcohol groups (OH) to ether groups (OCH_3) and the ability to hydrogen bond goes down, as does water solubility. However, the ability to dissolve in nonpolar carbon tetrachloride goes up, so B will be much more soluble in CCl_4.

Problem 14 (p 69) – Bile salts are released from your gall bladder when hydrophobic fats are eaten to allow your body to solubilize the fats, so that they can be absorbed and transported in the aqueous blood. The major bile salt glycolate, shown below, is synthesized from cholesterol. Explain the features of glycolate that makes it a good compromise structure that can mix with both the fat and aqueous blood. Use the 'rough' 3D drawings below to help your reasoning, or better yet, build models to see the structures for yourself (though it's a lot of work).

HO H H H cholesterol

synthesized in many,many steps in the body

HO H H H OH OH O N H O O glycolate (bile salt)

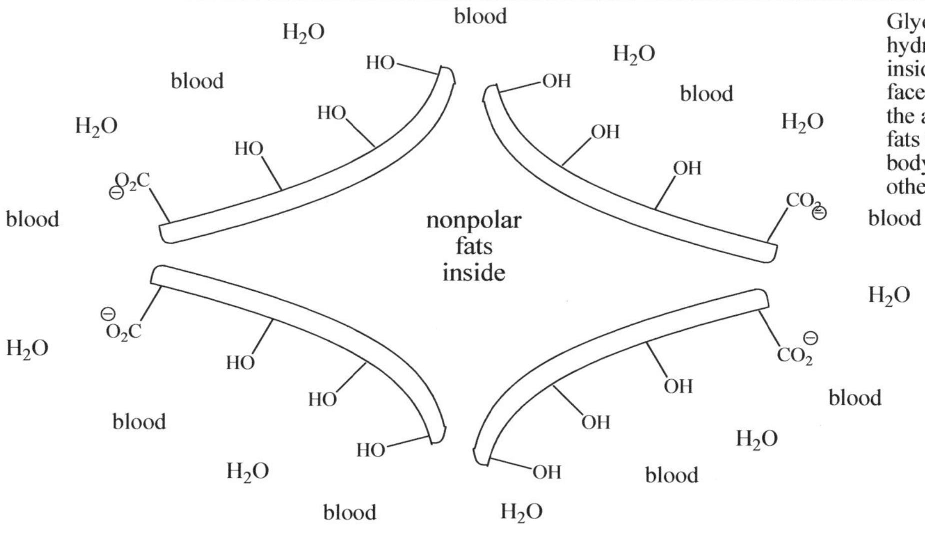

Glycolate has a nonpolar, hydrophobic face that can cover the inside of a fat ball and a hydorphilic face that can point outward toward the aqueous blood, which allows fats to be transported throughout the body to reach fat storage cells and other essential locations.

Chapter 3

Problem 1 (p 81) - Generate the 18 possible structural isomers of C_8H_{18}. (We'll save $C_{40}H_{82}$ for another lifetime. If you generated one isomer per second, it would take you about 2,000,000 years. We better make that several lifetimes.)

1. octane
2. 2-methylheptane
3. 3-methylheptane
4. 4-methylheptane
5. 2,2-dimethylhexane
6. 2,3-dimethylhexane
7. 2,4-dimethylhexane
8. 2,5-dimethylhexane
9. 3,3-dimethylhexane
10. 3,4-dimethylhexane
11. 3-ethylhexane
12. 2,2,3-trimethylpentane
13. 2,2,4-trimethylpentane
14. 2,3,3-trimethylpentane
15. 2,3,4-trimethylpentane
16. 2-methyl-3-ethylpentane
17. 3-methyl-3-ethylpentane
18. 2,2,3,3-tetramethylbutane

Problem 2 (p 82) – Draw all of the possible isomers of C_4H_9Br. Hint: There should be four. If you feel ambitious, try and draw all of the possible isomers of $C_6H_{13}F$. There should be about 17 of them.

1. 1-bromobutane
2. 2-bromobutane
3. 1-bromo-2-methylpropane
4. 2-bromo-2-methylpropane

1. 1-fluorohexane
2. 2-fluorohexane
3. 3-fluorohexane
4. 1-fluoro-2-methylpentane
5. 2-fluoro-2-methylpentane
6. 3-fluoro-2-methylpentane
7. 2-fluoro-4-methylpentane

8 1-fluoro-4-methylpentane

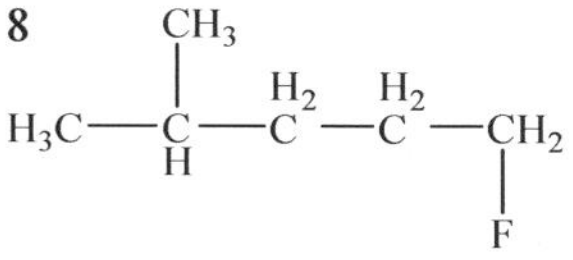

9 1-fluoro-3-methylpentane

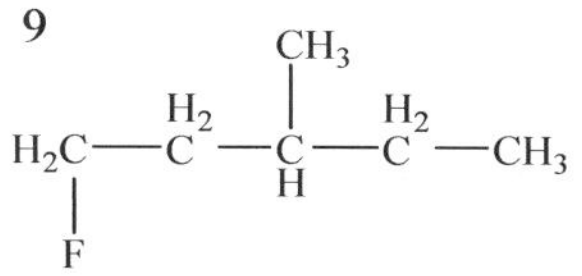

10 2-fluoro-3-methylpentane

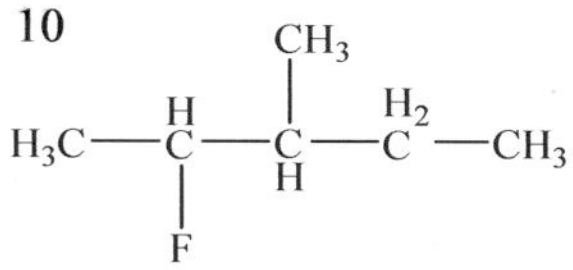

11 3-fluoro-3-methylpentane

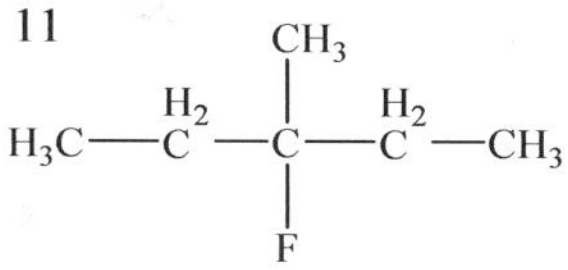

12 1-fluoro-2-ethylbutane

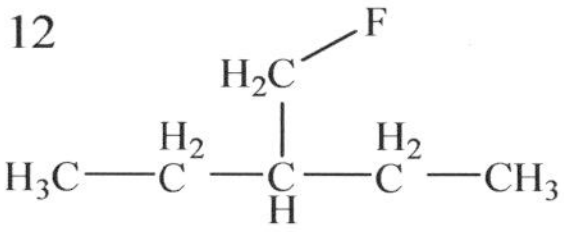

13 1-fluoro-3-ethylpentane

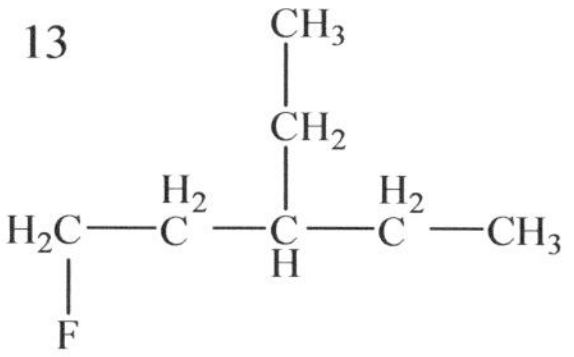

14 2-fluoro-3-ethylpentane

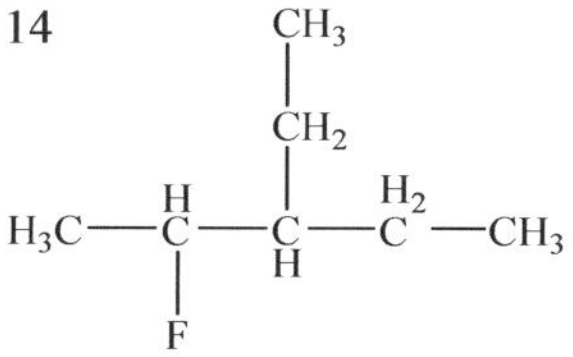

15 3-fluoro-3-ethylpentane

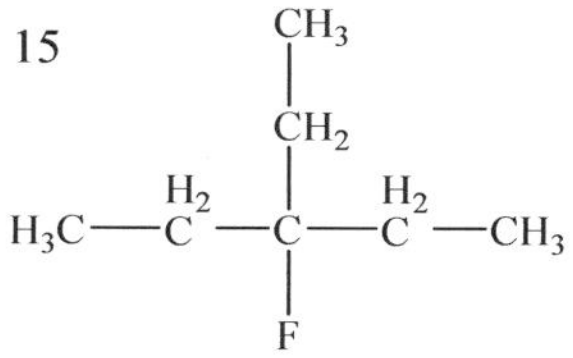

16 1-fluoro-2,2,3-trimethylbutane

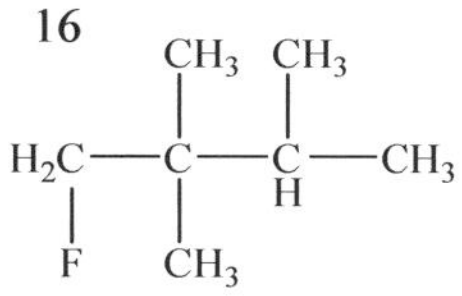

17 2-fluoro-2,3,3-trimethylbutane

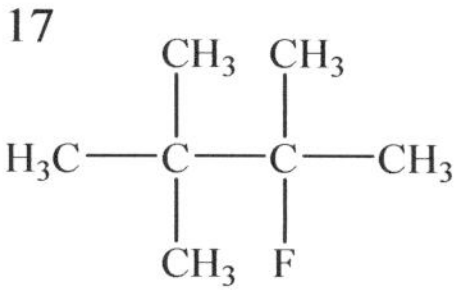

18 1-fluoro-2,3,3-trimethylbutane

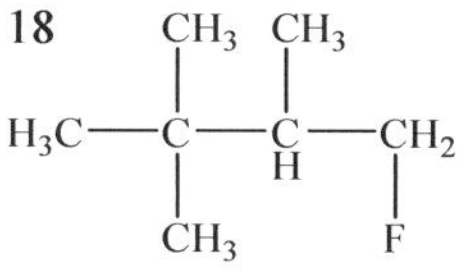

Problem 3 (p 84) – Draw all of the possible isomers of $C_4H_{10}O$. Hint: There should be seven. If you feel ambitious, try and draw all of the possible isomers of $C_6H_{14}O$. There should be about 32 of them (17 alcohols and 15 ethers, if I counted correctly).

alcohols

1 butan-1-ol

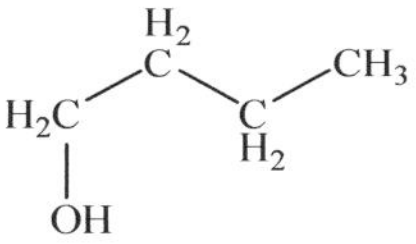

2 butan-2-ol

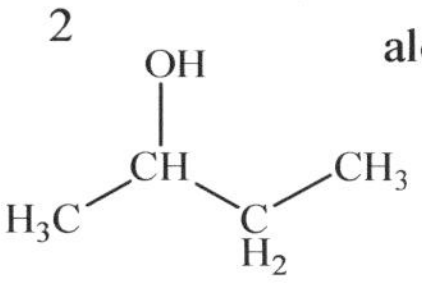

3 2-methylpropan-1-ol

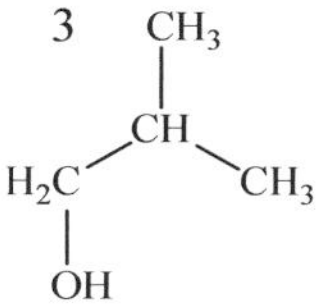

4 2-methylpropan-2-ol

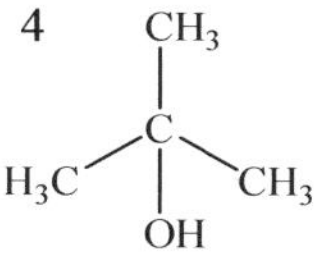

ethers

5 1-methoxypropane

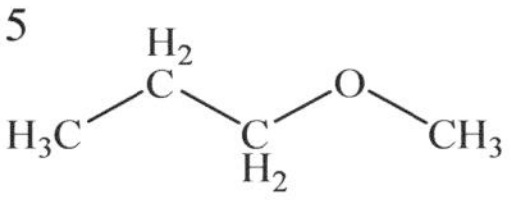

6 2-methoxypropane

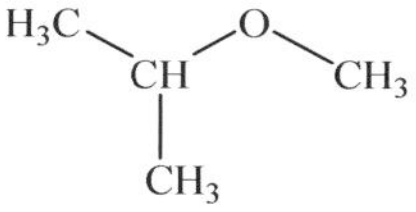

7 ethoxyethane

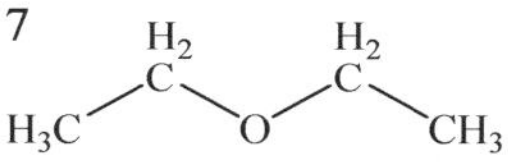

Alcohols

1 hexan-1-ol

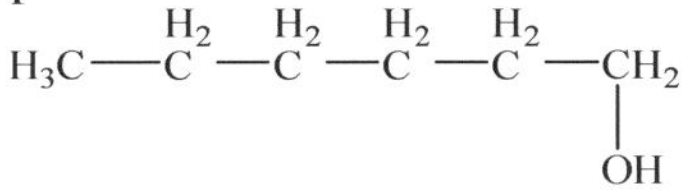

2 hexan-2-ol

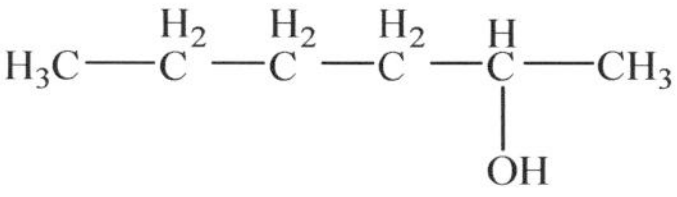

3 hexan-3-ol

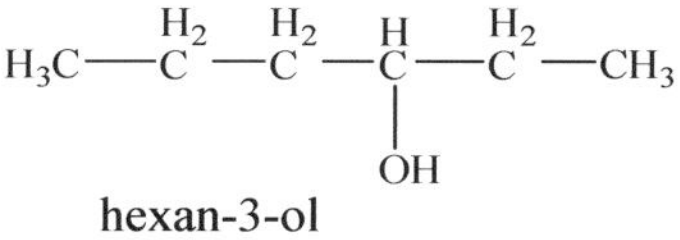

4 2-methylpentan-1-ol

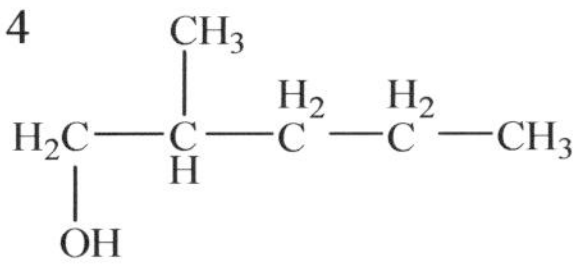

5 2-methylpentan-2-ol

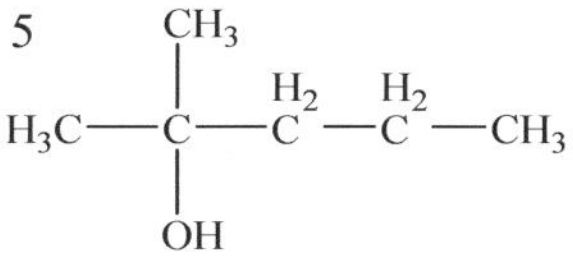

6 2-methylpentan-3-ol

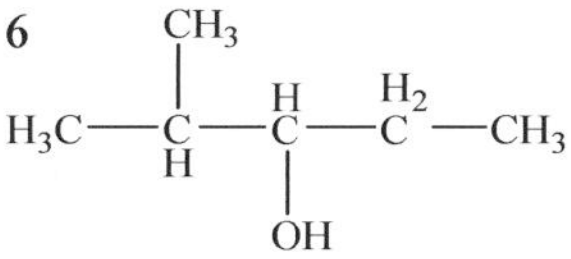

7 4-methylpentan-2-ol

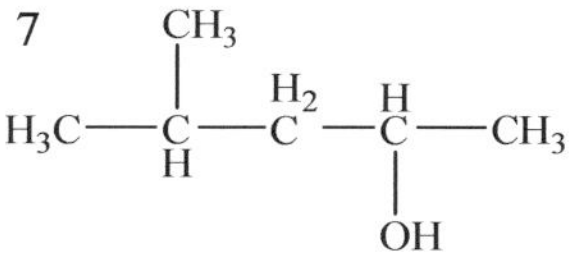

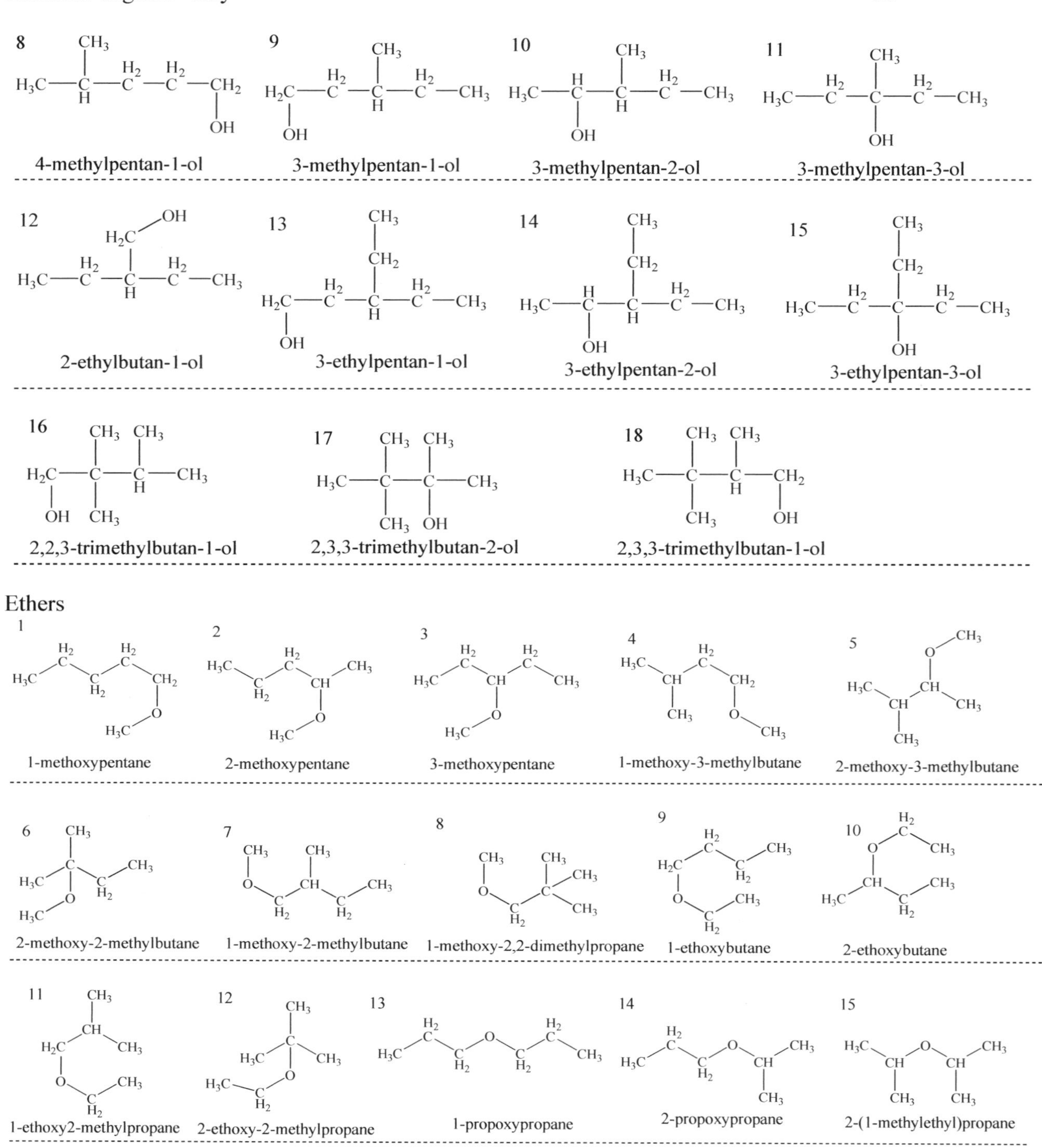

Problem 4 (p 85) – Draw all of the possible isomers of $C_4H_{11}N$. Hint: There should be eight. If you feel ambitious, try and draw all of the possible isomers of $C_6H_{15}N$. There should be about 39 of them. (Yikes!)

1. $H_3C{-}CH_2{-}CH_2{-}CH_2{-}NH_2$ — 1° amine — butan-1-amine
2. $CH_3CH(NH_2)CH_2CH_3$ — 1° amine — butan-2-amine
3. $(CH_3)_2CHCH_2NH_2$ — 1° amine — 2-methylpropan-1-amine
4. $(CH_3)_3C{-}NH_2$ — 1° amine — 2-methylpropan-2-amine
5. $CH_3CH_2CH_2{-}NH{-}CH_3$ — 2° amine — N-methylpropan-1-amine

6 — 2° amine — N-methylpropan-2-amine

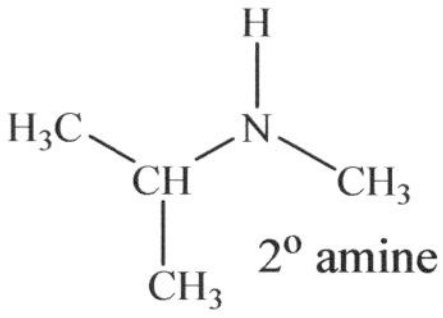

7 — 2° amine — N-ethylethanamine

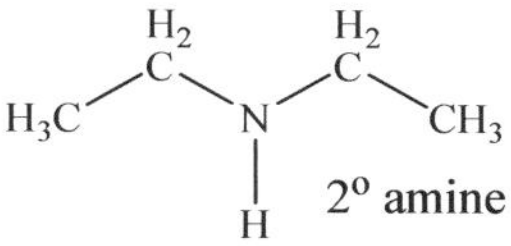

8 — 3° amine — N,N-dimethylethanamine

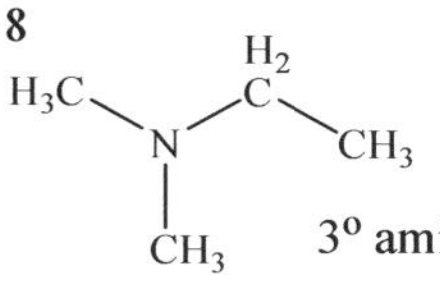

---

1 — 1° amine — hexan-1-amine

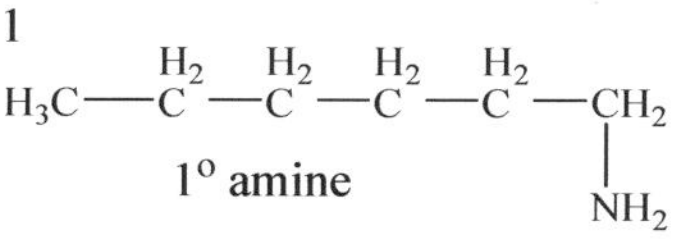

2 — 1° amine — hexan-2-amine

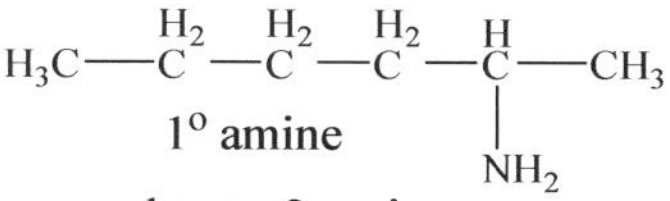

3 — 1° amine — hexan-3-amine

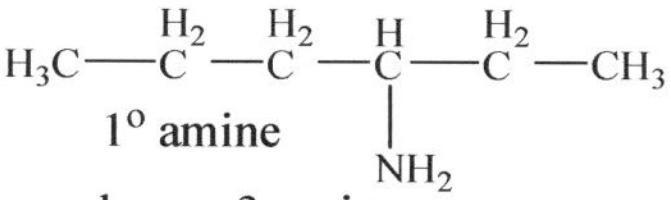

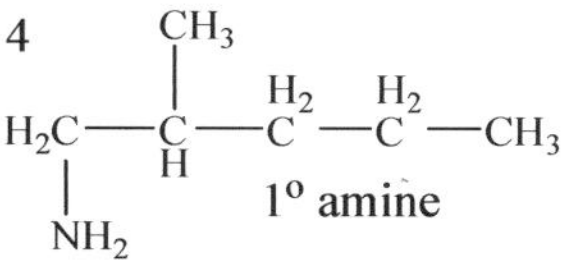

---

4 — 1° amine — 2-methylpentan-1-amine

5 — 1° amine — 2-methylpentan-2-amine

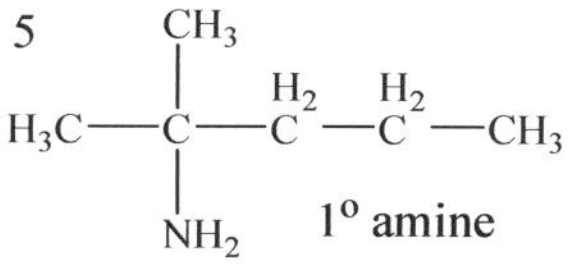

6 — 1° amine — 2-methylpentan-3-amine

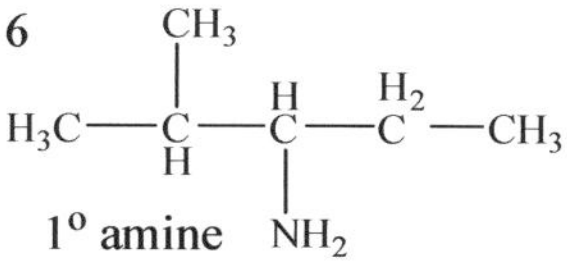

7 — 1° amine — 4-methylpentan-2-amine

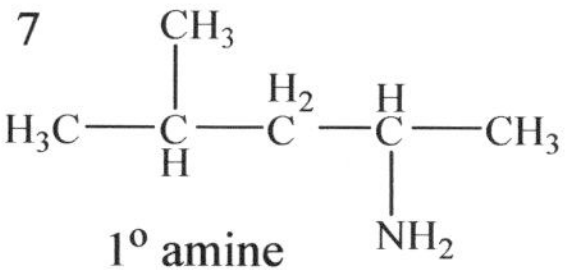

---

8 — 1° amine — 4-methylpentan-1-amine

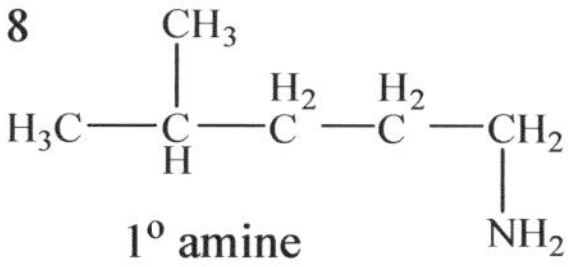

9 — 1° amine — 3-methylpentan-1-amine

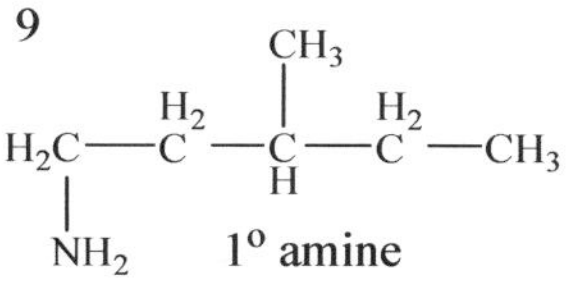

10 — 1° amine — 3-methylpentan-2-amine

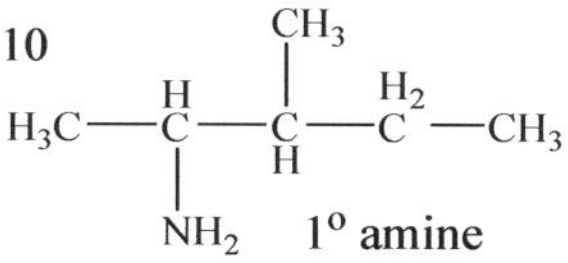

11 — 1° amine — 3-methylpentan-3-amine

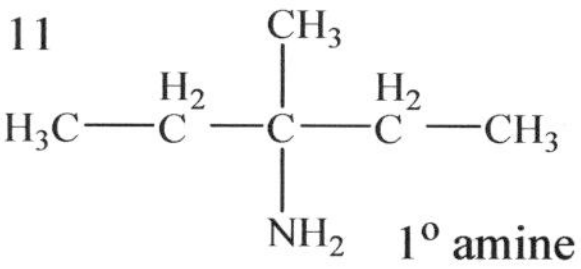

---

12 — 1° amine — 2-ethylbutan-1-amine

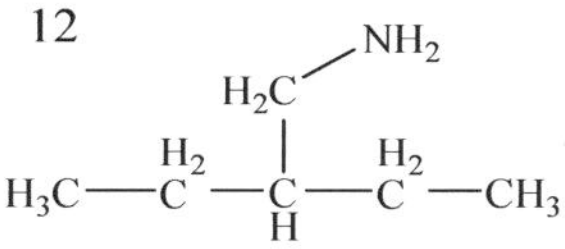

13 — 1° amine — 3-ethylpentan-1-amine

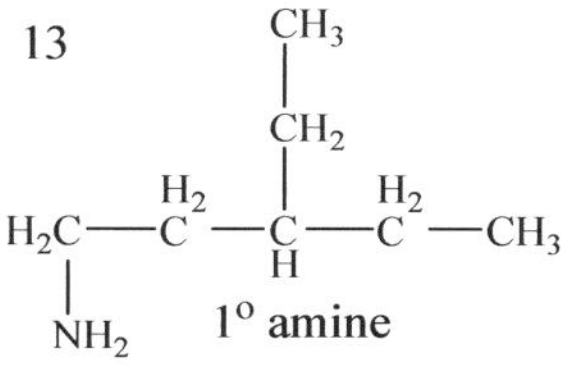

14 — 1° amine — 3-ethylpentan-2-amine

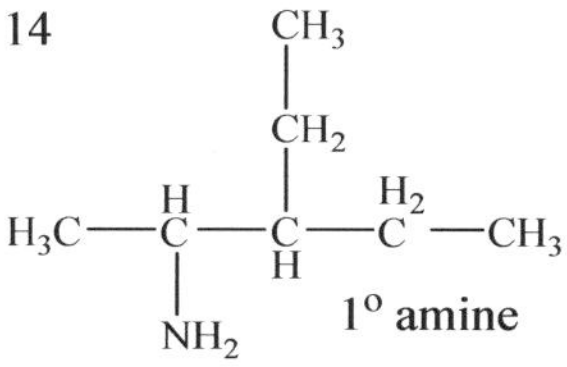

15 — 1° amine — 3-ethylpentan-3-amine

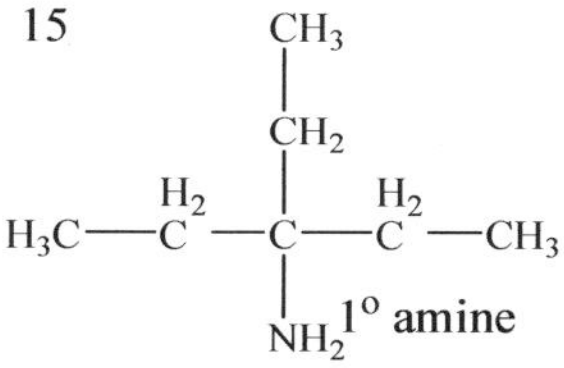

---

16 — 1° amine — 2,2,3-trimethylbutan-1-amine

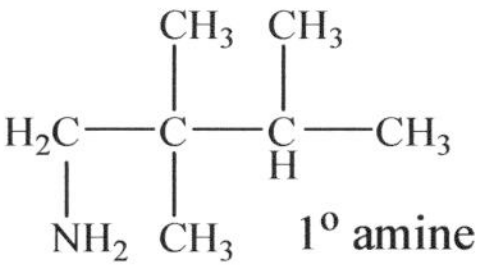

17 — 1° amine — 2,3,3-trimethylbutan-2-amine

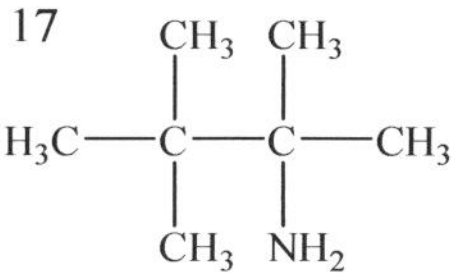

18 — 1° amine — 2,3,3-trimethylbutan-1-amine

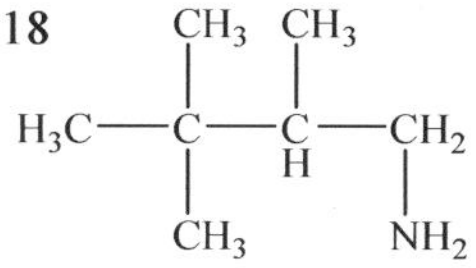

19 — 2° amine — N-methylpentan-1-amine

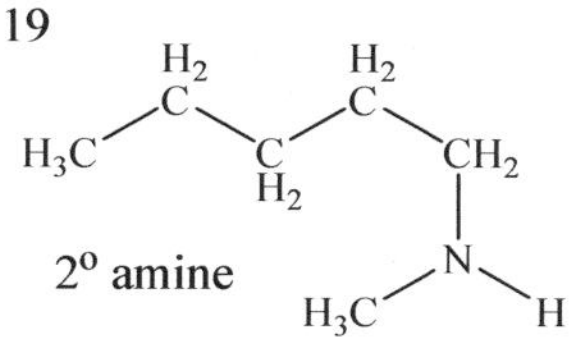

---

20 — 2° amine — N-methylpentan-1-amine

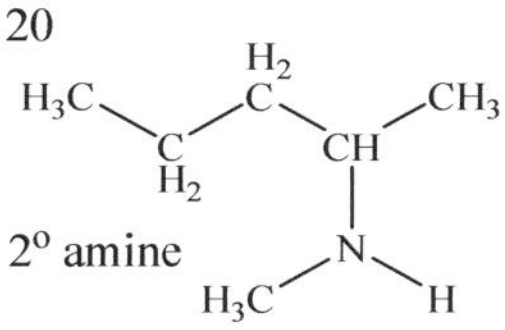

21 — 2° amine — N-methylpentan-3-amine

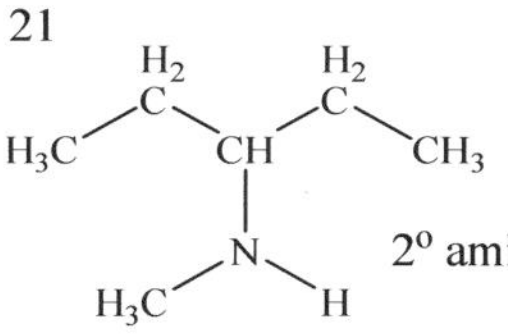

22 — 2° amine — N-methyl-3-methylbutan-1-amine

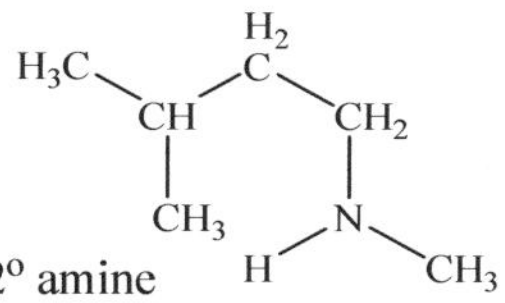

23 — 2° amine — N-methyl-3-methylbutan-2-amine

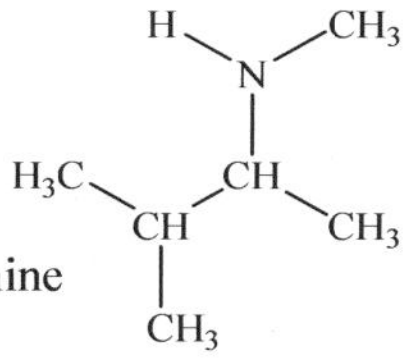

24 — 2° amine — N,2-dimethylbutan-2-amine

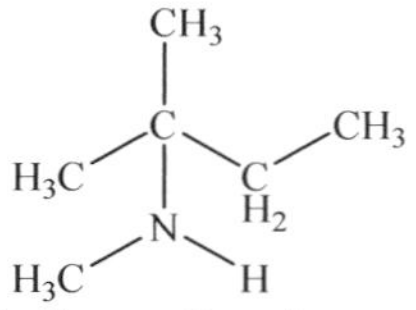

25 — 2° amine — N,2-dimethylbutan-1-amine

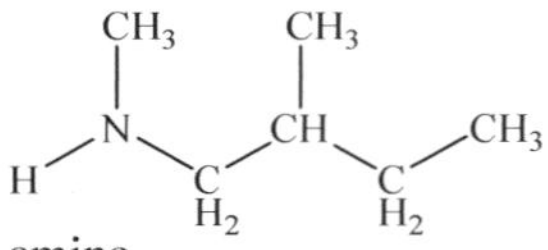

26 — 2° amine — N,2,2-trimethylpropan-1-amine

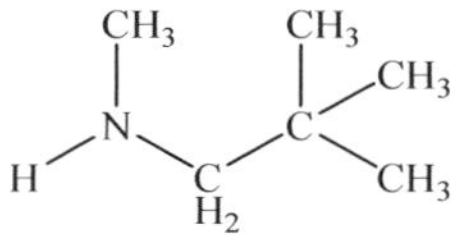

27 — 2° amine — N-ethylbutan-1-amine

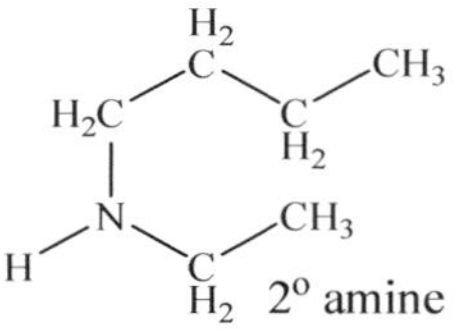

---

28 — 2° amine — N-ethylbutan-2-amine

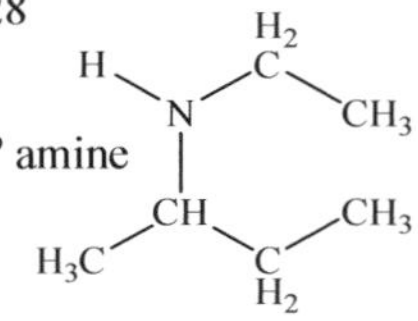

29 — 2° amine — N-ethyl-2-methylpropan-1-amine

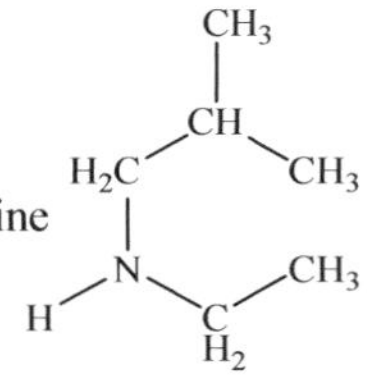

30 — 2° amine — N-ethyl-2-methylpropan-2-amine

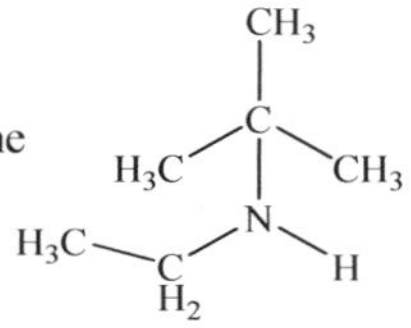

31 — 2° amine — N-propylpropan-1-amine

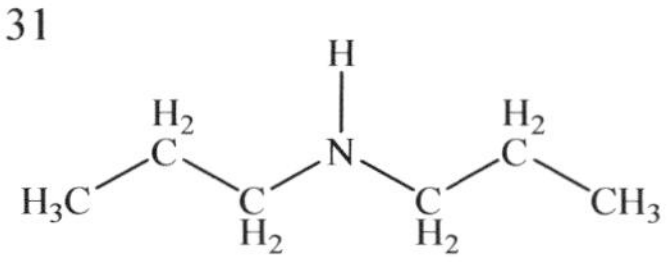

---

32 — 2° amine — N-(1-methylethyl)propan-1-amine

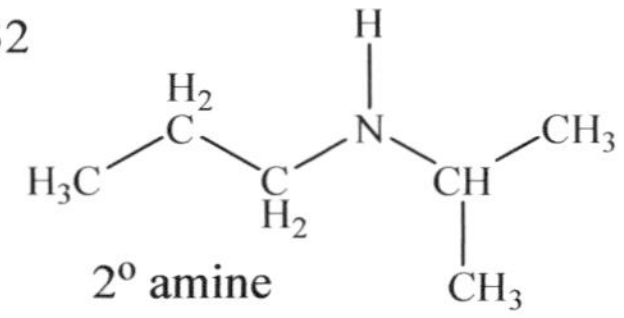

33 — 2° amine — N-(1-methylethyl)-1-methylethan-1-amine

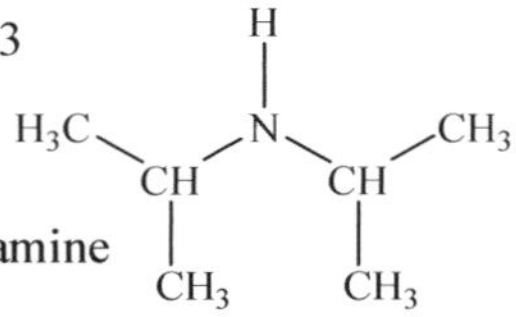

34 — 3° amine — N,N-dimethylbutan-1-amine

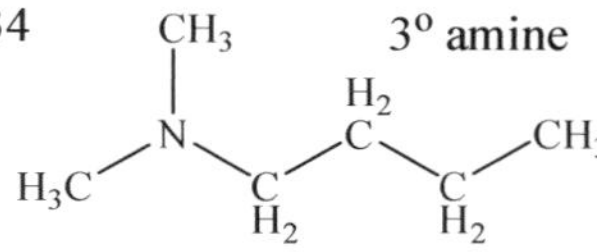

---

35 — 3° amine — N,N-dimethylbutan-2-amine

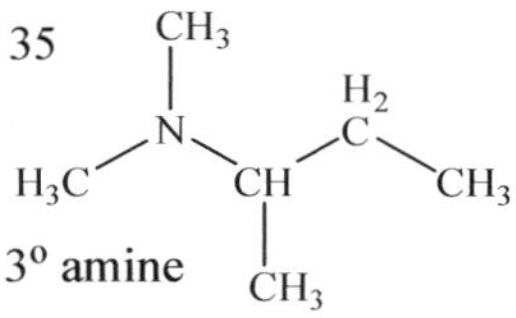

36 — 3° amine — N,N,2-trimethylpropan-1-amine

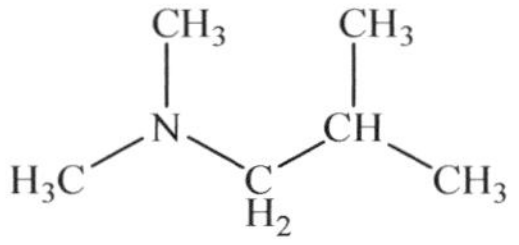

37 — 3° amine — N,N,1,1-tetramethylethan-1-amine

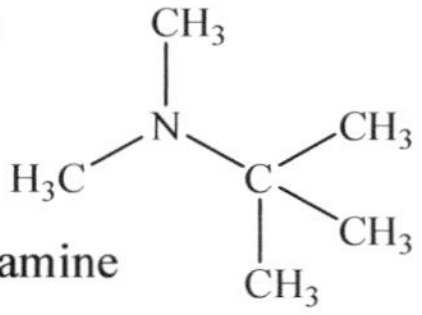

---

38 — 3° amine — N-ethyl-N-methylpropan-1-amine

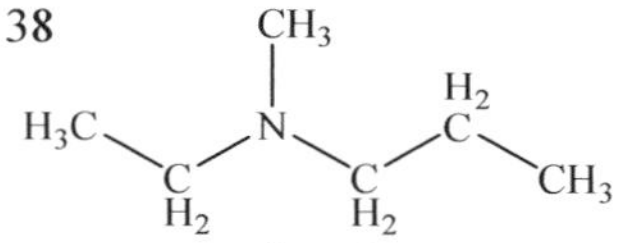

39 — 3° amine — N-ethyl-N-methylpropan-2-amine

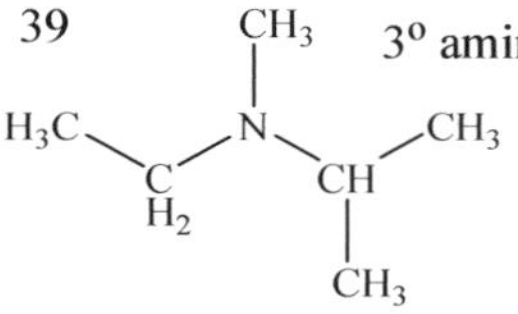

40 — 3° amine — triethylamine

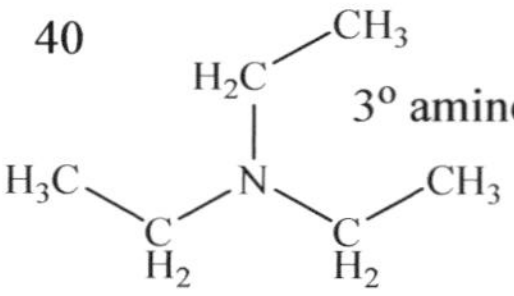

Problem 5 (p 88) – Calculate the degree of unsaturation in the hydrocarbon formulas below. Draw one possible structure.

a.

C_7H_8 $2(7) + 2 = 16$

$- 8$

$8 \div 2 = 4$ degrees of unsaturation

possible structure

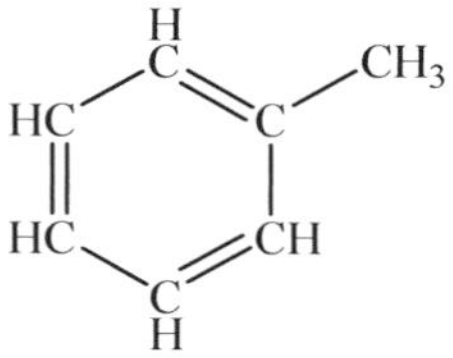

b

$C_{10}H_8$ $2(10) + 2 = 22$

$- 8$

$14 \div 2 = 7$ degrees of unsaturation

possible structure

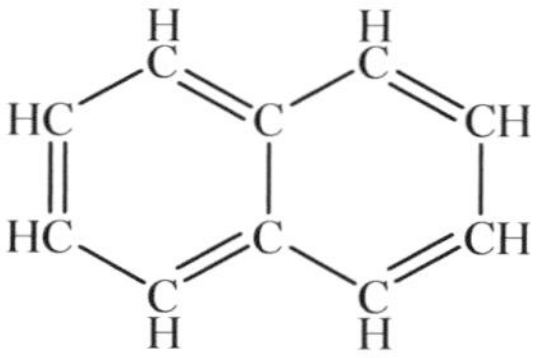

c.
C_8H_{10} $\quad 2(8) + 2 = 18$
$\quad -10$
$8 \div 2 = 4$ degrees of unsaturation

possible structure

---

d.
$C_{12}H_{26}$ $\quad 2(12) + 2 = 26$
$\quad -26$
$0 \div 2 = 0$ degrees of unsaturation

possible structure

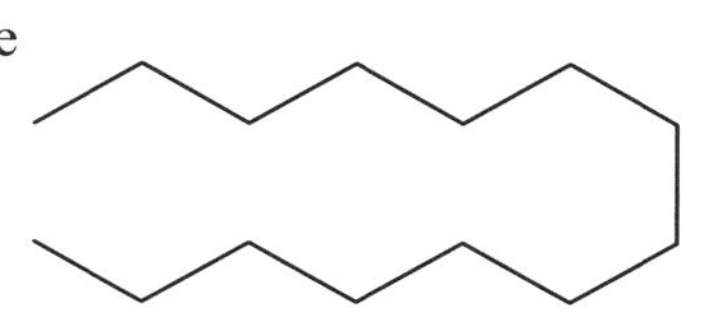

---

e.
C_6H_{12} $\quad 2(6) + 2 = 14$
$\quad -12$
$2 \div 2 = 1$ degrees of unsaturation

possible structures

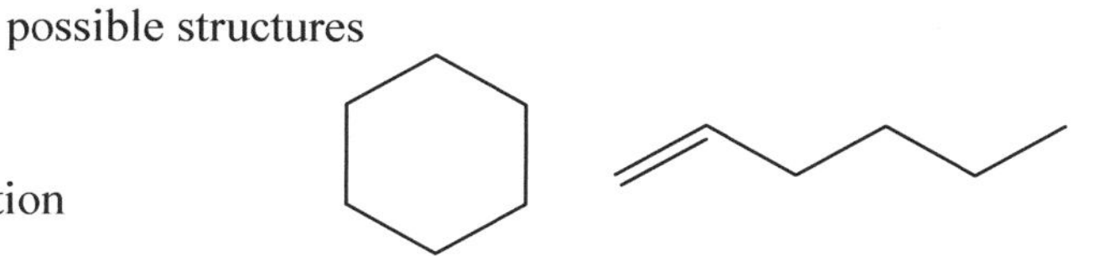

---

f.
C_6H_{10} $\quad 2(6) + 2 = 14$
$\quad -10$
$4 \div 2 = 2$ degrees of unsaturation

possible structures

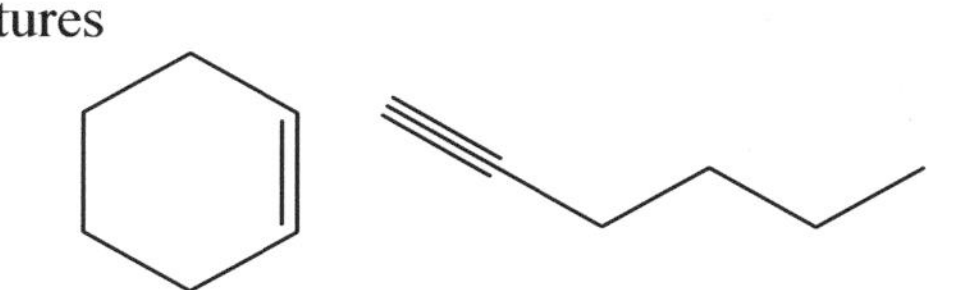

---

g.
C_6H_8 $\quad 2(6) + 2 = 14$
$\quad -8$
$6 \div 2 = 3$ degrees of unsaturation

possible structures

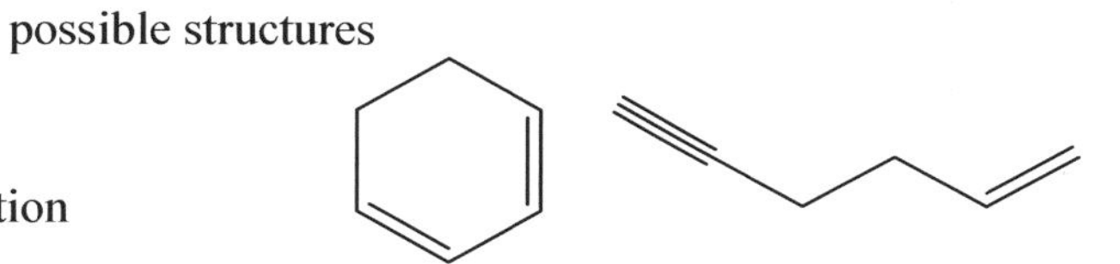

---

h.
C_6H_6 $\quad 2(6) + 2 = 14$
$\quad -6$
$8 \div 2 = 4$ degrees of unsaturation

possible structures

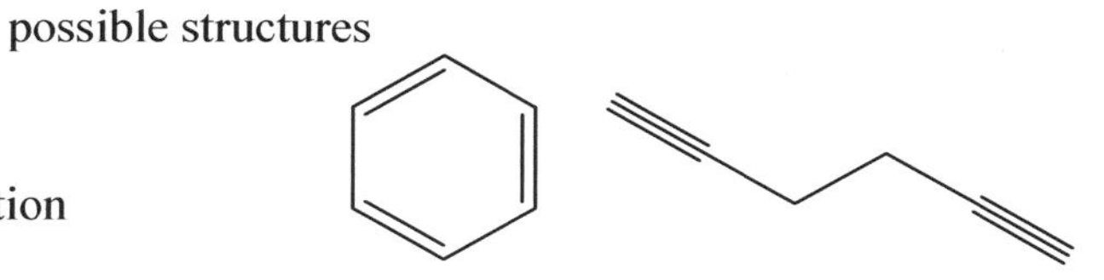

---

i.
C_6H_4 $\quad 2(6) + 2 = 14$
$\quad -4$
$10 \div 2 = 5$ degrees of unsaturation

possible structures

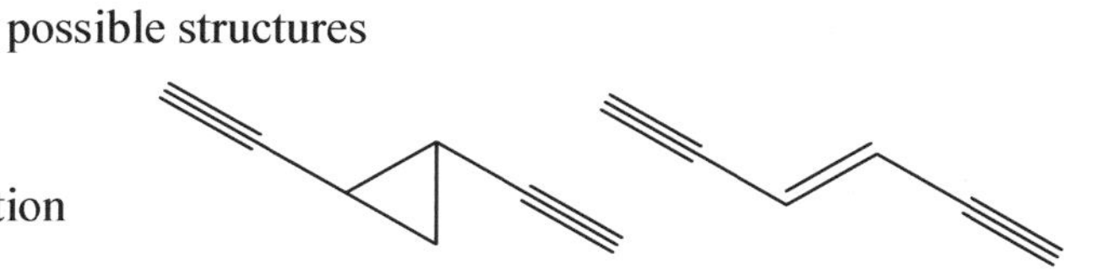

---

j.
C_6H_2 $\quad 2(6) + 2 = 14$
$\quad -2$
$12 \div 2 = 6$ degrees of unsaturation

possible structures

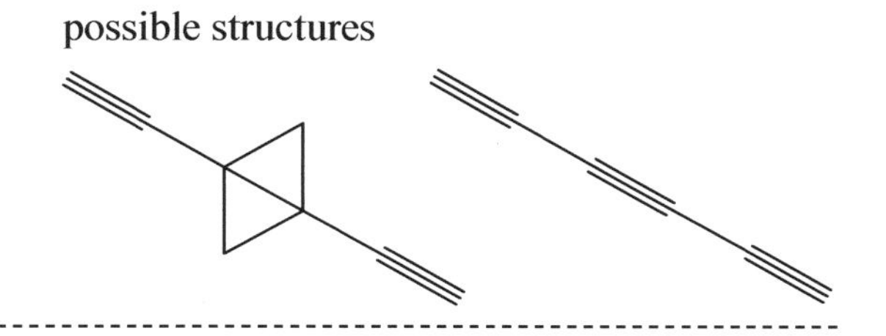

Problem 6 (p 88) – Draw all of the alkene and cycloalkane isomers of C_5H_{10}. I calculate that there should be six alkenes and six cycloalkanes. Start with the three alkane skeletal isomers.

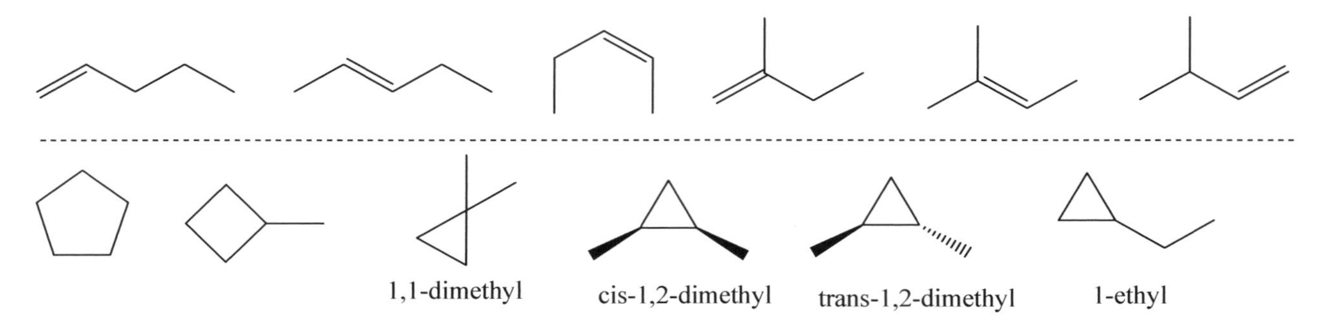

Problem 7 (p 89) – If you feel daring, try and draw some of the isomers of C_5H_8. I found four dienes (2 pi), three alkynes (2 pi), three allenes (2 pi), ten ring and pi bond isomers and five isomers with two rings. It's quite possible that I missed some. Why not shoot for two of each possibility.

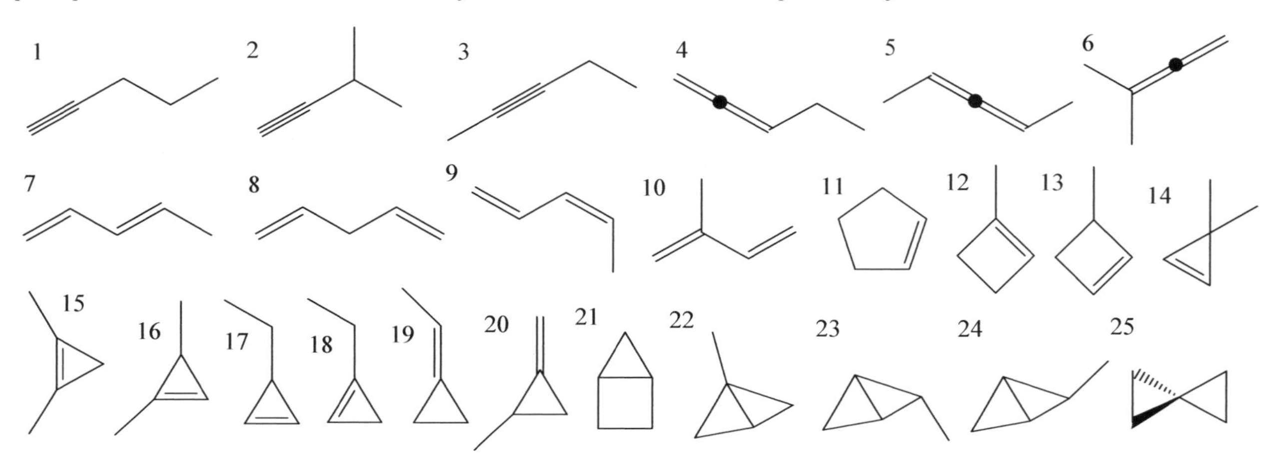

Problem 8 (p 90) – Calculate the degree of unsaturation in the formulas below. Draw one possible structure for each formula.

a.

C_7H_7FBrCl

$$2(7) + 2 = 16$$
$$-10$$
$$6 \div 2 = 3 \text{ degrees of unsaturation}$$

possible structure

Cl

Br F

b.

$C_{10}H_6F_2$

$$2(10) + 2 = 22$$
$$-\ 8$$
$$14 \div 2 = 7 \text{ degrees of unsaturation}$$

possible structure

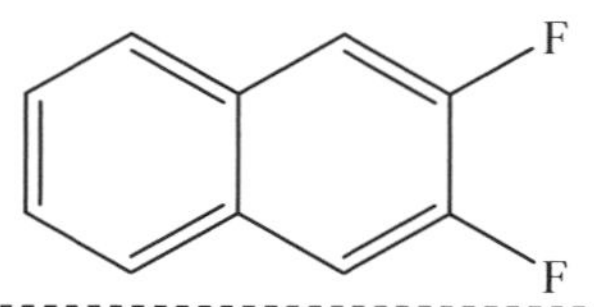

c.

C_5H_4FCl

$$2(5) + 2 = 12$$
$$-6$$
$$6 \div 2 = 3 \text{ degrees of unsaturation}$$

possible structure

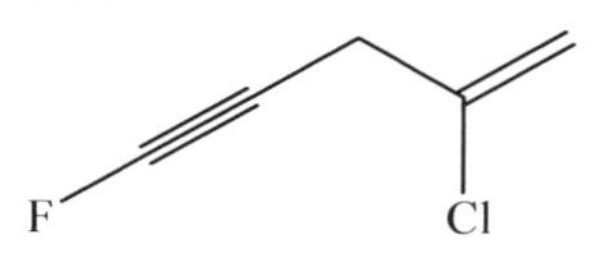

d.

C_6Br_6

possible structure

$2(6) + 2 = 14$
-6
$8 \div 2 = 4$ degrees of unsaturation

---

e.

$C_6H_{12}I_2$

possible structure

$2(6) + 2 = 14$
-14
$0 \div 2 = 0$ degrees of unsaturation

Problem 9 (p 88) – Calculate the degree of unsaturation in the formulas below. Draw one possible structure for each formula.

a.

C_2H_6O

possible structures

$2(2) + 2 = 6$
-6
$0 \div 2 = 0$ degrees of unsaturation

---

b.

C_3H_6O

possible structures

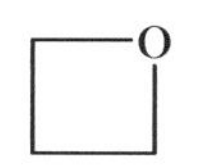

$2(3) + 2 = 8$
-6
$2 \div 2 = 1$ degree of unsaturation

---

c.

C_4H_5ClO

possible structures

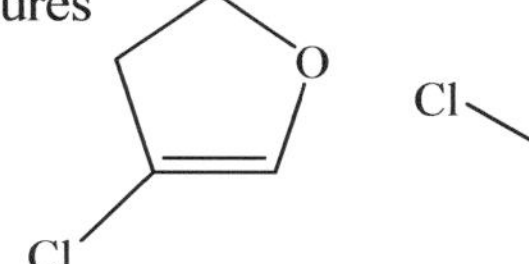

$2(4) + 2 = 10$
-6
$4 \div 2 = 2$ degrees of unsaturation

---

d.

$C_6H_4F_2O_2$

possible structure

$2(6) + 2 = 14$
-6
$8 \div 2 = 4$ degrees of unsaturation

---

e.

$C_6H_{12}O_6$

possible structures

$2(6) + 2 = 14$
-12
$2 \div 2 = 1$ degree of unsaturation

Problem 10 (p 91) – Calculate the degree of unsaturation in the formulas below. Draw one possible structure for each formula.

a.

C_2H_3N possible structure

$2(2) + 2 + 1 = 7$
$- 3$
$4 \div 2 = 2$ degrees of unsaturation

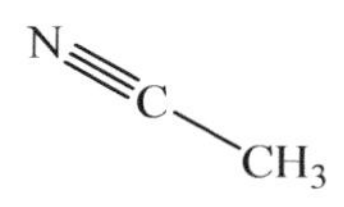

---

b.

C_5H_5N possible structure

$2(5) + 2 + 1 = 13$
$- 5$
$8 \div 2 = 4$ degrees of unsaturation

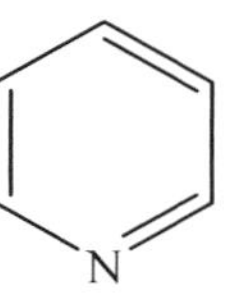

---

c.

$C_4H_4N_2$ possible structure

$2(4) + 2 + 2 = 12$
$- 4$
$8 \div 2 = 4$ degrees of unsaturation

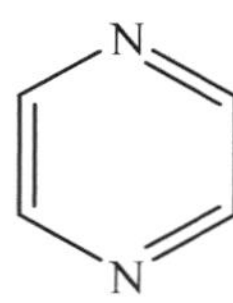

---

d.

$C_6H_6F_2N_2O$ possible structure

$2(6) + 2 + 2 = 16$
$- 8$
$8 \div 2 = 4$ degrees of unsaturation

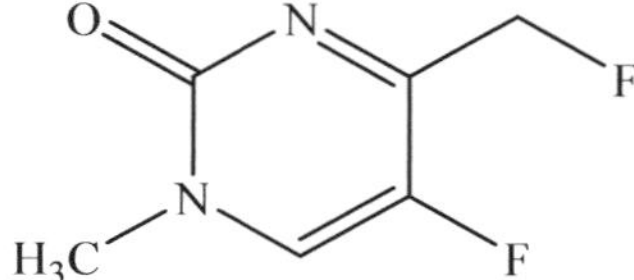

---

e.

$C_6H_{13}NO_2$ possible structure

$2(6) + 2 + 1 = 15$
$- 13$
$2 \div 2 = 1$ degrees of unsaturation

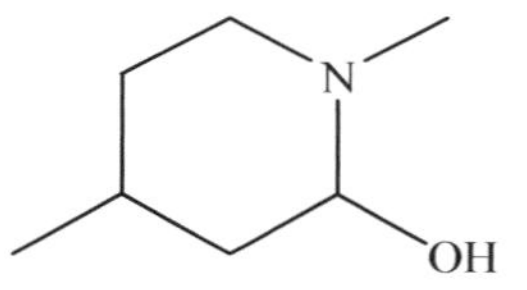

Problem 11 (p 91) – The following functional groups are unsaturated. Draw as many structures as you can for each of the examples below. Consider the alkane skeletons with the same number of carbon atoms as your starting points.

alkene isomers of C_5H_{10}, at least six, don't forget cis and tran possibilities

$2(5) + 2 = 12$
$- 10$
$2 \div 2 = 1$ degree of unsaturation

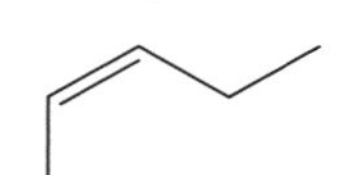
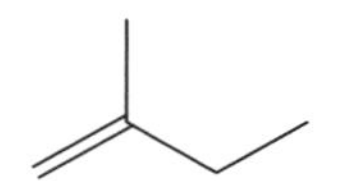
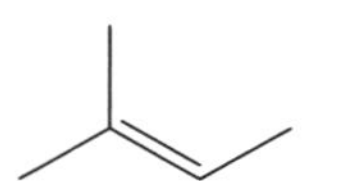
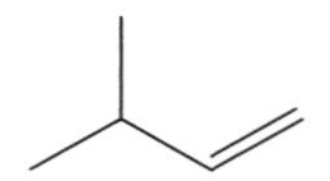

---

alkyne isomers of C_5H_8, at least three

$2(5) + 2 = 12$
$- 8$
$4 \div 2 = 2$ degrees of unsaturation

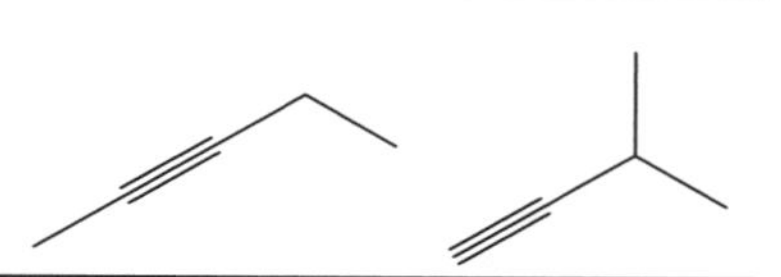

---

acid chloride isomers of C_5H_9OCl, at least four

$2(5) + 2 = 12$
$- 10$
$2 \div 2 = 1$ degree of unsaturation

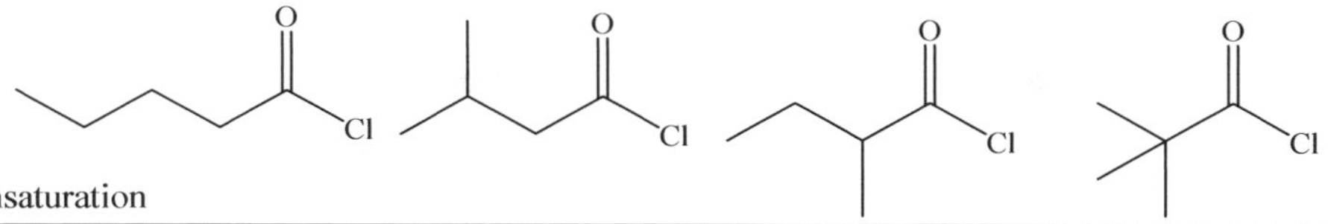

nitrile isomers of C_5H_9N, at least four

$2(5) + 2 + 1 = 13$
$- 9$
$4 \div 2 = 2$ degree of unsaturation

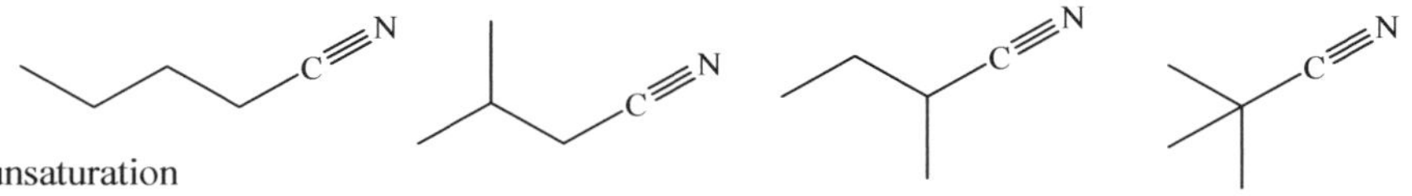

aldehyde and ketone isomers of $C_5H_{10}O$, at least seven

$2(5) + 2 = 12$
$- 10$
$2 \div 2 = 1$ degree of unsaturation

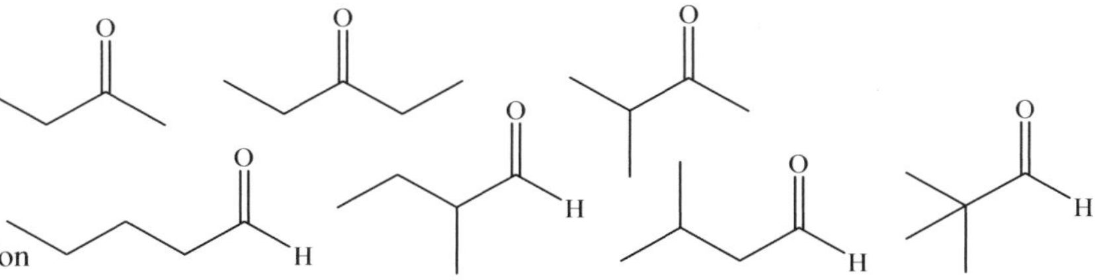

carboxylic acid and ester isomers of $C_5H_{10}O_2$, at least 13

$2(5) + 2 = 12$
$- 10$
$2 \div 2 = 1$ degree of unsaturation

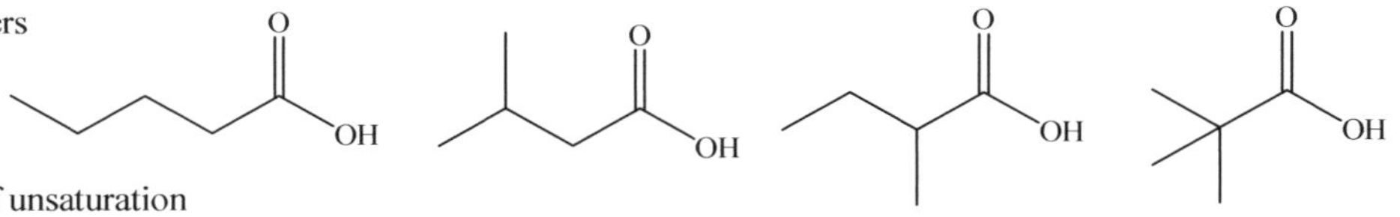

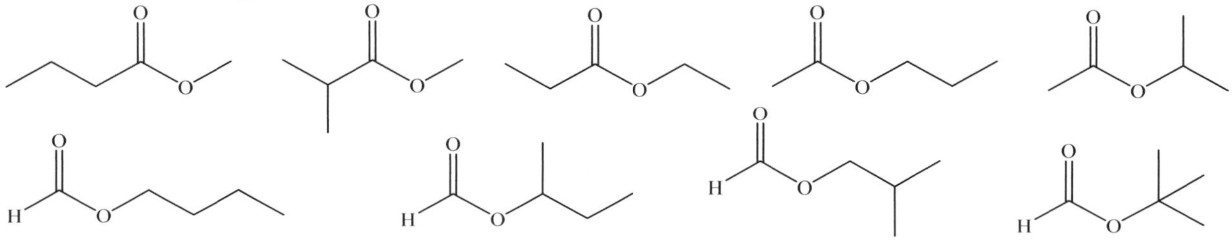

primary, secondary and tertiary amide isomers of $C_5H_{11}NO$, at least 17

$2(5) + 2 + 1 = 13$
$- 11$
$2 \div 2 = 1$ degree of unsaturation

1° amide 1° amide 1° amide 1° amide

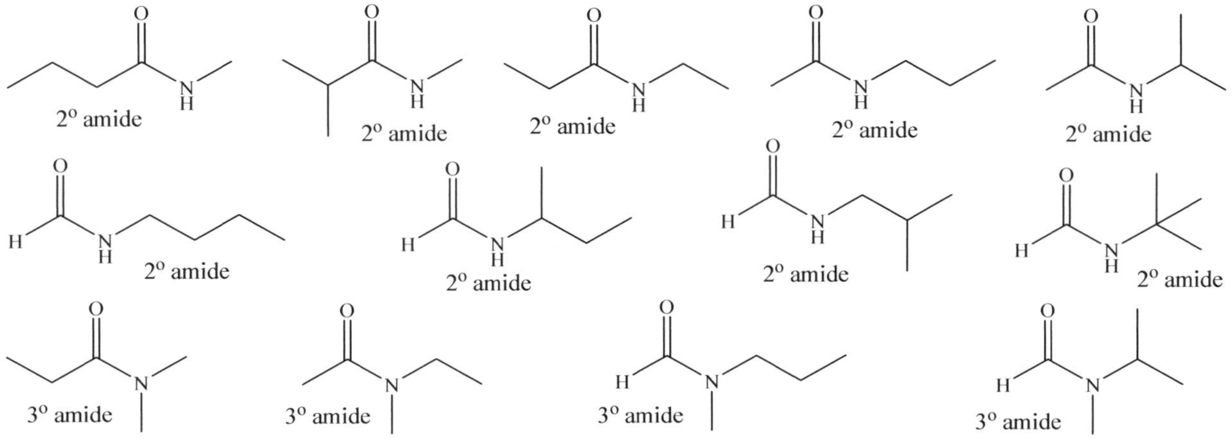

2° amide 2° amide 2° amide 2° amide 2° amide

2° amide 2° amide 2° amide 2° amide

3° amide 3° amide 3° amide 3° amide

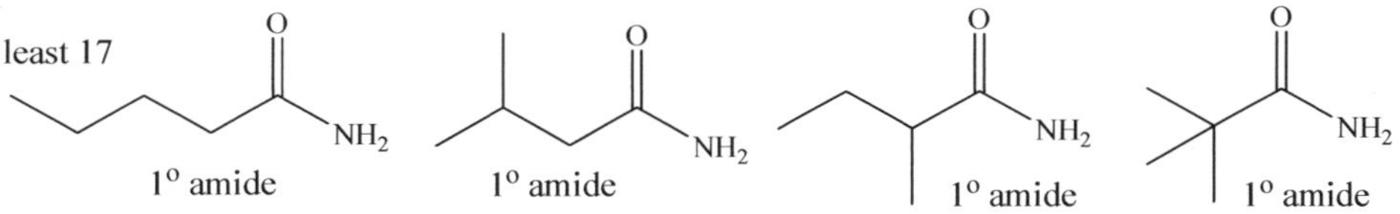

Problem 12 (p 91) – Draw several isomers of molecular formula = CH_3NO_2? (There are many.)

CH_3NO_2 $\quad 2(1) + 2 + 1 = 5$

$\quad - 3$

$\quad 2 \div 2 = 1$ degree of unsaturation

several possible structures

O, H, C, N, O, H, H (H–C(=O)–N(H)–O–H)

H–O–C(H)=N–O–H

H–O–C(H)(H)–N=O

H–O–C(=O)–N(H)–H

H–C(=O)–O–N(H)–H

O—N–H / H₂C—O (four-membered ring)

O—N–H / O—CH₂ (four-membered ring)

O / O—CH–NH₂ (three-membered ring)

O / HN—CH–OH (three-membered ring)

O / O—N–CH₃ (three-membered ring)

O=N–O–CH₃

H_3C–N(⊕)(=O)–O⊖

Chapter 4

Problem 1 (p 97) – Provide an acceptable name for the following structures.

a.

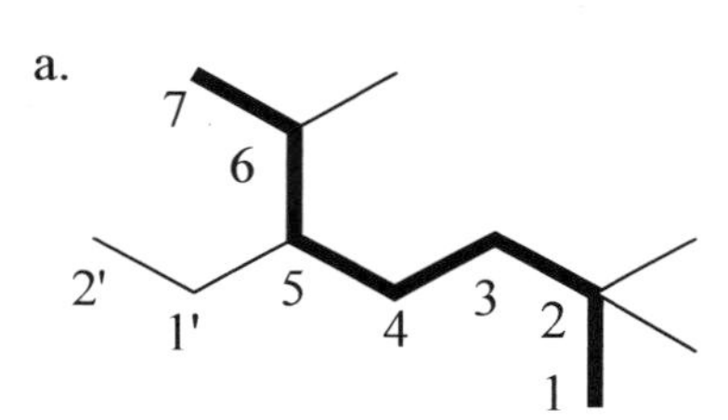

5-ethyl-2,2,6-trimethylheptane

b.

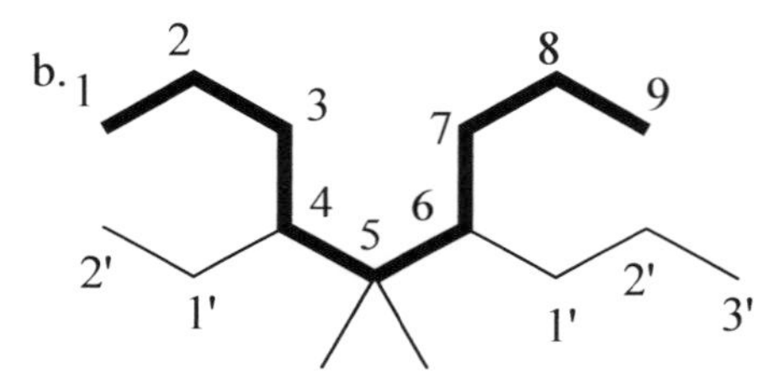

4-ethyl-5,5-dimethyl-6-propylnonane

c.

3 2 1

4

2' 1' 1' 5 6 7

2'

3'

4-ethyl-3-methyl-4-propylheptane

d.

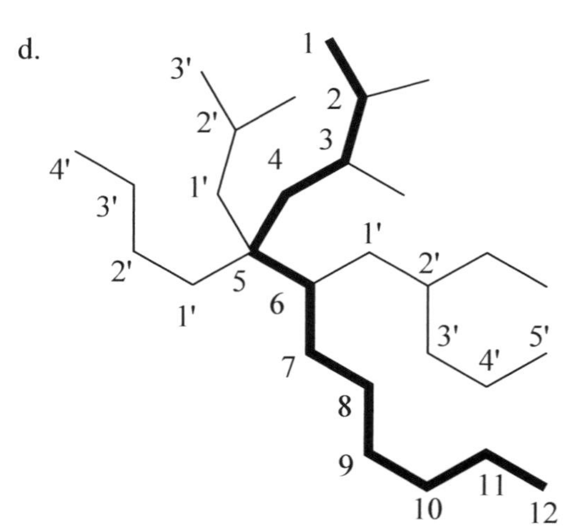

2,3-dimethyl-5-(2-methylpropyl)-5- butyl-6-(2-ethylpentyl)dodecane

Problem 2 (p 102) - Identify each of the substituent patterns below by its common name. Point out an example of a 1°, 2°, 3° and 4° carbon and nitrogen. Also, point out an example of a methyl, methylene and methine (methylidene) position.

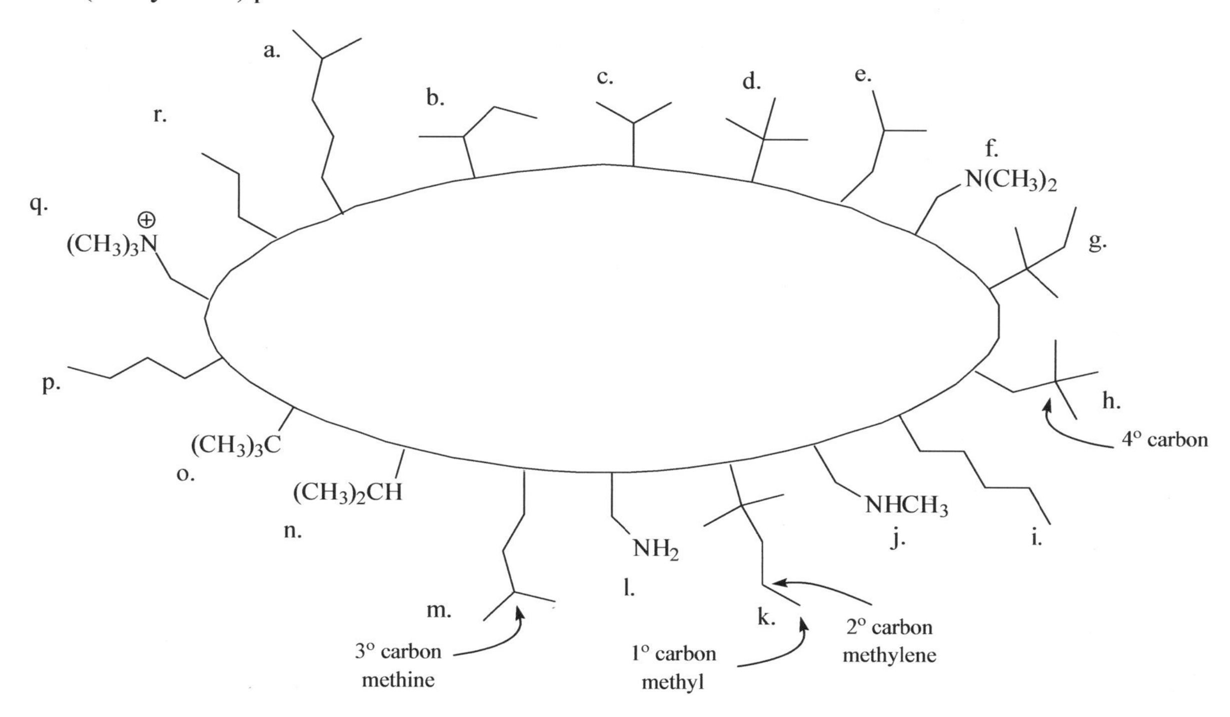

a isohexyl
b. sec-butyl
c. isopropyl
d. t-butyl
e. isobutyl
f. 3° amine (N,N-dimethylamino)
g t-pentyl
h. neopentyl
i. n-pentyl (or pentyl)
j. 2° amine (N-methylamino)
k. t-hexyl
l. 1° amine
m. isopentyl
n. isopropyl
o. t-butyl
p. n-butyl (or butyl)
q. 4° ammonium ion
r. n-propyl (or propyl)

Problem 3 (p 104) – Provide an acceptable name for each of the following structures.

a.
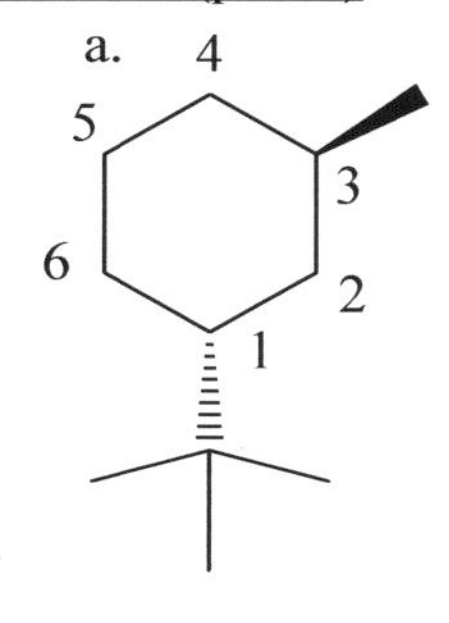

b.

c.
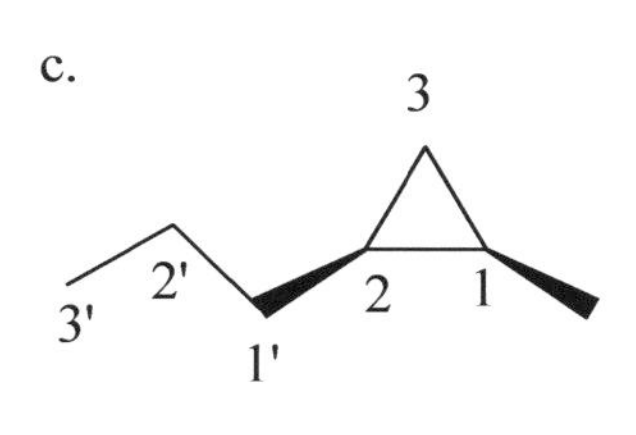

d.
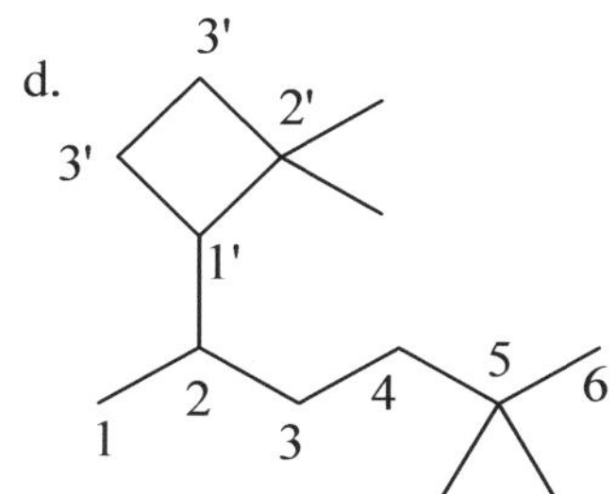

a. trans-1-(1,1-dimethylethyl)-3-methylcyclohexane

b. 1-ethyl-1-methylcycloheptane

c. cis-1-methyl-2-propylcyclopropane

d. 2-(2,2-dimethylcyclobutyl)-5,5-dimethylhexane

Problem 4 (p 107) - Classify the order of priorities among each group below (1 = highest). The "*" specifies attachment to a stereogenic center.

a.

1	2	3	4
* —NH_2	* —CH_2F	* —CH_2CH_2Cl	* —$CH_2CH_2CH_2I$
N (H,H)	C_1 (F,H,H)	C_1 (C,H,H) C_2 (Cl,H,H)	C_1 (C,H,H) C_2 (C,H,H) C_3 (I,H,H)

b.

2	1	4	3
H * —C—F H	* —O—H	H H * —C—C—C(=O)OH H H	CH_3 * —C—CH_3 CH_3
C_1 (F,H,H)	O (H)	C_1 (C,H,H) C_2 (C,H,H) C_3 (O,O,O)	C_1 (C,C,C) C_2 (C,H,H)

c.

1	3	5	4	2
* —F	H * —C—Cl H	* —C≡N	* —C(=O)OH	* —C(=O)Cl
F	C_1 (Cl,H,H)	C_1 (N,N,N)	C_1 (O,O,O)	C_1 (Cl,O,O)

d.

5	4	2	1	3
* —C(H)=C(H)—F	* —C(H)=C(H)—Cl	* —C≡C—H	* —C≡C—CH_3	CH_3 * —C—CH_3 CH_3
C_1 (C,C,H) C_2 (F,C,H)	C_1 (C,C,H) C_2 (Cl,C,H)	C_1 (C,C,C) C_2 (C,C,H)	C_1 (C,C,C) C_2 (C,C,C)	C_1 (C,C,C) C_2 (H,H,H)

e.

3	5	1	2	4
* —C_6H_4 (ring C—H, H, H, H; ortho H_3C)	* —C_6H_5 (ring C—H: H, H, H, H, H)	* —C_6H_4 (ring C—H, H, H, H; ortho HO—CH_2)	* —C_6H_4 (ring C—H, H, H, H; ortho C≡N)	* —C_6H_4 (ring C—H, H, H, H; F)
C_1 (C,C,C) C_2 (C,C,C) C_3 (C,C,H)	C_1 (C,C,C) C_2 (C,C,H) C_3 (C,C,H)	C_1 (C,C,C) C_2 (C,C,C) C_3 (O,H,H)	C_1 (C,C,C) C_2 (C,C,C) C_3 (N,N,N)	C_1 (C,C,C) C_2 (C,C,H) C_3 (F,C,C)

Problem 5 (p 107) - Classify each alkene, below, as E, Z or no stereochemistry present.

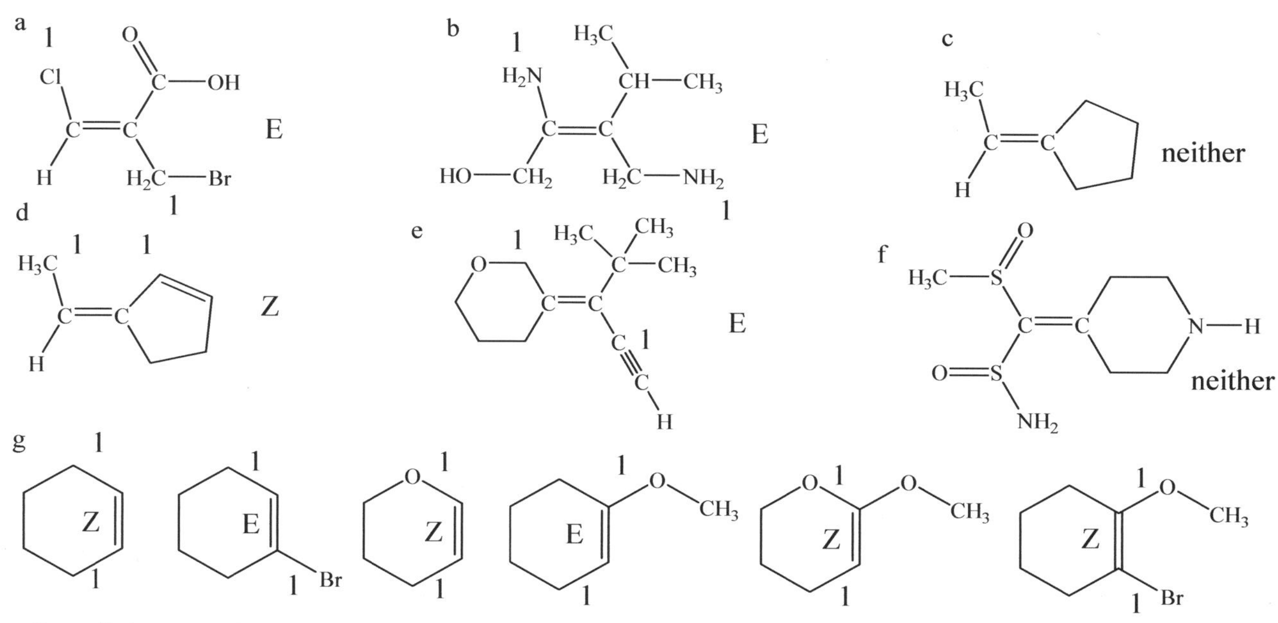

All small rings are 'cis' inside the ring, but not all are Z. The E/Z terms are part of a priority classification system. If you change any atom in a structure you may alter the priorities and change the absolute configuration (E or Z).

Problem 6 (p 110) – Provide an acceptable name for each of the following structures.

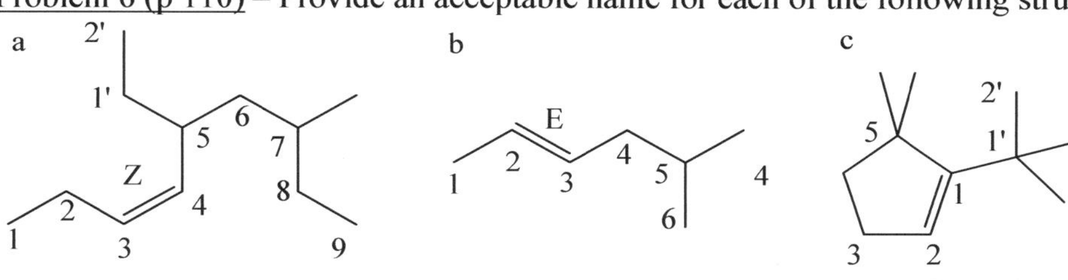

a 5-ethyl-7-methylnon-3Z-ene

b 5-methylhex-2E-ene

c 1-(1,1-dimethylethyl)-5,5-dimethylcyclopent-1-ene or...1-t-butyl....

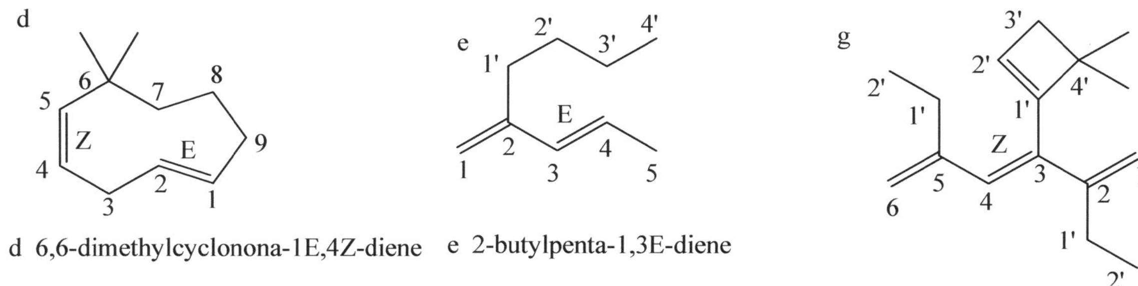

d 6,6-dimethylcyclonona-1E,4Z-diene

e 2-butylpenta-1,3E-diene

g 2,5-diethyl-3-(4,4-dimethylcyclobut-1-enyl)hexa-1,3Z,5-triene

f 1-methyl-3-(1-methylethyl)-8-(prop-2-enyl)cycloocta-1,3,5,7-tetraene
...3-isopropyl... ...8-allyl...

Problem 7 (p 112) - Provide an acceptable name for each of the following structures. (Know how to do all of these.)

a

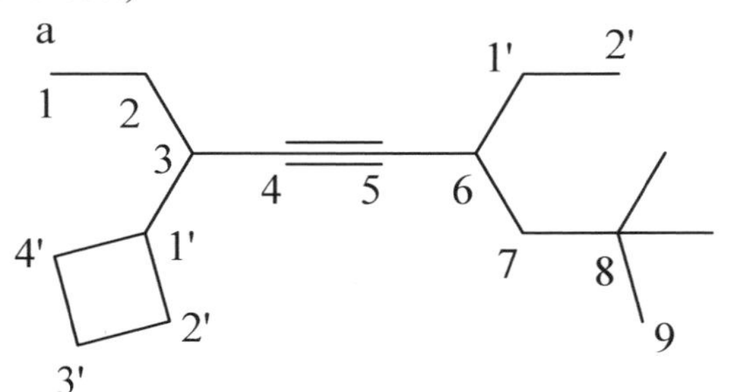

b

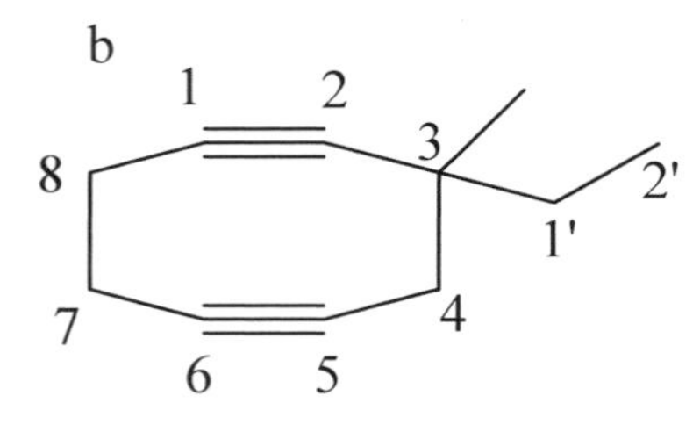

b 3-ethyl-3-methylcycloocta-1,5-diyne

a 3-cyclobutyl-6-ethyl-8,8-dimethylnon-4-yne

c

d

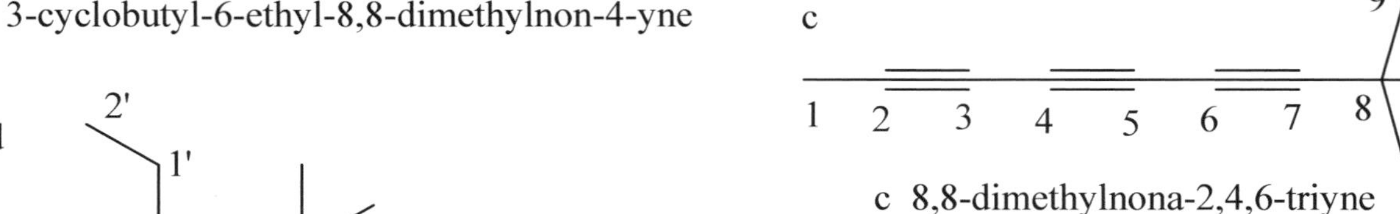

c 8,8-dimethylnona-2,4,6-triyne

d 3-ethyl-5,5-dimethylhex-1-yne

e

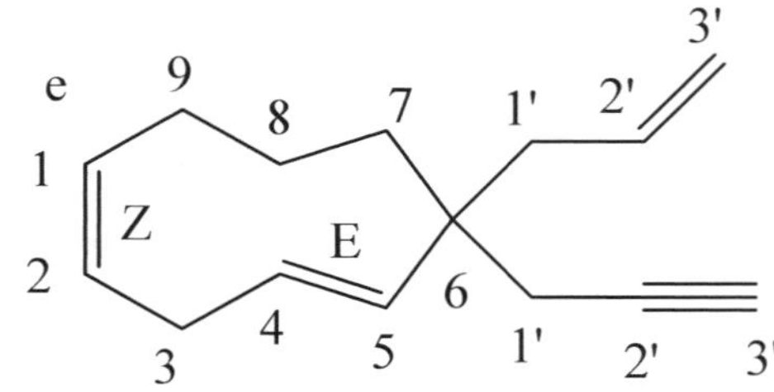

e 6-(prop-2-enyl)-6-(prop-2-ynyl)cyclonona-1Z,4E-diene
...6-allyl... ...6-propargyl...

Problem 8 (p 114) - Provide an acceptable name for each of the following structures.

a.

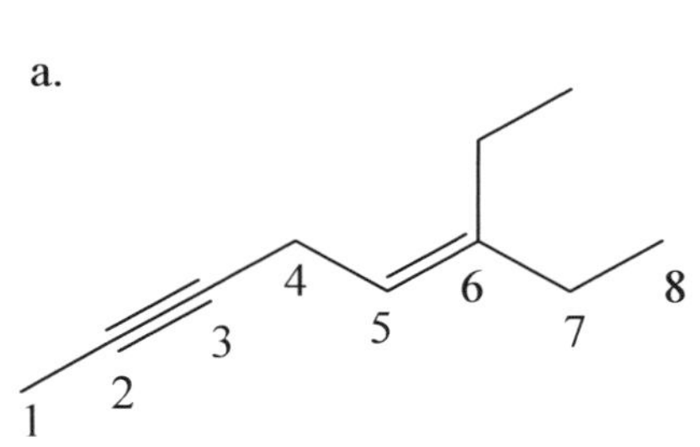

a 6-ethyloct-5-en-2-yne

b.

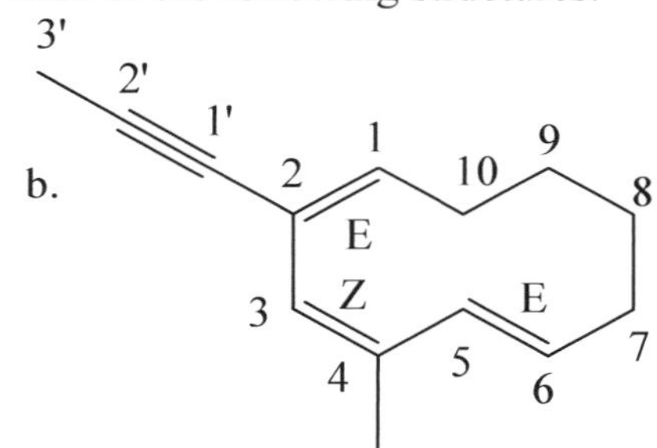

b 2-(prop-1-ynyl)-4-methylcyclodeca-1E,3Z,5E-triene

---

c.

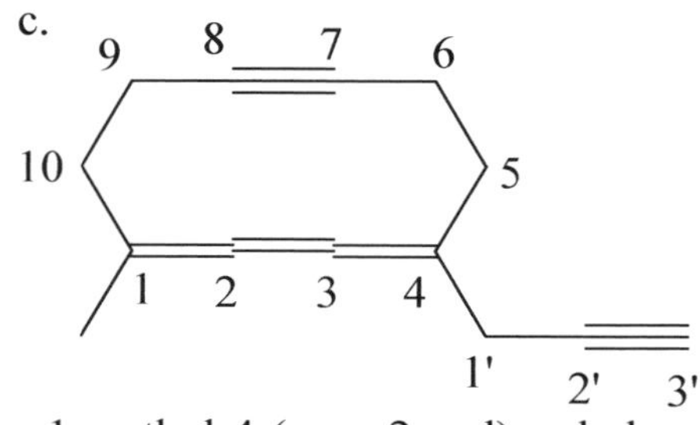

c 1-methyl-4-(prop-2ynyl)cyclodeca-1,2,3-trien-7-yne
...4-propargyl...

d.

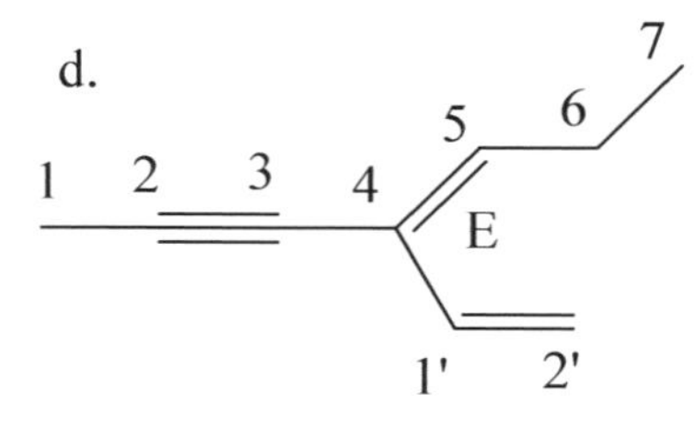

d 4-ethenylhept-4E-en-2-yne
...4-vinyl...

---

e.

e 1-(prop-2-enyl)-3-(2-methylbut-2E-enyl)cyclodec-1Z-en-6-yne
...1-allyl...

f.

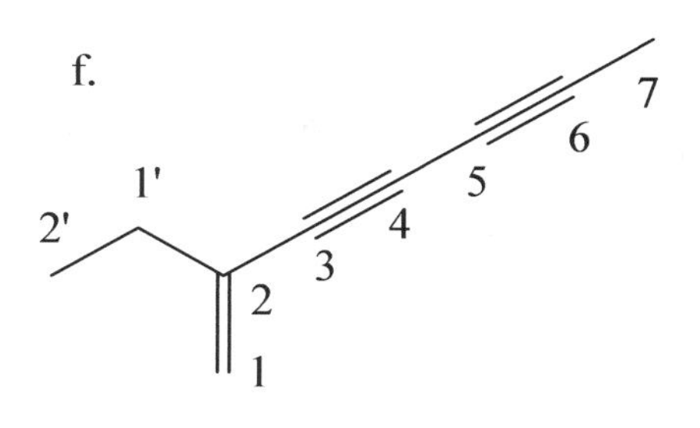

f 2-ethylhept-1-en-3,5-diyne

---

g.

g 4-(5,5-dimethylcyclopent-2-enyl)oct-2E-en-5-yne

Problem 9 (p 138) – Provide an acceptable name for each of the following structures.

a

2-hydroxy-4-phenylnon-3Z-enoic acid

b

-5-isobutyl-

4-benzyl-5-(2-methylpropyl)undec-10-en-3-one

c

isopropyl

1-methylethyl 2-methoxy-5-cyanooct-6Z-enoate

d

9-amino-7-methyl-5-mercaptodec-6E-en-3-ol

e

2-chlorohept-6-ynoyl chloride

f

2-methyl-3-ethyl-9-bromodecan-4-amine

g

2-aminohexanamide

h

-4-isopropyl

ethanoic 2-methyl-3-ethoxy-4-(1-methylethyl)nonanoic anhydride

i

2-oxo-4-methylpentanal

j

-2-t-butyl- -7-sec-butyl-

2-(1,1-dimethylethyl)-7-(1-methylpropyl)dodec-2E-en-9-ynenitrile

Problem 10 (p 139) – Provide an acceptable name for each of the following structures.

a

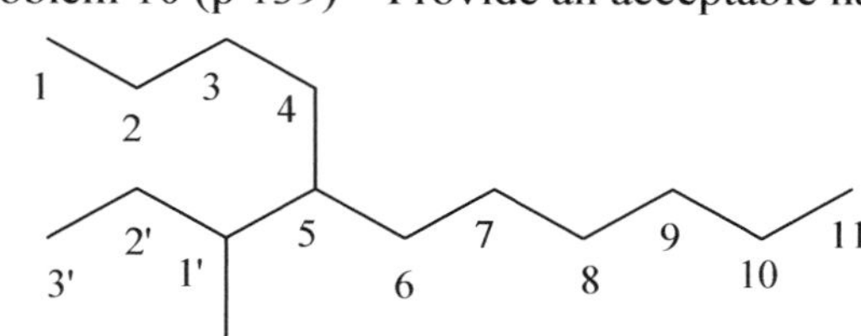

5-(1-methylpropyl)undecane

b

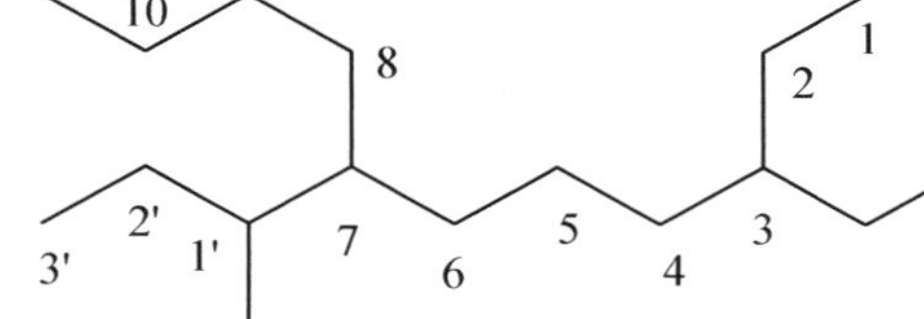

7-(1-methylpropyl)-3-ethylundecane

c

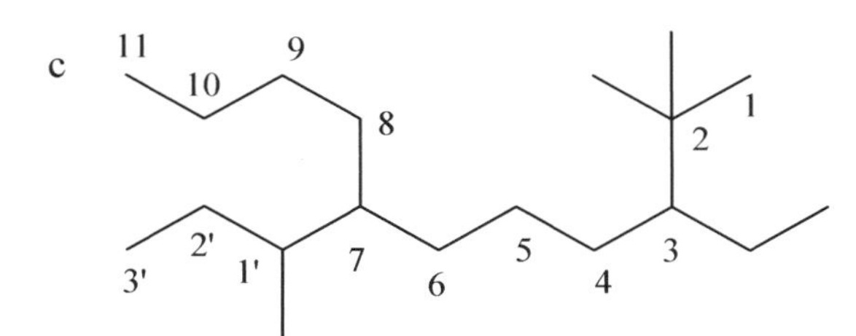

2,2-dimethyl-7-(1-methylpropyl)-3-ethylundecane

---

d

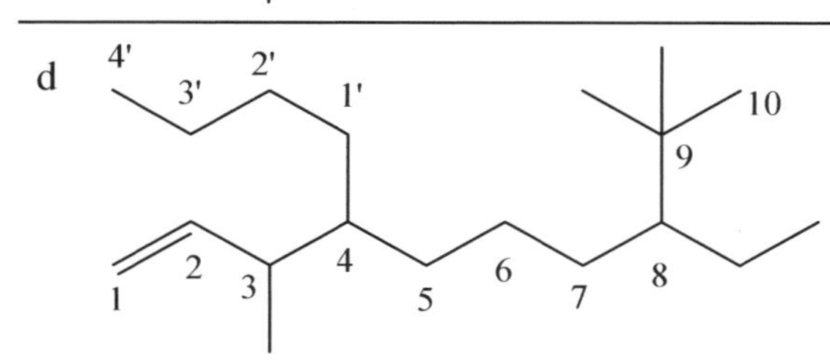

3,9,9-trimethyl-4-butyl-8-ethyldec-1-ene

---

e

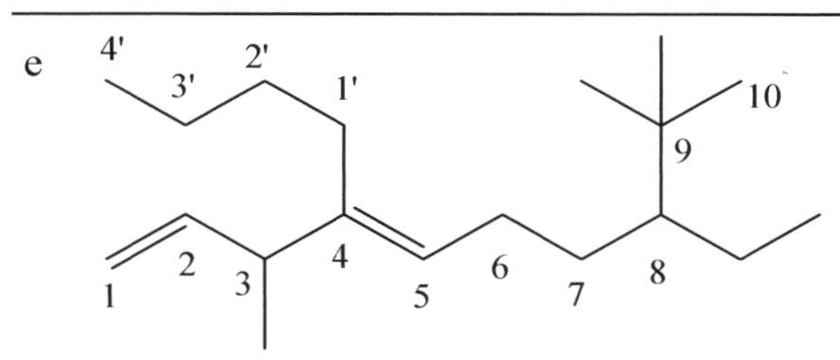

3,9,9-trimethyl-4-butyl-8-ethyldeca-1,4E-diene

---

f

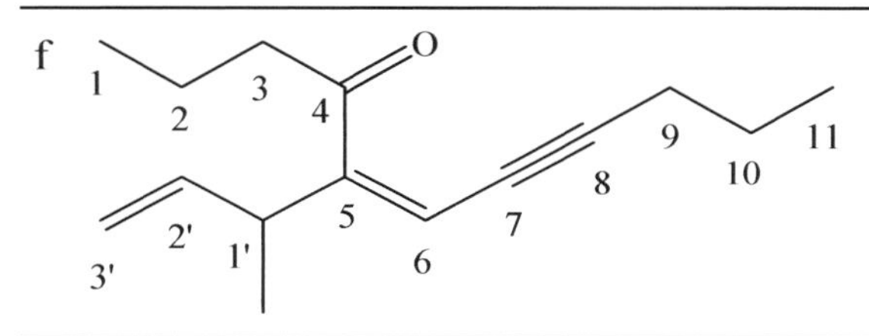

5-(1-methylprop-2-enyl)undec-5Z-en-7-yn-4-one

---

g

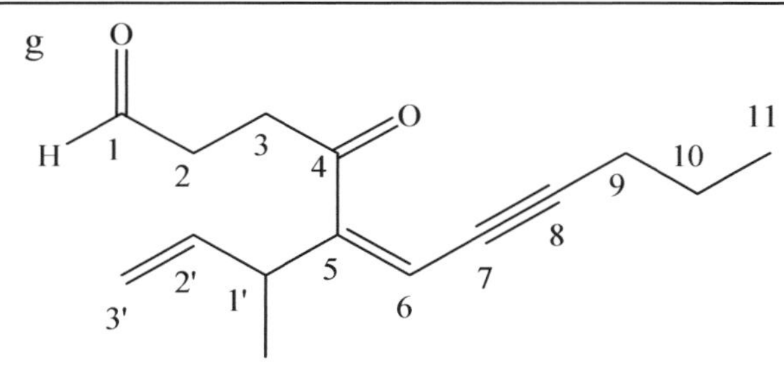

5-(1-methylprop-2-enyl)-4-oxoundec-5Z-en-7-ynal

---

h

10-(N-methyl-N-ethylamino)-5-(1-methylprop-2-enyl)-4-oxoundec-5Z-en-7-ynal

---

i

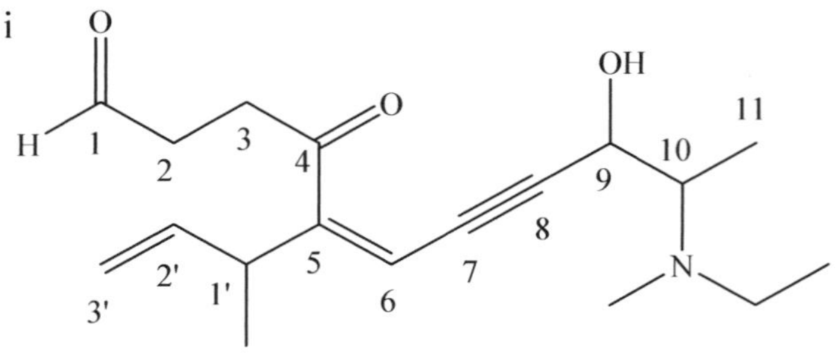

O
OH
H 1 3 O 11
2 4 10
9
8
2' 5 1' N
3' 6 7

10-(N-methyl-N-ethylamino)-9-hydroxy-5-(1-methylprop-2-enyl)-4-oxoundec-5Z-en-7-ynal

j

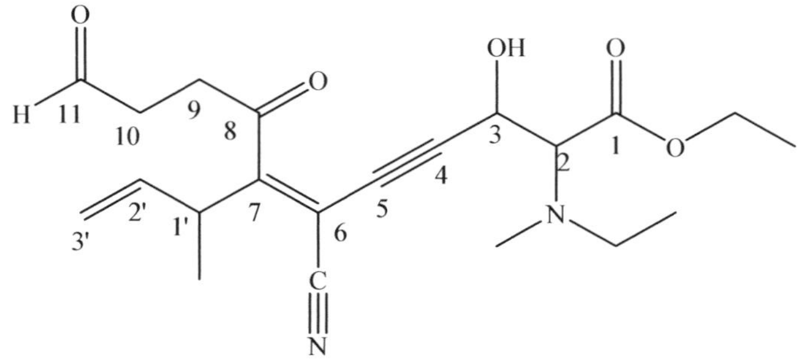

O
OH O
H 11 9 O
10 8 3 1 O
2
4
2' 1' 7 5 N
3' 6
C
N

ethyl 2-(N-methyl-N-ethylamino)-3-hydroxy-6-cyano-7-(1-methylprop-2-enyl)-8,11-dioxoundec-6E-en-4-ynoate

k

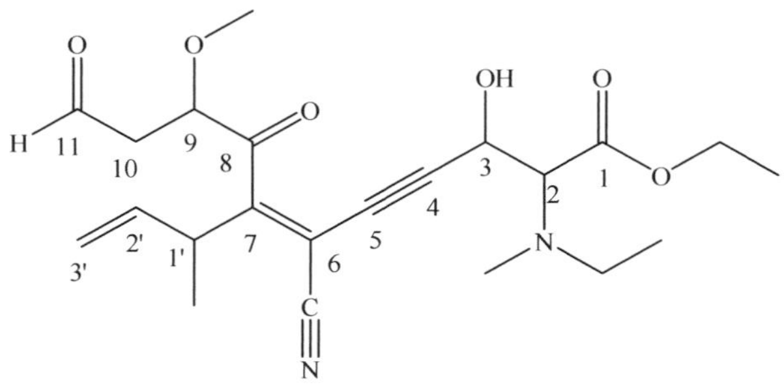

O O
OH O
H 11 9 O
10 8 3 1 O
2
4
2' 1' 7 5 N
3' 6
C
N

ethyl 2-(N-methyl-N-ethylamino)-3-hydroxy-6-cyano-7-(1-methylprop-2-enyl)-8,11-dioxo-9-methoxyundec-6E-en-4-ynoate

l

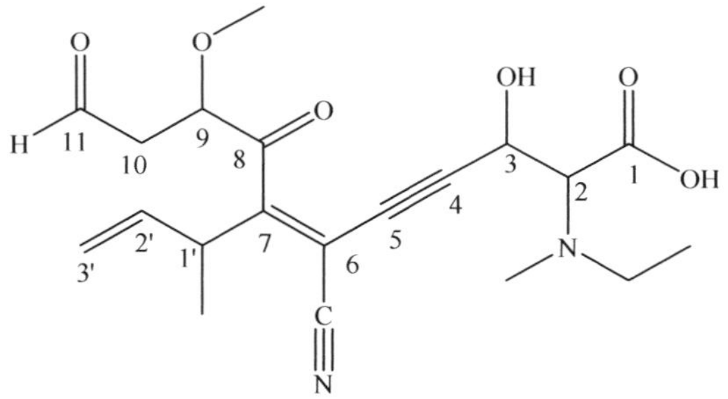

O O
OH O
H 11 9 O
10 8 3 1 OH
2
4
2' 1' 7 5 N
3' 6
C
N

2-(N-methyl-N-ethylamino)-3-hydroxy-6-cyano-7-(1-methylprop-2-enyl)-8,11-dioxo-9-methoxyundec-6E-en-4-ynoic acid

m

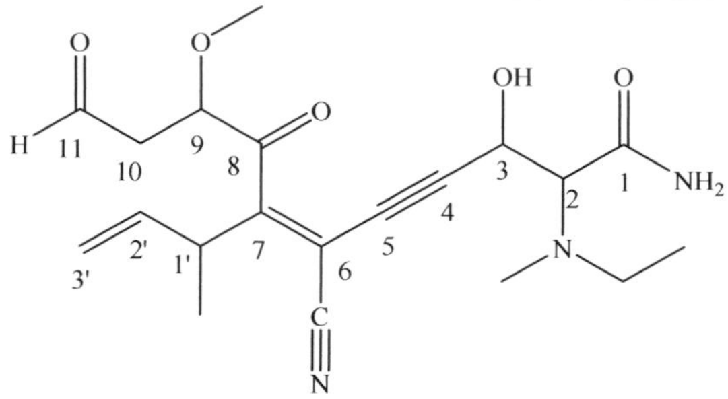

O O
OH O
H 11 9 O
10 8 3 1 NH_2
2
4
2' 1' 7 5 N
3' 6
C
N

2-(N-methyl-N-ethylamino)-3-hydroxy-6-cyano-7-(1-methylprop-2-enyl)-8,11-dioxo-9-methoxyundec-6E-en-4-ynamide

n

2-(N-methyl-N-ethylamino)-3-hydroxy-6-cyano-7-(1-methylprop-2-enyl)-8,11-dioxo-9-methoxyundec-6E-en-4-ynoyl chloride

---

o

2-(N-methyl-N-ethylamino)-3-hydroxy-6-formyl-7-(1-methylprop-2-enyl)-8,11-dioxo-9-methoxyundec-6E-en-4-ynenitrile

---

Key for lecture problems to work in class.

1

start here

3,3-dimethyl-4-propyl-8-(1-ethylpropyl)dodecane

No priority functional groups so lowest number for alkyl branch decides which end to number from.

2

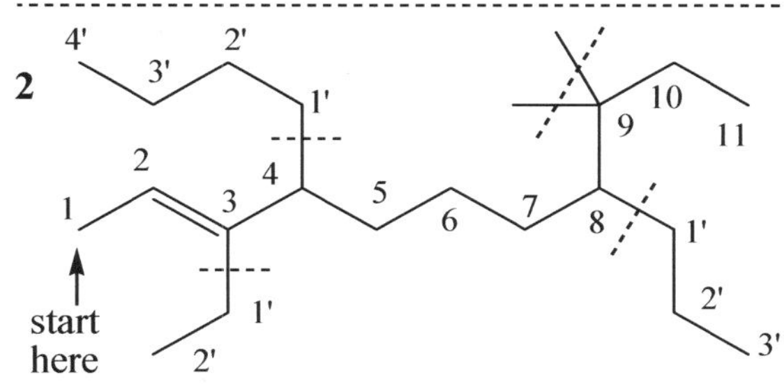

3-ethyl-4-butyl-8-propyl-9,9-dimethylundec-2E-ene

"-ene" only named as a suffix

3

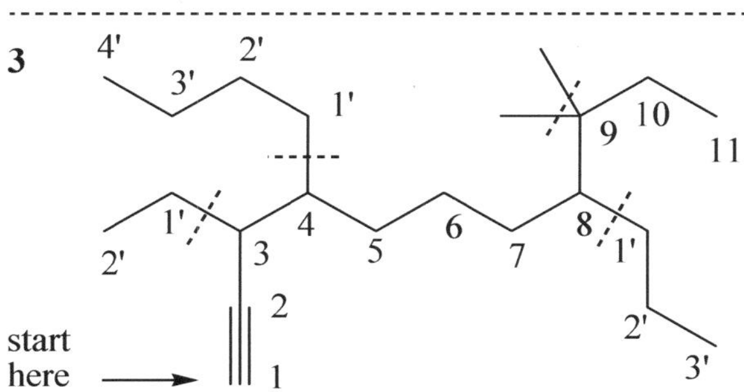

e b p m

3-ethyl-4-butyl-8-propyl-9,9-dimethylundec-1-yne

"-yne" only named as a suffix

4

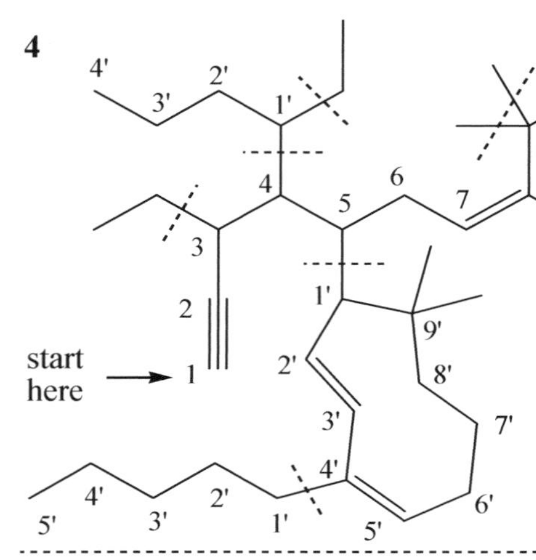

3-ethyl-4-(1-ethylbutyl)-5-(4-pentyl-9,9-dimethylcyclonona-2E,4Z-dienyl)-8-propyl-9,9-dimethylundec-7Z-en-1-yne

---

5 start here

5-(1-ethylprop-2-ynyl)-6-(4-methylcyclopent-2-enyl)-9-(1,1-dimethylpropyl)-10-mercapto-11-(1-methylethoxy)-12-aminododecan-3-ol

---

6

start here

3-hydroxy-5-(1-ethylprop-2-ynyl)-6-(4-methylcyclopent-2-enyl)-8-oxo-9-(1,1-dimethylpropyl)-10-mercapto-11-(1-methylethoxy)-12-aminododecanal

---

7

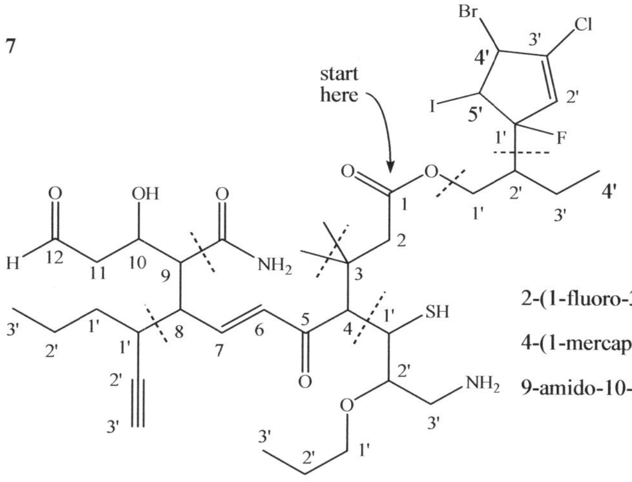

start here

2-(1-fluoro-3-chloro-4-bromo-5-iodocyclopent-2-enyl)butyl 3,3-dimethyl-4-(1-mercapto-2-propoxy-3-aminopropyl)-5,12-dioxo-8-(1-propylprop-2-ynyl)-9-amido-10-hydroxydodec-6E-enoate

---

8

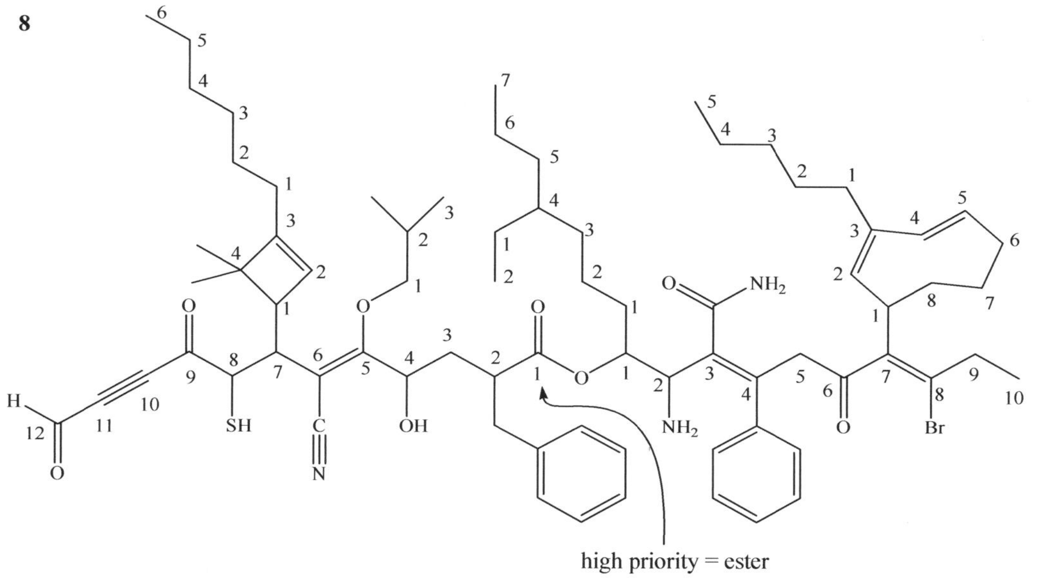

high priority = ester

1-(4-ethylheptyl)-2-amino-3-amido-4-phenyl-6-oxo-7-(3-pentylcycloocta-2Z,4E-dienyl)-8-bromodec-3E,7Z-dienyl 2-benzyl-4-hydroxy-5-(2-methylpropoxy)-6-cyano-7-(3-hexyl-4,4-dimethylcyclobut-2-enyl)-8-mercapto-9,12-dioxododec-5E-en-10-ynoate

---

9 Extra problem with functional group prefixes.

9

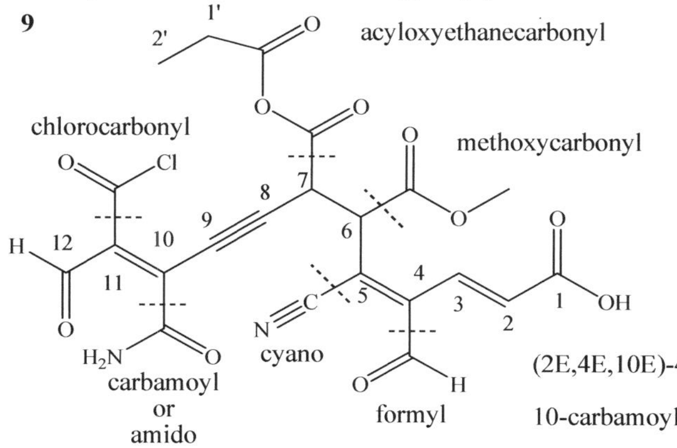

acyloxyethanecarbonyl

chlorocarbonyl

methoxycarbonyl

cyano

carbamoyl or amido

formyl

(2E,4E,10E)-4-formyl-5-cyano-6-methoxycarbonyl-7-acyloxyethanecarbonyl-10-carbamoyl-11-chlorocarbonyl-12-oxododeca-2,4,10-trien-8-ynoic acid

---

Chapter 5

Problem 1 (p 152) - If chlorine atoms (Cl) are substituted for all hydrogen atoms in ethane (perchloroethane), the energy barrier to rotation is 10.8 kcal/mole. The increase in the energy barrier is thought to be, in part, due to the polar nature of the C-Cl bond (repulsion from the similar bond dipoles). Draw Newman projections and show the potential energy changes associated with them for a complete 360° rotation cycle. Label each conformation as staggered or eclipsed. Calculate the approximate percent of staggered versus eclipsed conformations.

$$K = 10^{\frac{-\Delta G}{2.3RT}} \approx 10^{\frac{-\Delta H}{2.3RT}} = 10^{\frac{-(10{,}800\ \text{cal/mol})}{2.3(2\ \text{cal/mol-K})(298\text{K})}} = 10^{-7.82} = \frac{1}{6.7\text{x}10^{7}} = \frac{\text{eclipsed}}{\text{staggered}}$$

$\Delta H = +10.8 \frac{\text{kcal}}{\text{mole}}$

uphill in energy

staggered ≈ 100%

eclipsed ≈ 0%

$$K = \frac{\text{eclipsed}}{\text{staggered}} = \frac{1}{6.7\text{x}10^{7}} = \frac{1}{67{,}000{,}000}$$

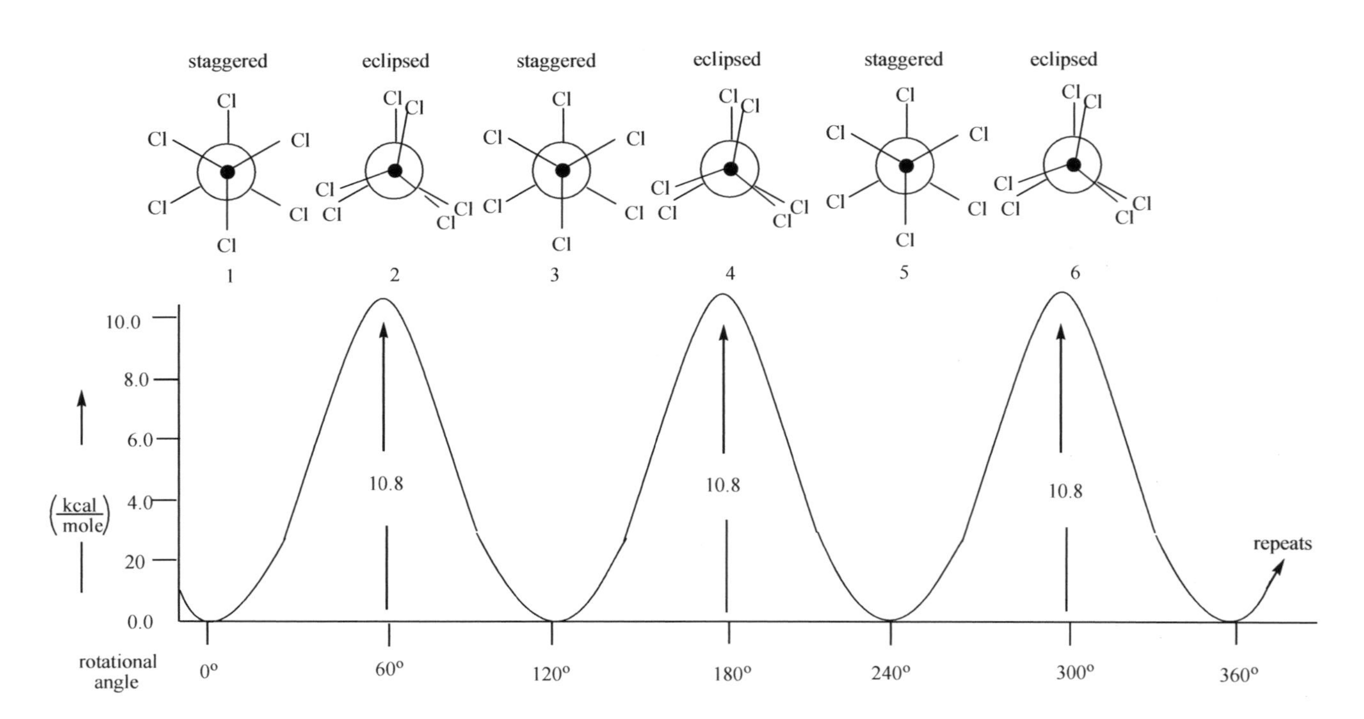

Problem 2 (p 158) – Draw Newman projections for all staggered and eclipsed conformations of 2-methylpropane. Sight down the C_1-C_2 bond. The energy barrier to rotation is 3.9 kcal/mole. What value does this suggest for a CH_3/H eclipsing interaction? Is this reasonable from the values calculated earlier using butane or propane?

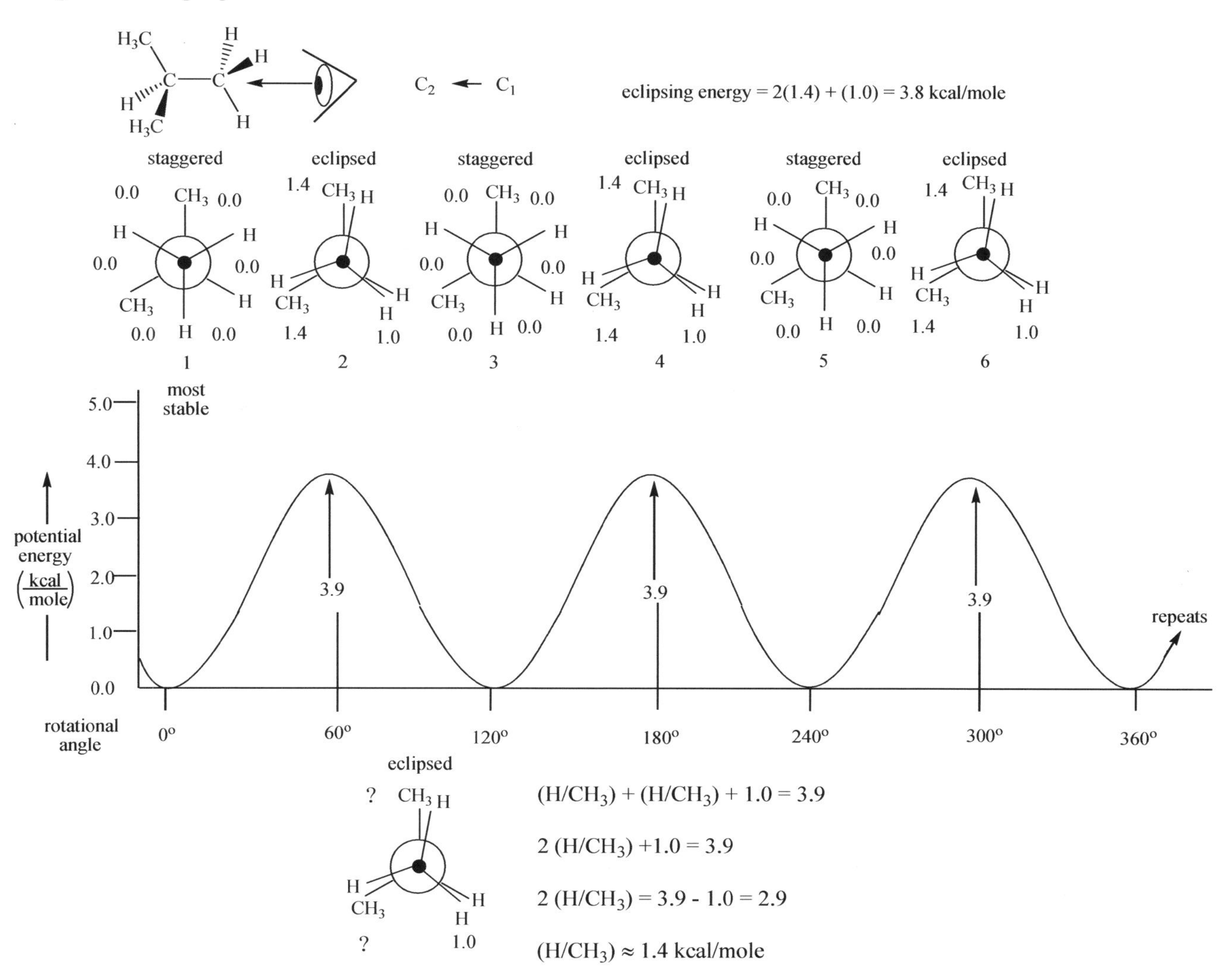

Problem 3 (p 159) – Use the $C_1 \rightarrow C_2$ bond to draw a Newman projection for 2-methylbutane. What are the three additional groups on C_1? What are the three additional groups on C_2? Do the same using the $C_2 \rightarrow C_3$ bond. For each part, start with the most stable conformation and rotate through 360°. Point out all gauche, anti and syn methyl interactions and provide an estimate of what the relative energy for each conformation is (use the butane example as your guide). Start your energy scale at the lowest energy calculated. Assume an ethyl/H eclipsing interaction is 1.5 kcal/mole and a methyl/H eclipsing interaction is 1.4 kcal/mole and an ethyl/H gauche is 0.1 kcal/mole. Estimate the equilibrium ratio between the high and low energy conformations in the first example. In the second example estimate the equilibrium ratio between the lowest two staggered conformations.

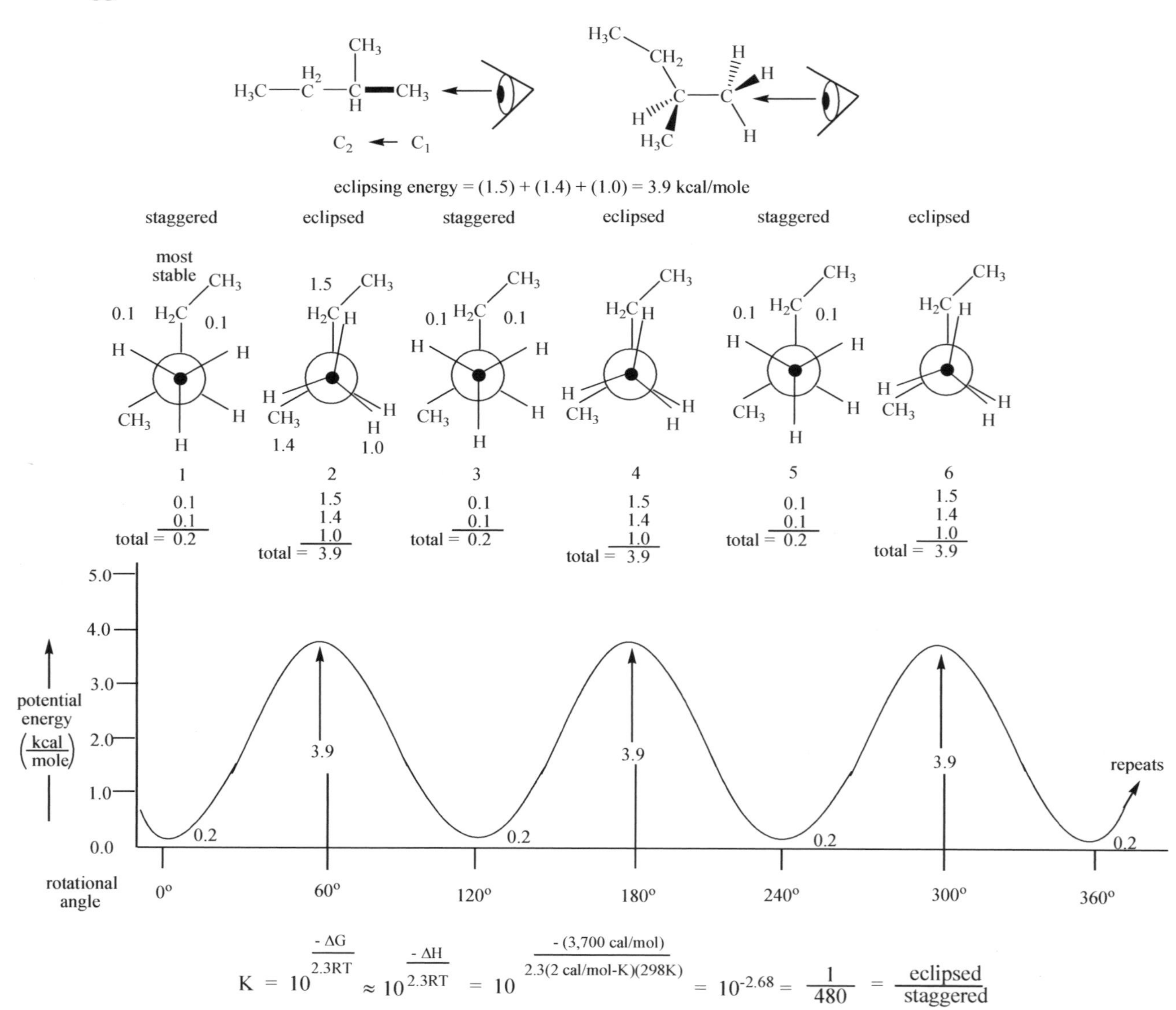

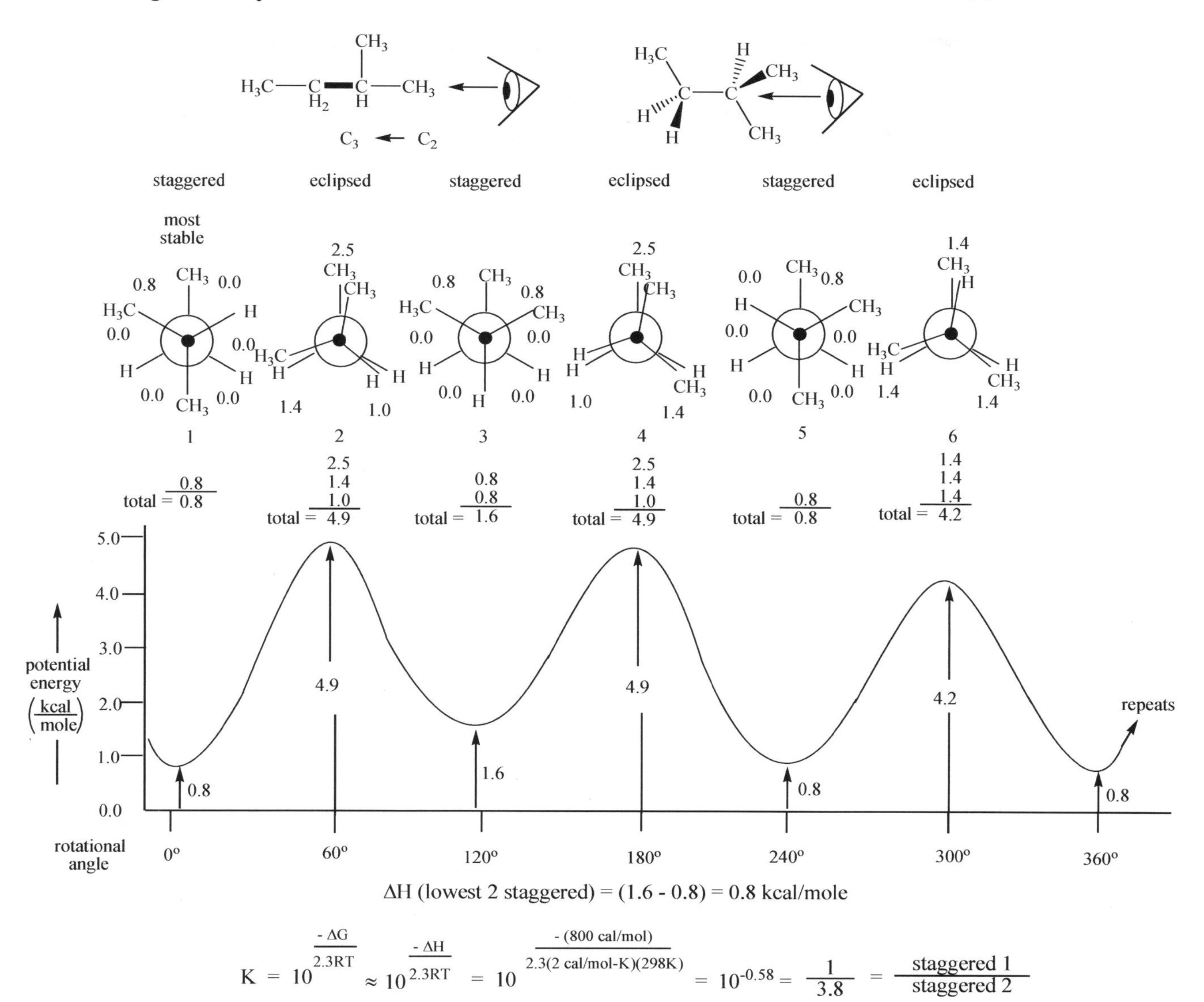

ΔH (lowest 2 staggered) = (1.6 - 0.8) = 0.8 kcal/mole

$$K = 10^{\frac{-\Delta G}{2.3RT}} \approx 10^{\frac{-\Delta H}{2.3RT}} = 10^{\frac{-(800\ \text{cal/mol})}{2.3(2\ \text{cal/mol-K})(298\text{K})}} = 10^{-0.58} = \frac{1}{3.8} = \frac{\text{staggered 1}}{\text{staggered 2}}$$

Problem 4 (p 159) – Methanamine (CH_3NH_2) has rotational barriers of 1.98 kcal/mole and methanol (CH_3OH) has rotational barriers of 1.07 kcal/mole. Write out the staggered and eclipsed conformational structures along with a potential energy versus angle of rotation diagram using Newman projections. What are the 3 additional groups on "N" and "O" in the Newman projections? Do your diagrams suggest that a lone pair of electrons has more or less torsional strain than a sigma bond with a hydrogen atom? Fill in the energy data in the table provided. Estimate the equilibrium ratios between the high and low energy conformations.

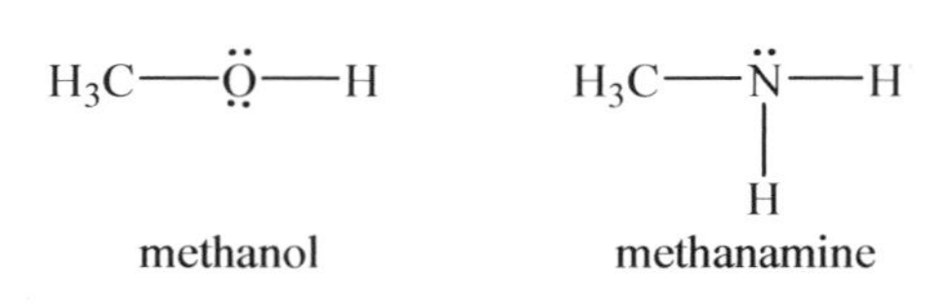

Eclipsing Groups	Increase in Potential Energy
H / H	1.0
CH_3 / H	1.4
CH_3 / CH_3	2.5
lone pair / H	0.0
gauche CH_3's	0.8

most stable

staggered 0.0 ⇌ eclipsed 2.0

total eclipsing energy = 2 (H/H) + (H/lone pair) = 2 kcal/mole
= 2 (1.0) + (H/lone pair) = 2 kcal/mole
(H/lone pair) = (2 - 2) kcal/mole = 0 kcal/mole

$$K \approx 10^{\frac{-\Delta H}{2.3RT}} = 10^{\frac{-(2{,}000\ \text{cal/mol})}{2.3(2\ \text{cal/mol-K})(298\text{K})}} \approx 10^{-1.45} \approx \frac{1}{28} \approx \frac{\text{eclipsed}}{\text{staggered}}$$

most stable

staggered 0.0 ⇌ eclipsed 1.0

total eclipsing energy = 1 (H/H) + 2 (H/lone pair) = 1 kcal/mole
= 1 (1.0) + 2 (H/lone pair) = 1 kcal/mole
(H/lone pair) = (1 - 1)/2 kcal/mole = 0 kcal/mole

$$K \approx 10^{\frac{-\Delta H}{2.3RT}} = 10^{\frac{-(1{,}000\ \text{cal/mol})}{2.3(2\ \text{cal/mol-K})(298\text{K})}} \approx 10^{-0.72} \approx \frac{1}{5.3} \approx \frac{\text{eclipsed}}{\text{staggered}}$$

Problem 5 (p 159) – a. Draw the Newman projections and dash/wedge 3D structures to show the possible rotations about the C_2-C_3 bond of 2,3-dimethylbutane. (Does it make any difference which perspective you use?) Rotate through 60° increments and start with the most stable conformation. Estimate the relative potential energies of the conformations using the energy values from the previous problem.

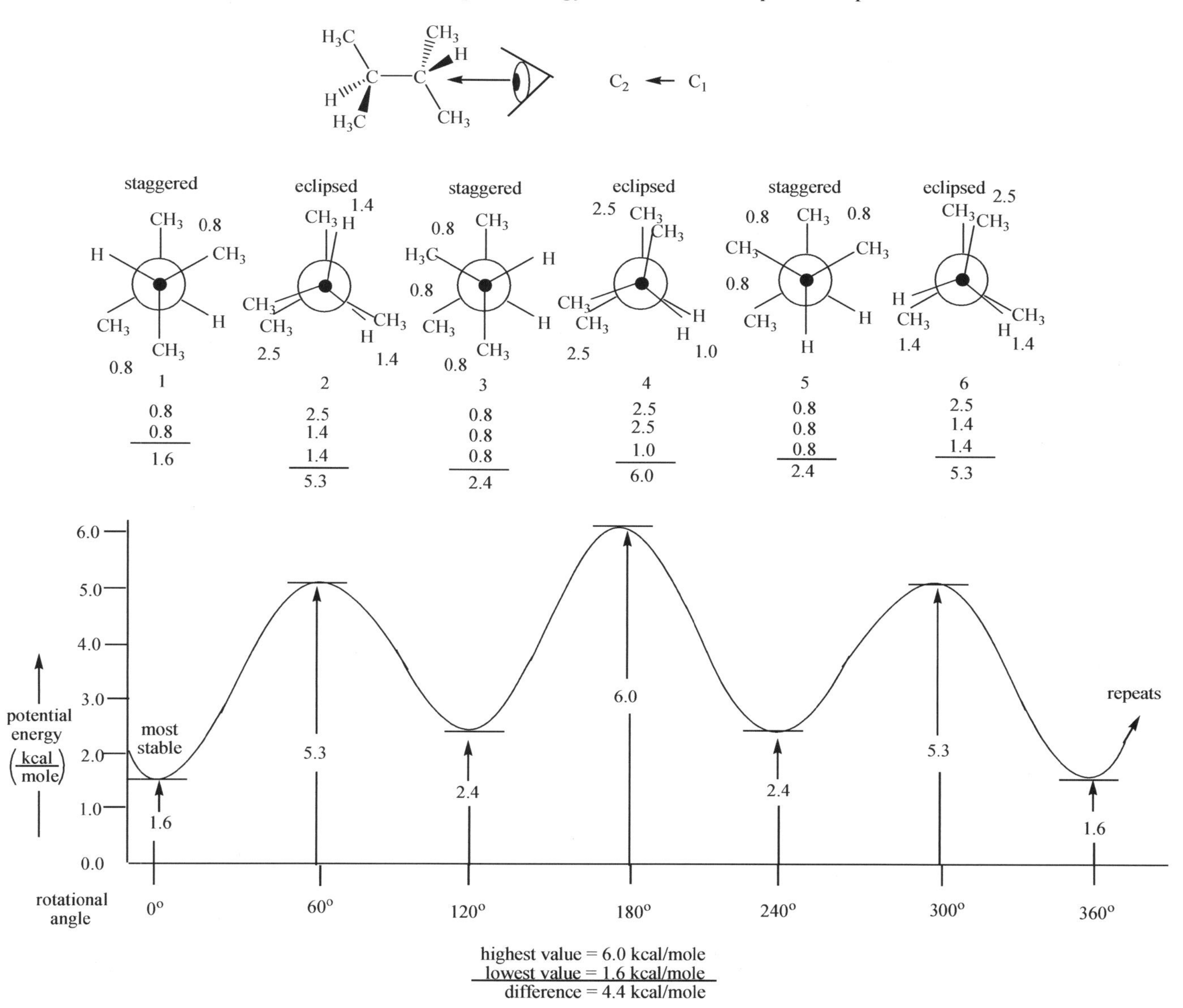

highest value = 6.0 kcal/mole
lowest value = 1.6 kcal/mole
difference = 4.4 kcal/mole

b. The highest energy barrier for rotation of 2,3-dimethylbutane is experimentally about 4.3 kcal/mole above the lowest energy conformation. Reevaluate your Newman projections for complete 360° rotation and state whether this seems reasonable with energy values above. (What is the energy difference between the most stable and least stable conformations?)

From above potential energy diagram: (highest value) – (lowest value) = 6.0 – 1.6 = 4.4 kcal/mole (very close)

We will use the following table of eclipsing and gauche energies to calculate energy differences and predict relative amounts in conformational isomer problems.

Approximate Eclipsing Energy Values (kcal/mole)

	H	Me	Et	i-Pr	t-Bu	Ph
H	1.0	1.4	1.5	1.6	3.0	1.7
Me	1.4	2.5	2.7	3.0	8.5	3.3
Et	1.5	2.7	3.3	4.5	10.0	3.8
i-Pr	1.6	3.0	4.5	7.8	13.0	8.1
t-Bu	3.0	8.5	10.0	13.0	23.0	13.5
Ph	1.7	1.7	3.8	8.1	13.5	8.3

$$\Delta G \approx \Delta H$$

$$K_{eq} = 10^{\frac{-\Delta H}{2.3RT}}$$

Approximate Gauche Energy Values (kcal/mole)

	H	Me	Et	i-Pr	t-Bu	Ph
H	0	0	0.1	0.2	0.5	0.2
Me	0	0.8	0.9	1.1	2.7	1.4
Et	0.1	0.9	1.1	1.6	3.0	1.5
i-Pr	0.2	1.1	1.6	2.0	4.1	2.1
t-Bu	0.5	2.7	3.0	4.1	8.2	3.9
Ph	0.2	1.4	1.5	2.1	3.9	2.3

Problem 6 (p 160) – Calculate the relative conformation energies of the conformations of 2,4-dimethylhexane shown below. Draw an energy pictures of the rotations. Use the $C_4 \rightarrow C_3$ perspective.

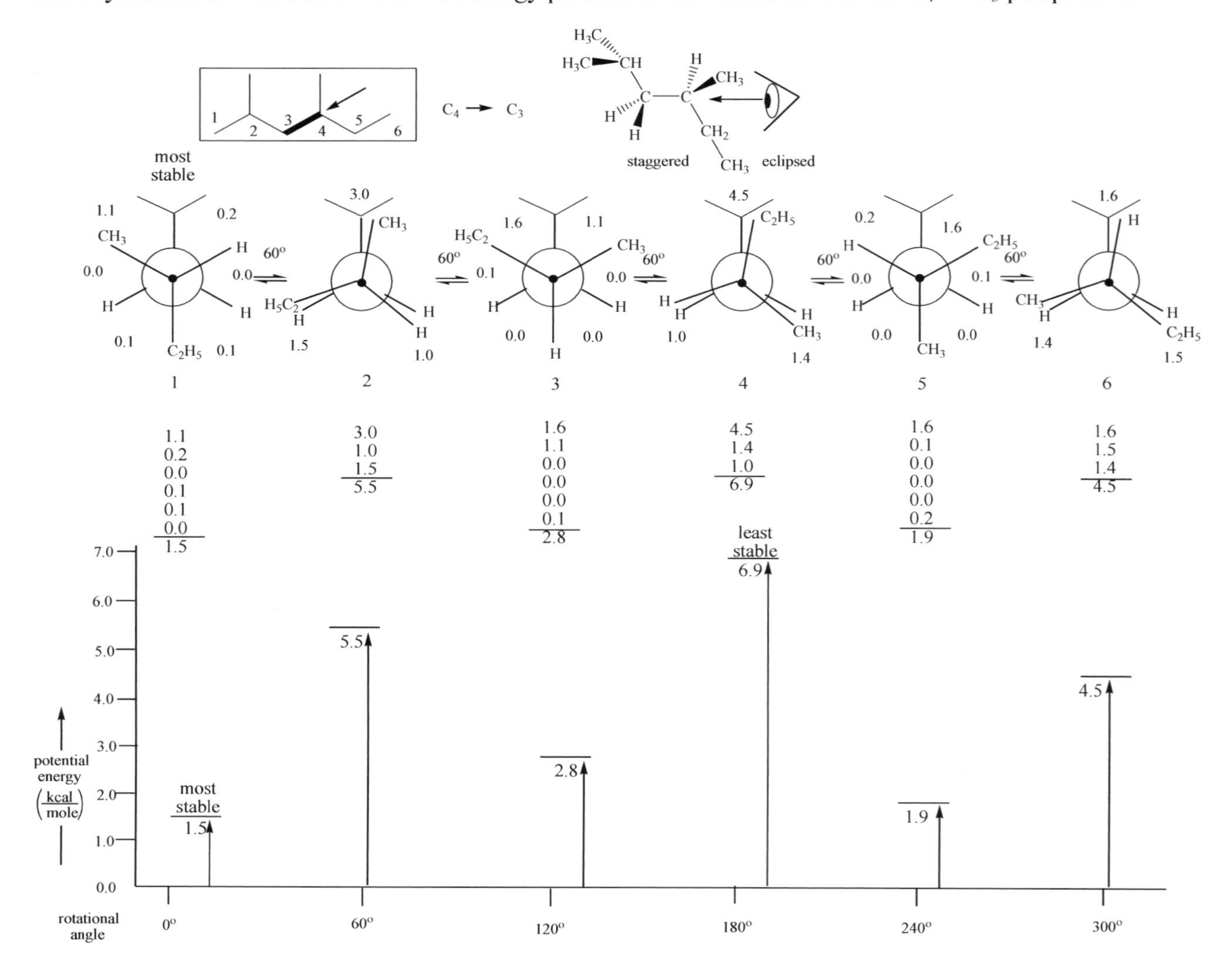

Problem 7 (p 160) – Calculate the relative conformation energies of the conformations of 2,2-dimethyl-4-phenylhexane shown below. Draw an energy pictures of the rotations. Use the $C_3 \rightarrow C_4$ perspective.

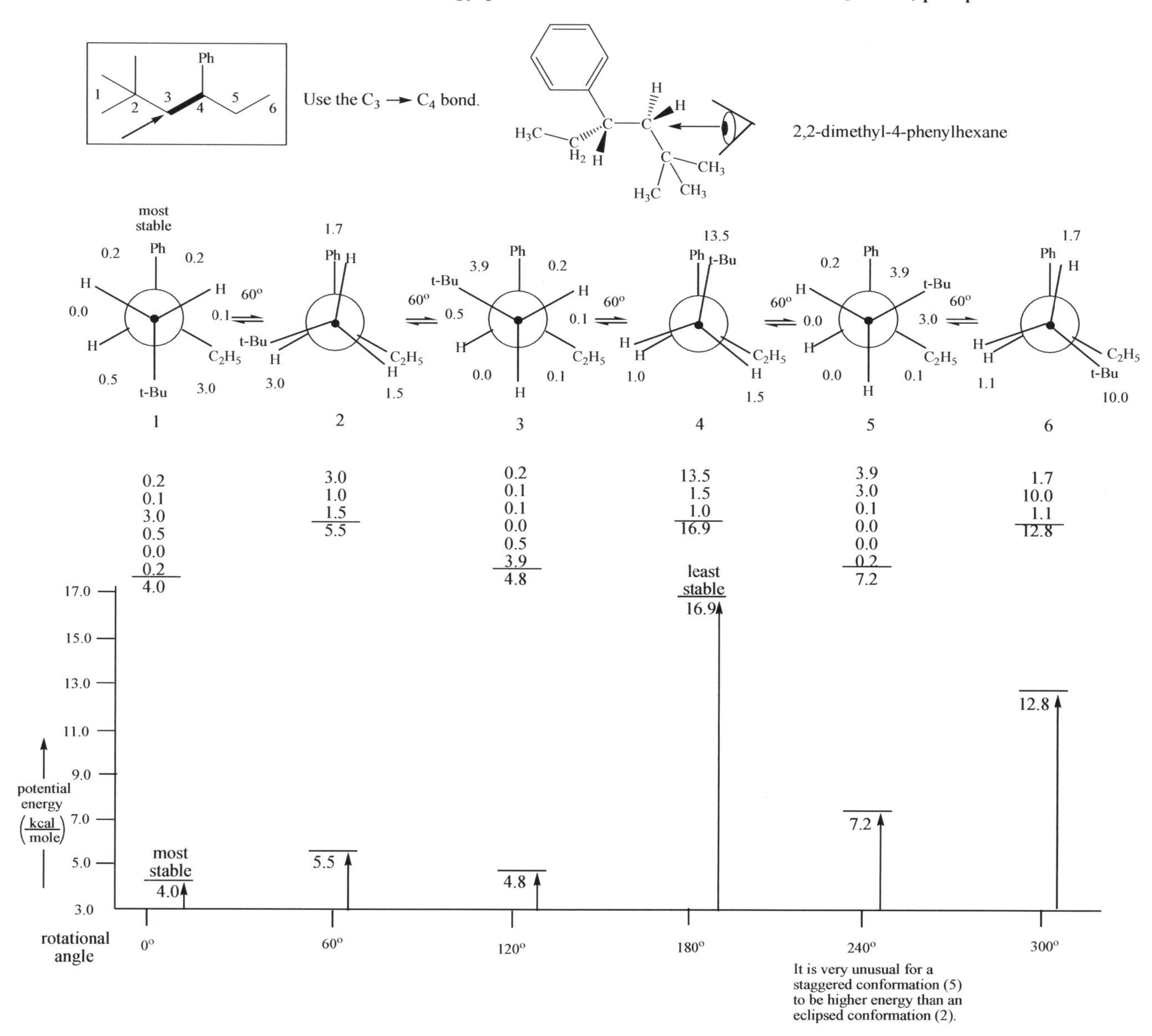

It is very unusual for a staggered conformation (5) to be higher energy than an eclipsed conformation (2).

Chapter 6

Problem 1 (p 164) - How many possible Newman perspectives of chairs are possible on the numbered cyclohexane below? Consider all numbers and front/back views. Write them as, C1→C2, C5→C4, etc.)

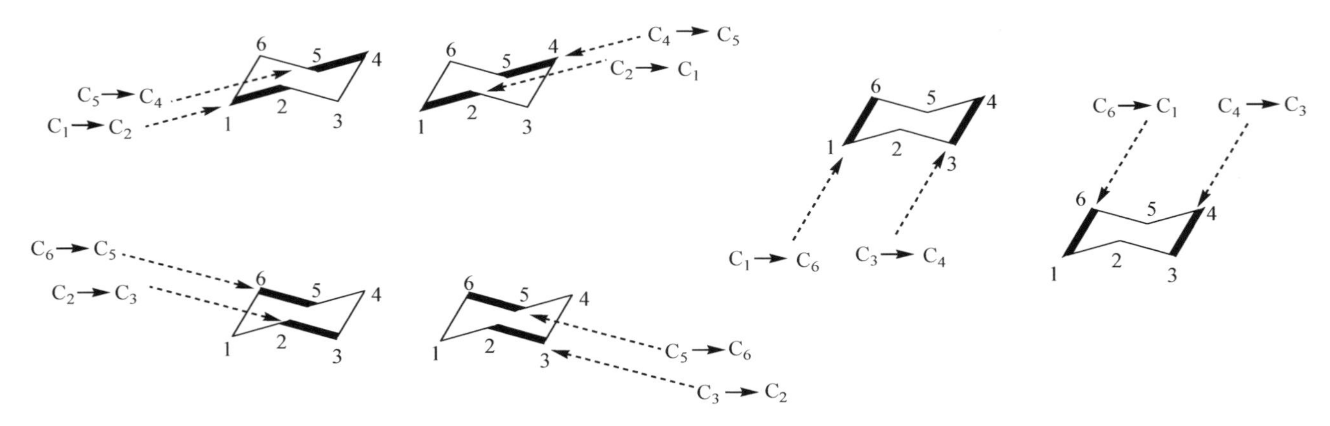

Problem 2 (p 172) - Which boat conformation would be a more likely transition state in interconverting the two chair conformations of methylcyclohexane...or are they equivalent? Explain your answer.

boat 1

chair 1
(more stable)

boat 2

chair 2
(less stable)

Boat 2 raises the potential energy more than boat 1 because the larger 'R' group is pointing into the middle of the ring causing greater electron-electron repulsion from the crowded positions. Boat 1 keeps the larger 'R' group pointing out away from the rest of the ring until the final flip puts it in the less stable axial position.

Problem 3 (p 174) – a. Propose an explanation for why ethenyl (-CH=CH$_2$) has a larger preference for the equatorial position than ethynyl (-CCH).

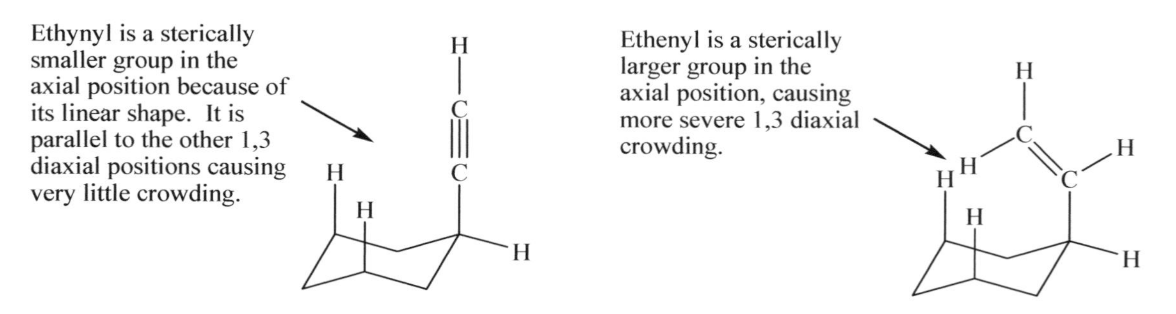

Ethynyl is a sterically smaller group in the axial position because of its linear shape. It is parallel to the other 1,3 diaxial positions causing very little crowding.
Ethenyl is a sterically larger group in the axial position, causing more severe 1,3 diaxial crowding.

b. Propose an explanation for why ethenyl ($-CH=CH_2$) has a smaller preference for the equatorial position than phenyl ($-C_6H_5$).

less strained with hydrogen pointing to middle of ring

rotate axial group

Even when ethenyl is axial it has the option of turning a hydrogen in towards the middle of the ring, greatly reducing the axial crowding.

rotate axial group

Nothing changes when the phenyl substituent rotates 180o. It is sterically crowded in both positions, so will have larger axial strain (energy).

c. Benzyl would seem to be a larger group than phenyl, but has a smaller A value. Propose a possible explanation.

less strained with hydrogen pointing to middle of ring

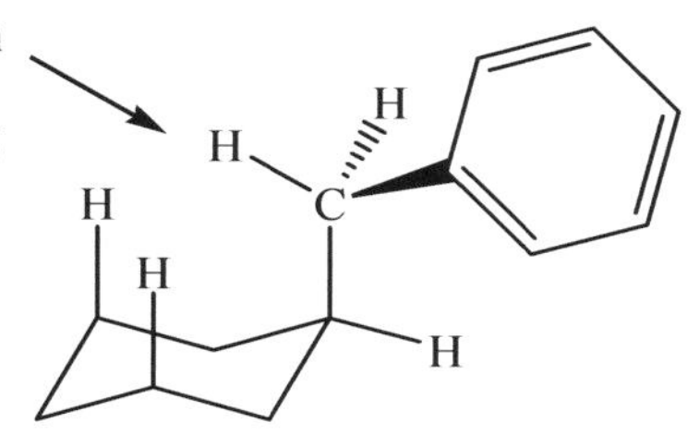

Because the attached carbon is sp3 hybridized, benzyl has an option of turning a hydrogen in towards the middle of the ring, greatly reducing the axial crowding.

As seen above, phenyl does not have any options for the relief of A strain (1,3 diaxial interactions).

Problem 4 (p 182) – Draw all isomers of dimethylcyclohexane and evaluate their relative energies and estimate an equilibrium distribution for the two chair conformations. Use the given energy values for substituted cyclohexane rings.

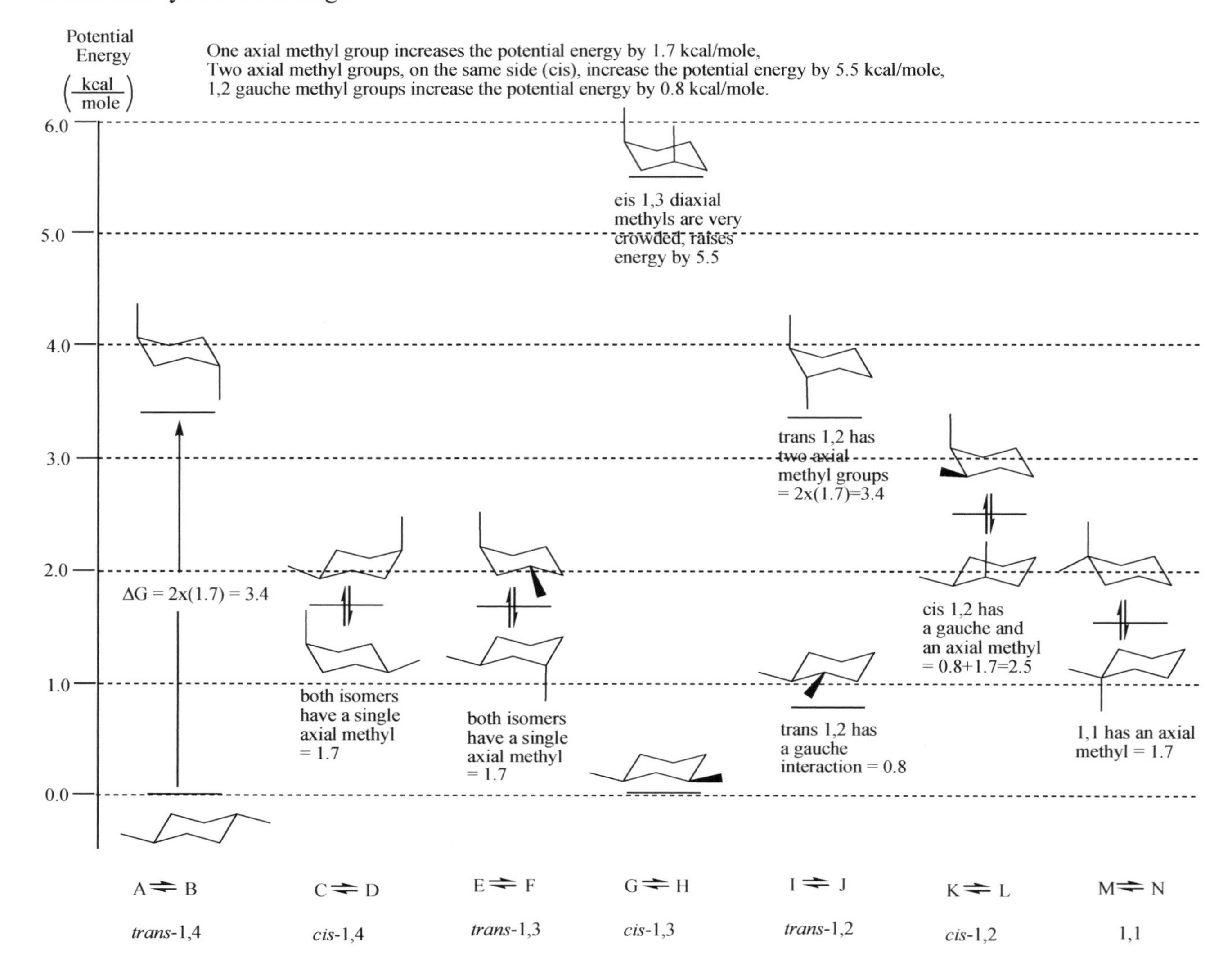
Potential Energy
(kcal/mole)
One axial methyl group increases the potential energy by 1.7 kcal/mole,
Two axial methyl groups, on the same side (cis), increase the potential energy by 5.5 kcal/mole,
1,2 gauche methyl groups increase the potential energy by 0.8 kcal/mole.
6.0
5.0
4.0
3.0
2.0
1.0
0.0
eis 1,3 diaxial methyls are very crowded, raises energy by 5.5
ΔG = 2x(1.7) = 3.4
trans 1,2 has two axial methyl groups = 2x(1.7)=3.4
both isomers have a single axial methyl = 1.7
both isomers have a single axial methyl = 1.7
cis 1,2 has a gauche and an axial methyl = 0.8+1.7=2.5
trans 1,2 has a gauche interaction = 0.8
1,1 has an axial methyl = 1.7
A ⇌ B
C ⇌ D
E ⇌ F
G ⇌ H
I ⇌ J
K ⇌ L
M ⇌ N
trans-1,4
cis-1,4
trans-1,3
cis-1,3
trans-1,2
cis-1,2
1,1

Problem 5 (p 182) - Both *cis* and *trans* 1-bromo-3-methylcyclohexane can exist in two chair conformations. Evaluate the relative energies of the two conformations in each isomer (use the energy values from the table presented earlier). Estimate the relative percents of each conformation at equilibrium using the difference in energy of the two conformations from you calculations. Draw each chair conformation in 3D bond line notation and as a Newman projection using the $C_1 \rightarrow C_6$ and $C_3 \rightarrow C_4$ bonds to sight down.

1-bromo-3-methylcyclohexane

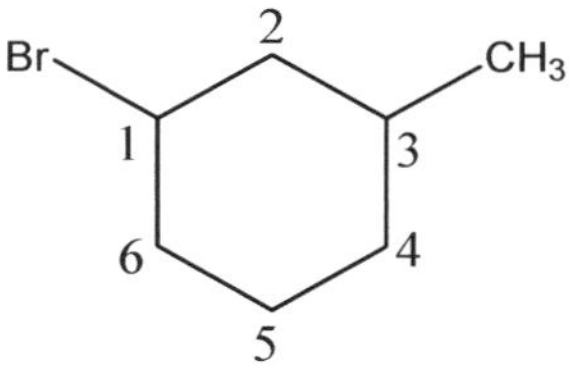

ΔH = -1.2 kcal/mole

axial Me = 1.7 axial Br = 0.5

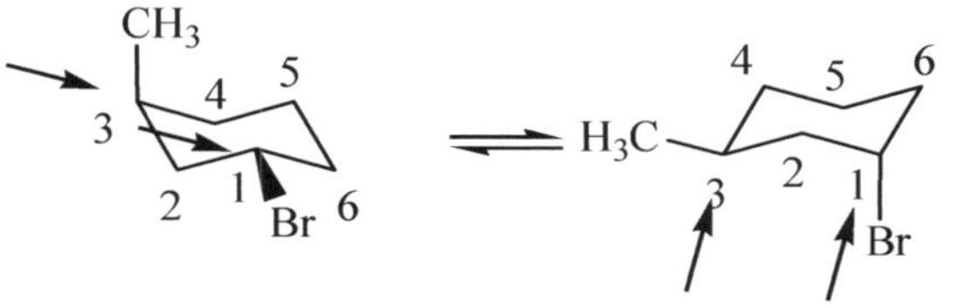

trans-1-bromo-3-methylcyclohexane

$$K = \frac{\text{chair 2}}{\text{chair 1}} = 10^{\frac{-\Delta G}{2.3RT}}$$

$$= 10^{\frac{-(-1{,}200 \text{ cal/mole})}{(2.3)(2 \text{ cal/mol-K})(298 \text{ K})}}$$

$$= 10^{+0.9} = \frac{7.4}{1.0} = \frac{88\%}{12\%}$$

ΔH = 2.2 kcal/mole

zero axial strain axial Me + Br = 1.7 + 0.5 = 2.2

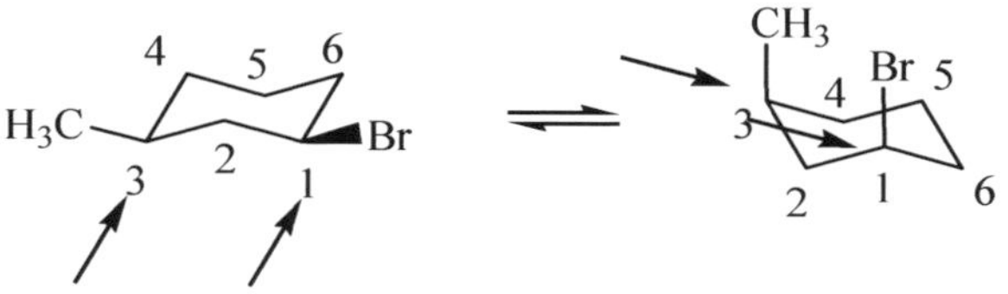

cis-1-bromo-3-methylcyclohexane

$$K = \frac{\text{chair 2}}{\text{chair 1}} = 10^{\frac{-\Delta G}{2.3RT}}$$

$$= 10^{\frac{-(2{,}200 \text{ cal/mole})}{(2.3)(2 \text{ cal/mol-K})(298 \text{ K})}}$$

$$= 10^{-1.6} = \frac{1}{39} = \frac{2\%}{98\%}$$

1.7 (Shows gauche on one side, but there is a second gauche on the other side.)

Probably a little higher because the CH_3 and Br are both axial on the same side.

1.7 0.5

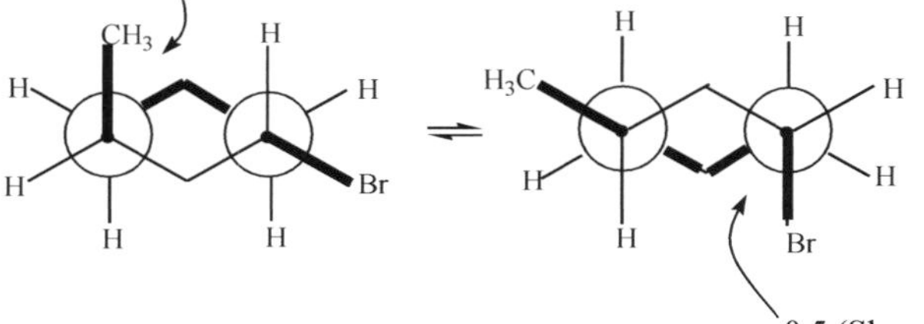

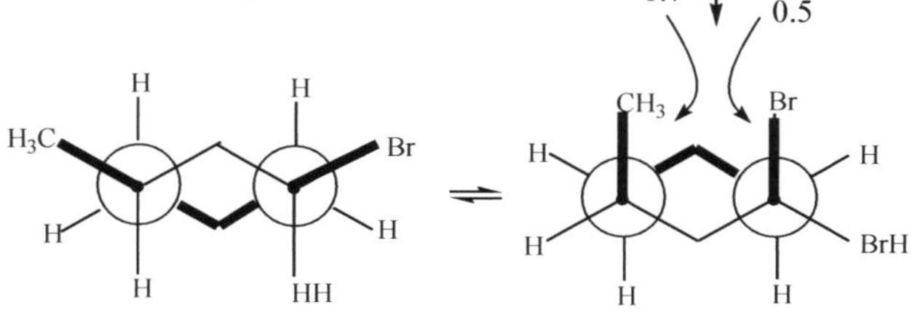

0.5 (Shows gauche on one side, but there is a second gauche on the other side.)

Problem 6 (p 183)– Estimate a value for the strain energy of the trimethylcyclohexane structures provided. Use the energy values below.

One axial methyl group increases the potential energy by 1.7 kcal/mole,
Two axial methyl groups, on the same side (cis), increase the potential energy by 5.5 kcal/mole,
Three axial methyl groups, on the same side, increase the potential energy by 12.9 kcal/mole and
1,2 gauche methyl groups increase the potential energy by 0.8 kcal/mole.

1. Reference compound (no features with strain) ΔH = 0 + 0 + 0 = 0	2 ΔH = 1x(ax)+2x(g) = 3.3	3 ΔH = 2x(ax)+1x(g) = 4.2	4 ΔH = 1x(ax) = 1.7
5 ΔH = 2x(ax) = 5.5	6 ΔH = 1x(ax) = 1.7	7 ΔH = 2x(ax) = 3.4	8 ΔH = 2x(g) = 1.8
9 ΔH = 1x(ax)+2x(g) = 3.3	10 ΔH = 1x(ax)+2x(g) = 3.3	11 ΔH = 2x(ax)+1x(g) = 4.2	12 ΔH = 2x(ax)+2x(g) = 7.1
13 ΔH = 3x(ax) = 7.2	14 ΔH = 1x(g) = 0.8	15 ΔH = 1x(ax)+1x(g) = 2.5	16 ΔH = 1x(ax)+1x(g) = 2.5
17 ΔH = 1x(ax)+1x(g) = 2.5	18 ΔH = 1x(ax)+1x(g) = 6.3	19 ΔH = 2x(ax)= 3.4	20 ΔH = 2x(ax)+1x(g) = 4.2
21 ΔH = 3x(ax) = 7.2	22 ΔH = 1x(ax)) = 1.7	23 ΔH = 2x(ax) = 5.5	24 ΔH = 3x(ax) = 12.9

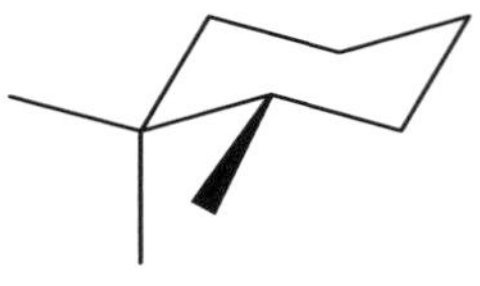

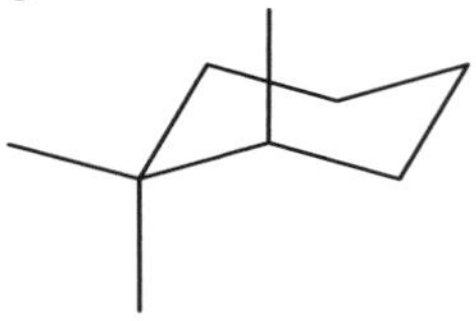

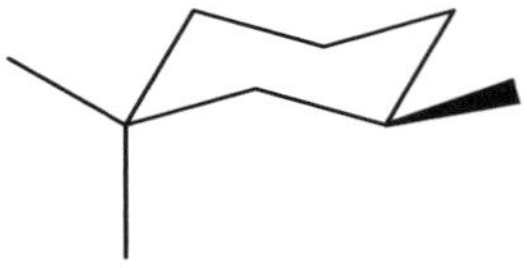

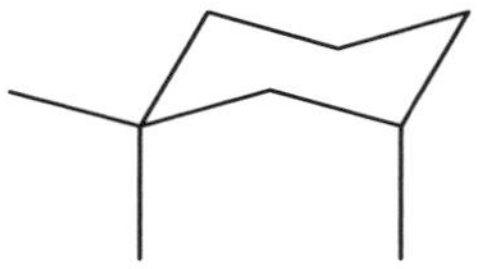

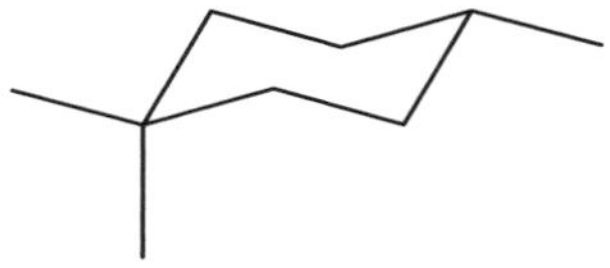

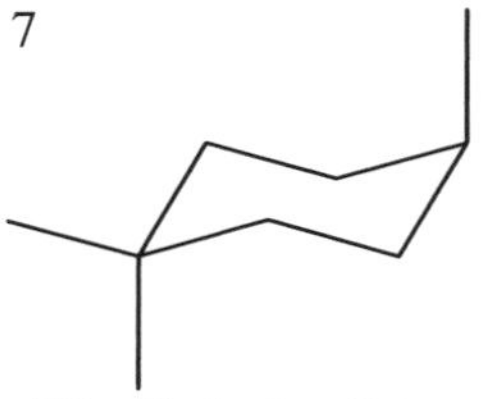

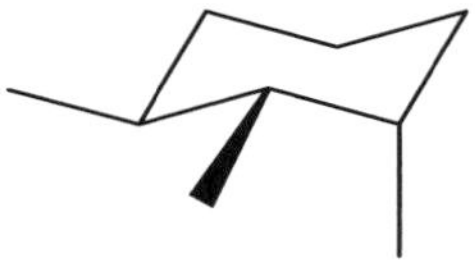

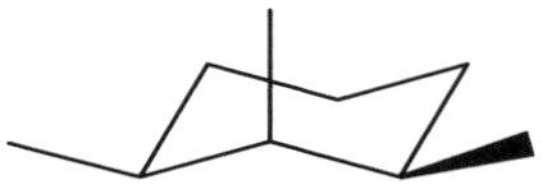

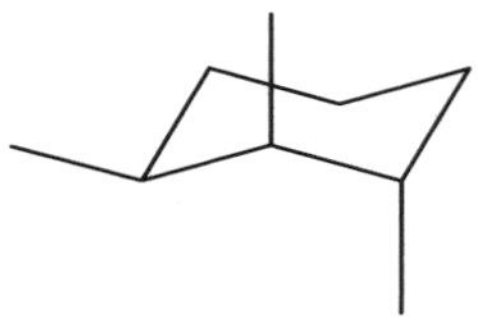

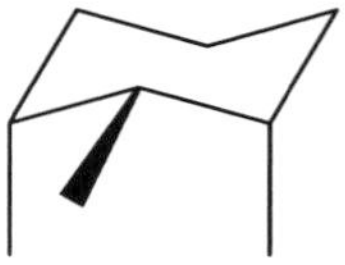

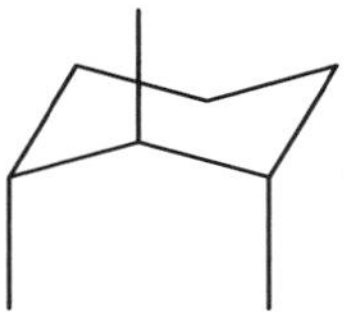

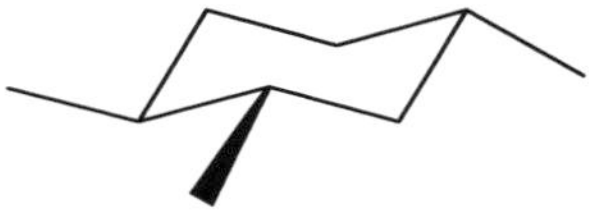

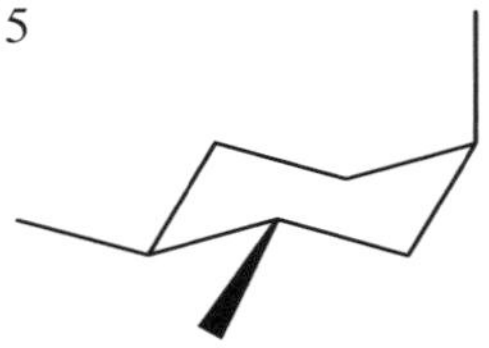

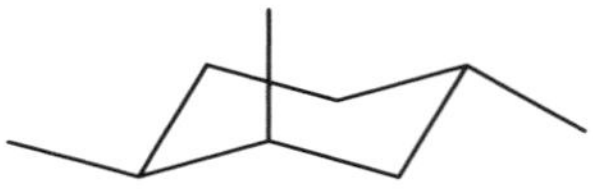

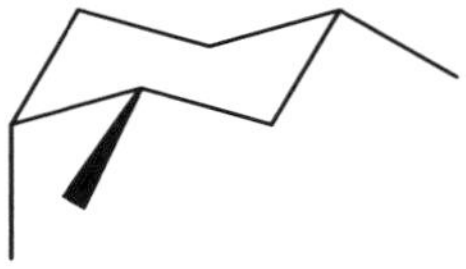

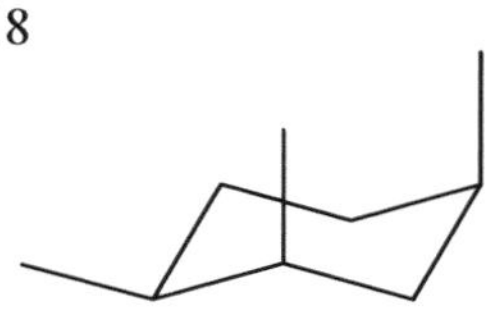

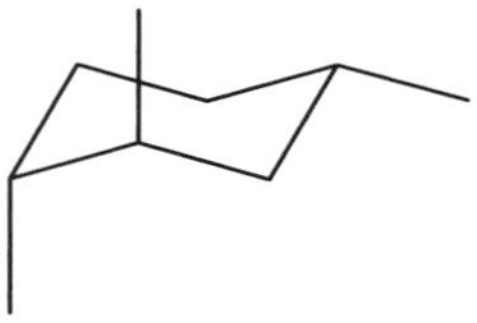

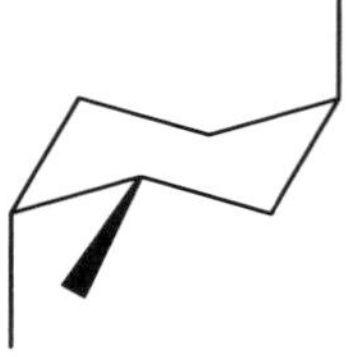

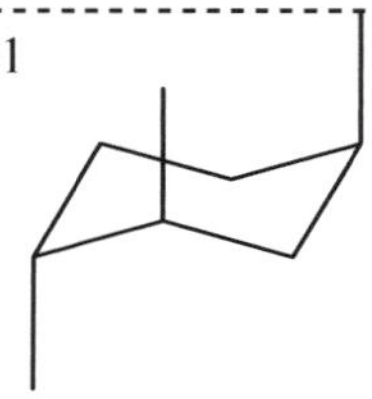

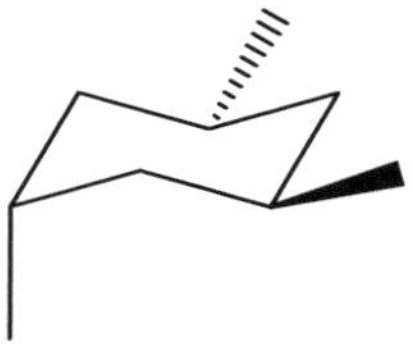

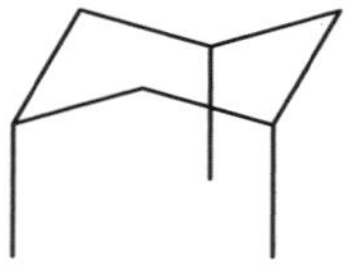

Chapter 7

Problem 1 (p 197) - Classify the absolute configuration of all chiral centers as R or S in the molecules below. Use hands (or model atoms) to help you see these configurations whenever the low priority group is facing towards you (the wrong way). Find the chiral centers, assign the priorities and make your assignments.

a. 1 Cl, 4 H, 3, 2, R

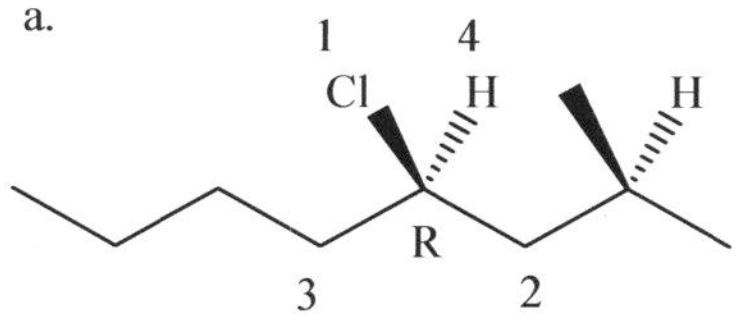

b. HO, 3 H_3C, 4 H, 1, 2, S

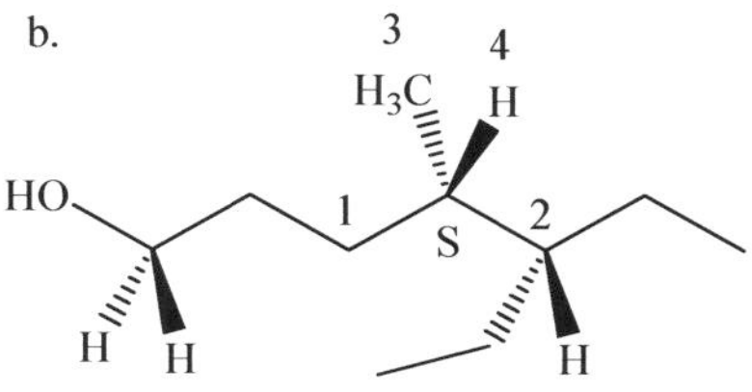

c. H_3C, R, H, Cl, S, CH_3

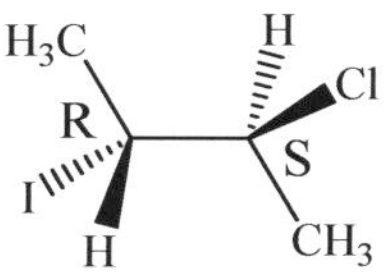

d. no chiral centers

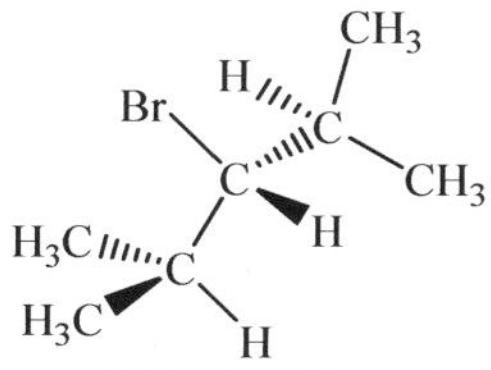

e. O, S, R, H_3C, Cl

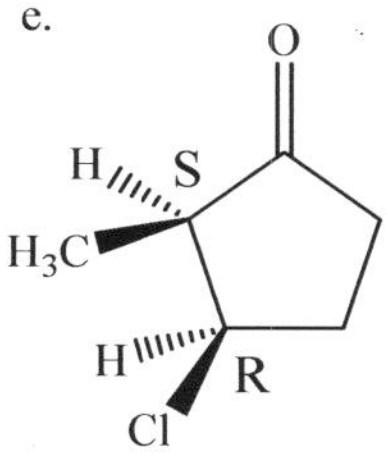

f. no chiral centers

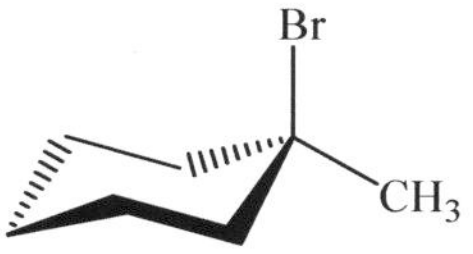

g. OH, H, R, CH_3, H_3C

h. Br, CH_3, S, H_3C, H

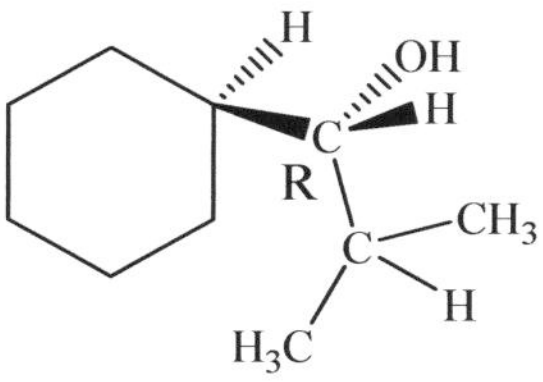

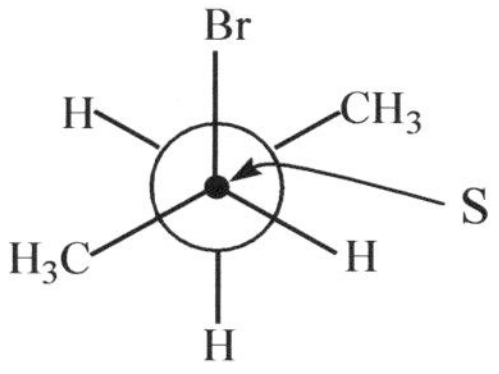

i. CH_2CH_3, H, Br, R, CH_3

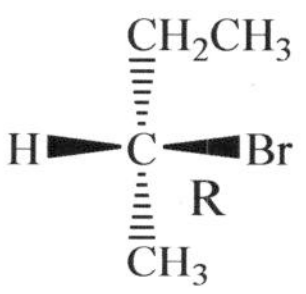

j. Cl, S, H, R

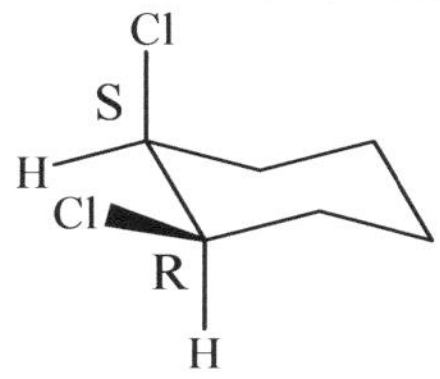

k. H_3CH_2C, Br, CH_3, R, D, H

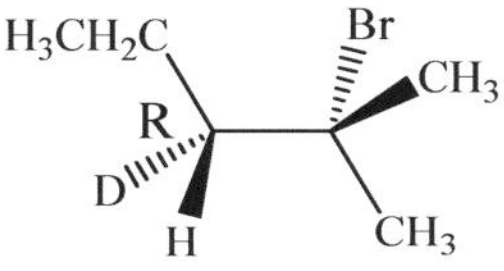

l. $\oplus$ S, CH_3, $\ominus$ O, S

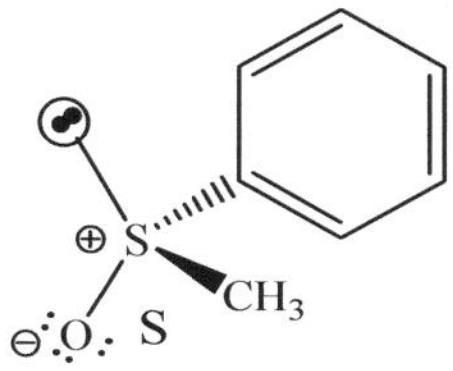

Problem 2 (p 198) - Evaluate the order of priority in each part from highest (= 1) to lowest (= 4).

a * = path to chiral center

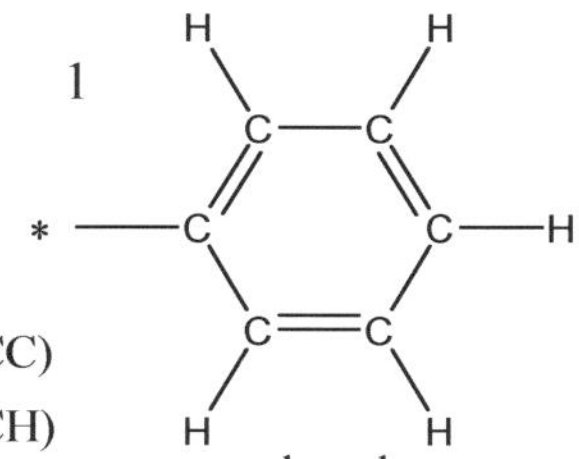

2 ethynyl: $C_1(CCC)$, $C_2(CCH)$, $C_3(none)$

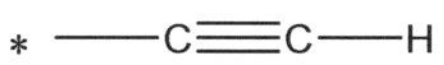

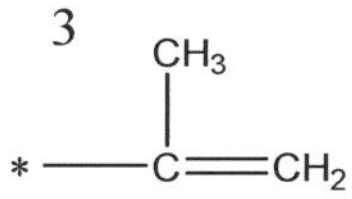

1 phenyl: $C_1(CCC)$, $C_2(CCH)$, $C_3(CCH)$

3 2-propenyl: $C_1(CCC)$, $C_2(CHH)$, $C_3(none)$

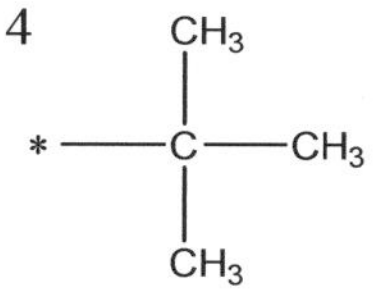

4 t-butyl: $C_1(CCC)$, $C_2(HHH)$, $C_3(none)$

b

3: $C_1(CHH)$, $C_2(OOH)$

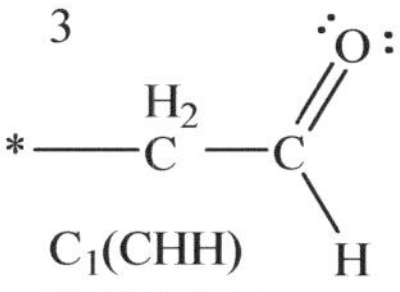

2: $C_1(CHH)$, $C_2(FHH)$

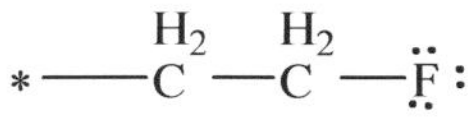

4: $C_1(CHH)$, $C_2(NNN)$

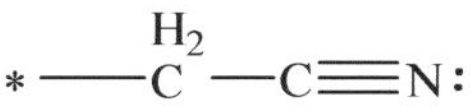

1: $C_1(OCC)$, $C_2(I\ HH)$

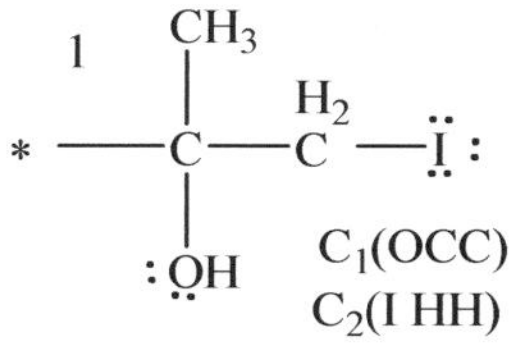

c

4: $C_1(OOH)$, $O_2(H)$

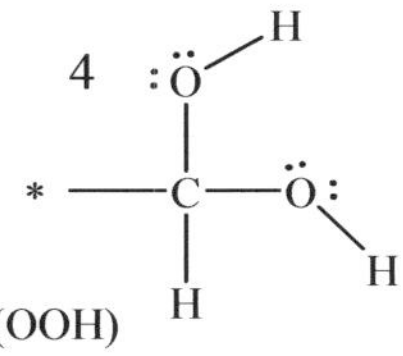

3: $C_1(OOH)$, $O_2(C)$

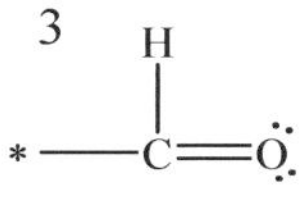

2: $C_1(OOH)$, $O_2(C)$, $C_3(HHH)$

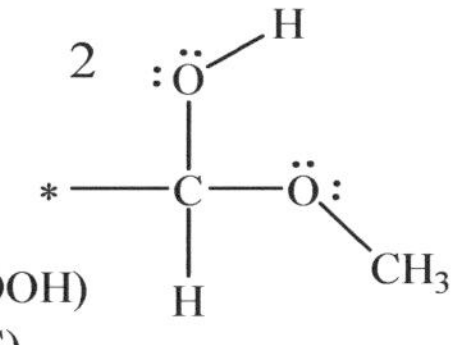

1: $C_1(OOH)$, $O_2(C)$ (twice), $C_3(HHH)$

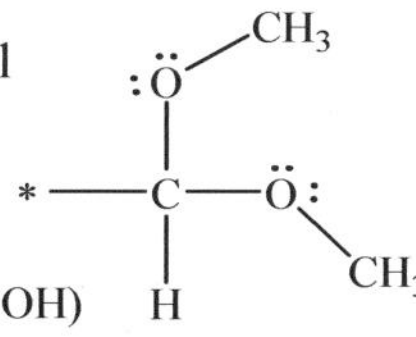

d

4	1	2	3
$C_1(CCC)$	$C_1(CCC)$	$C_1(CCC)$	$C_1(CCC)$
$C_2(CCH)$	$C_2(CCC)$	$C_2(CCH)$	$C_2(CCH)$
$C_3(CCH)$	$C_3(CCH)$	$C_3(C,C,Cl)$	$C_3(CCH)$
$C_4(CCH)$			$C_4(CCBr)$

Problem 3 (p 207) – For the following set of Fischer projections answer each of the questions below by circling the appropriate letter(s) or letter combination(s). Hint: Redraw the Fischer projections with the longest carbon chain in the vertical direction and having similar atoms in the top and bottom portion. Classify all chiral centers in the first structure as R or S absolute configuration. (15 pts)

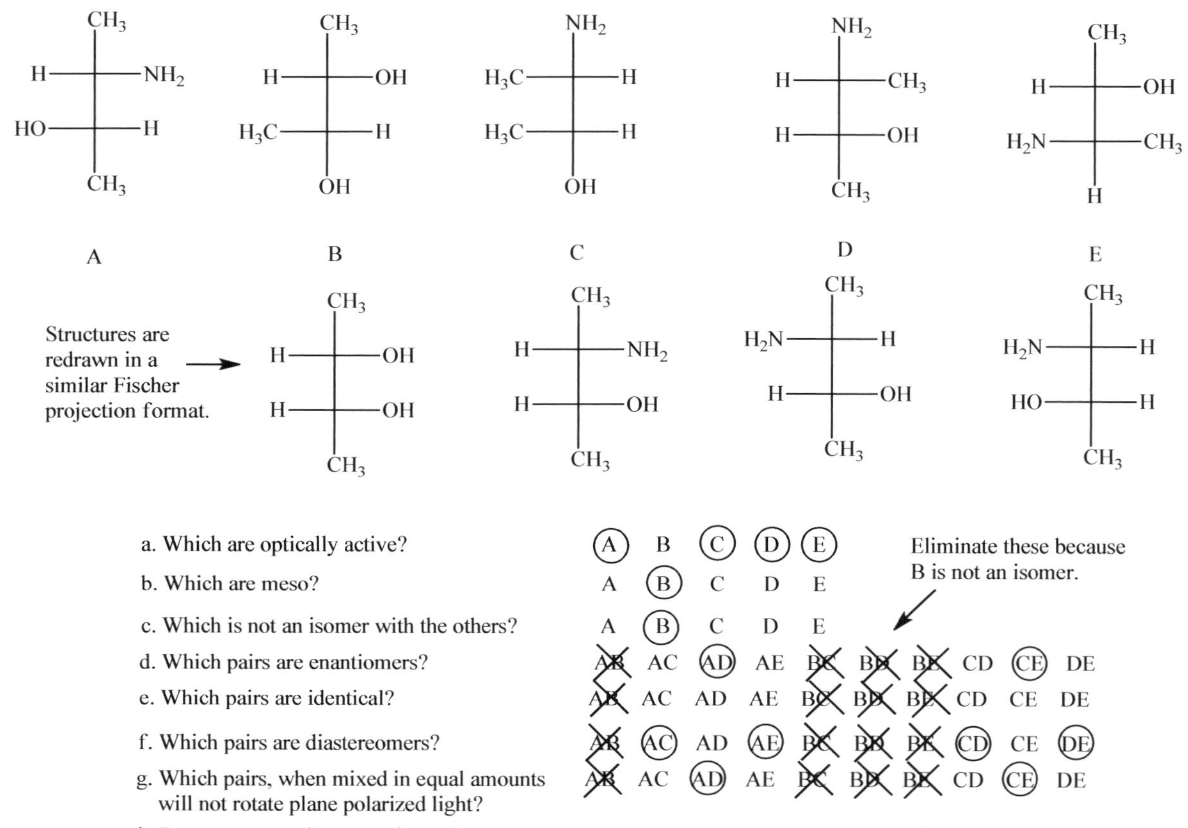

A
B
C
D
E
Structures are redrawn in a similar Fischer projection format.
a. Which are optically active?
b. Which are meso?
c. Which is not an isomer with the others?
d. Which pairs are enantiomers?
e. Which pairs are identical?
f. Which pairs are diastereomers?
g. Which pairs, when mixed in equal amounts will not rotate plane polarized light?
Eliminate these because B is not an isomer.

h. Draw any stereoisomers of 3-amino-2-butanol as Fischer projections, which are not shown above. If there are none, indicate this. All are shown.

i. Would anything change if, in compound C, the NH_2 was replaced with a OH group? It would be like B.

j. Circle all chiral centers in a recently discovered Costa Rican fungal compound showing antibacterial properties against vancomycin resistant bacteria. How many stereoisomers are possible with that many chiral centers?

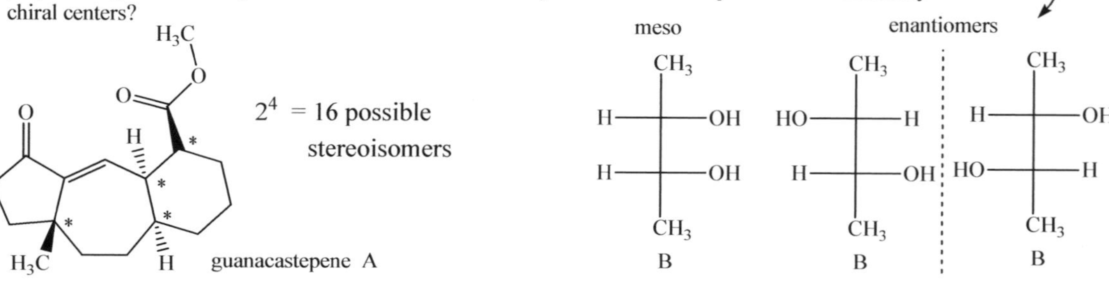

Problem 4 (p 208) - For the following set of Fischer projections answer each of the questions below by circling the appropriate letter(s) or letter combination(s). Hint: Redraw the Fischer projections having the longest carbon chain in the vertical direction and having similar atoms in the top and bottom portion. Classify all chiral centers in the first structure as R or S absolute configuration. (15 pts)

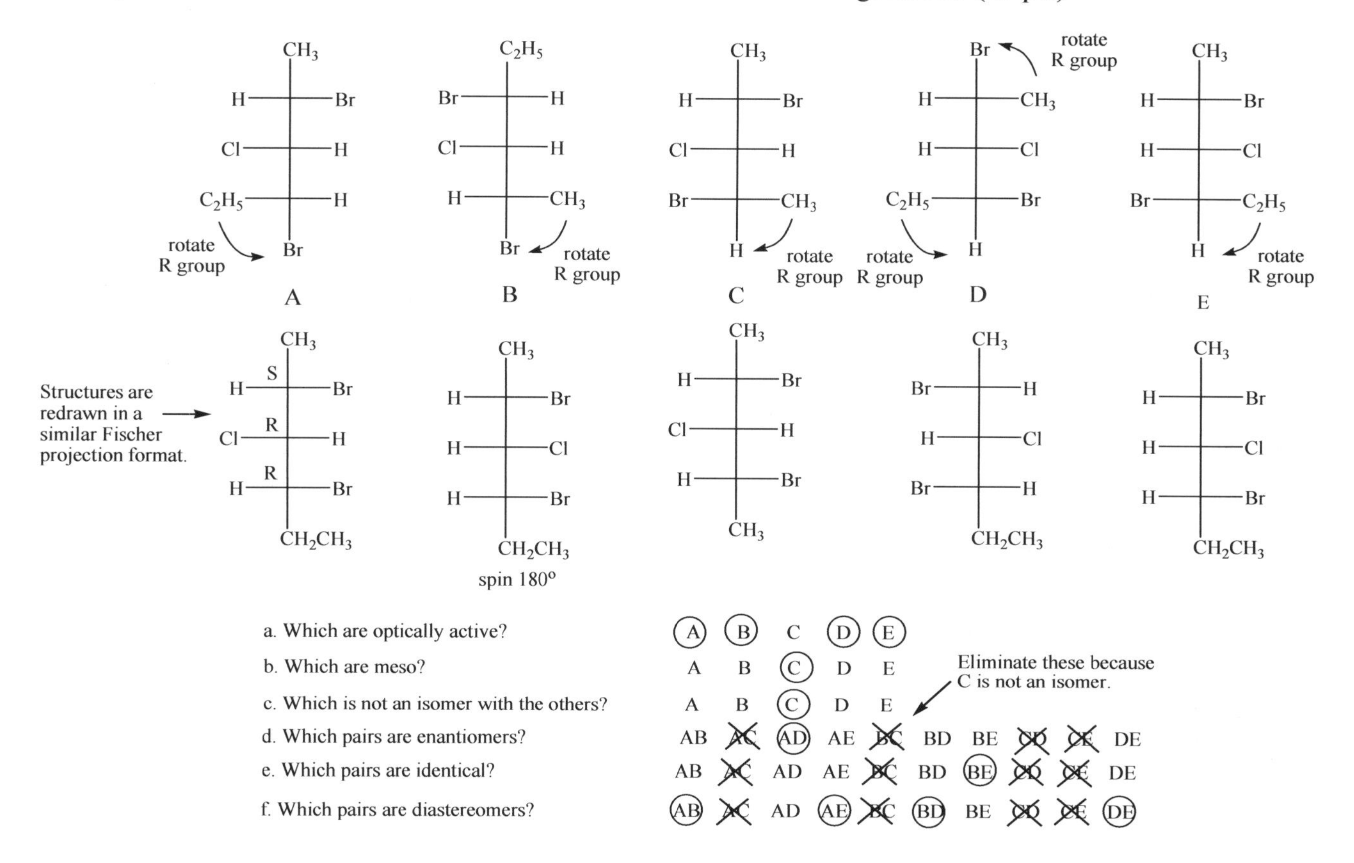

e. Draw Fischer projections of any stereoisomers of "A" which are not shown above. If there are none, indicate this. (8 pts)

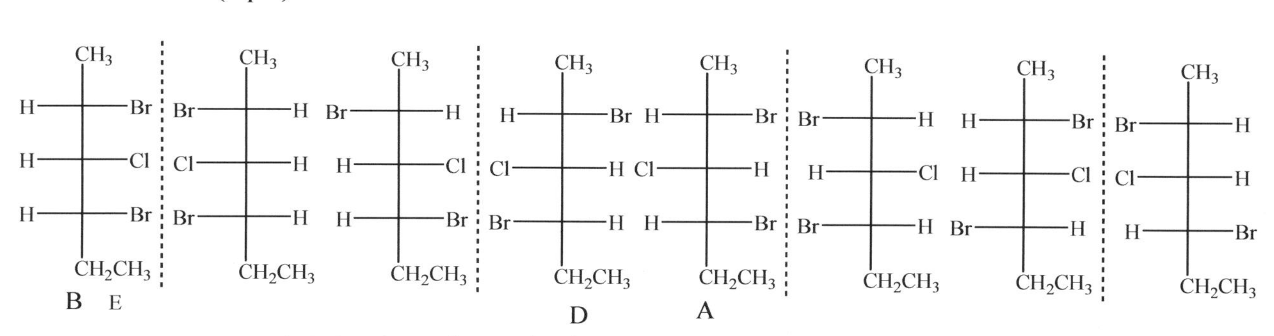

A, B, D and E are shown. The others were not shown above.

f. Derivatives of the antitumor steroidal saponin were recently prepared. The are highly potent and selective anticancer compounds. They inhibit Na^+/Ca^{+2} exchange leading to higher Ca^{+2} in the cytosol and mitochrondria causing cell death (apotosis) (Org. Lett. ASAP, 2014). Circle all chiral centers and any other stereogenic features in the partial structure below, and calculate the maximum number of stereoisomers possible. (4 pts)

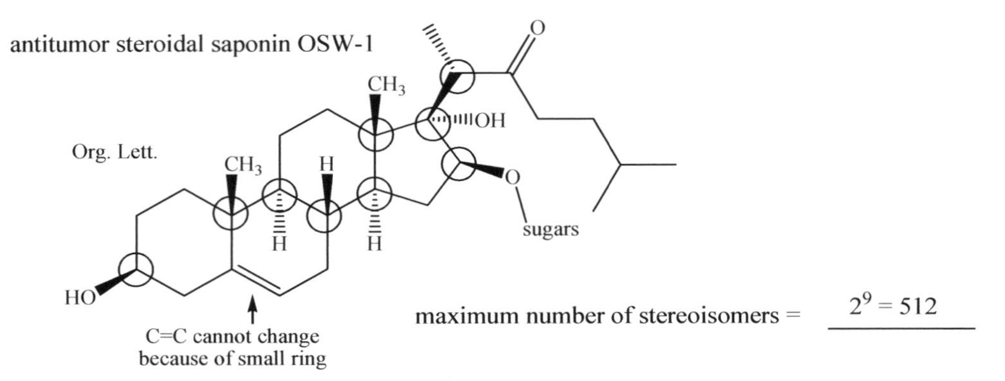

maximum number of stereoisomers = $2^9 = 512$

Problem 5 (p 209) – For the following set of Fischer projections answer each of the questions below by circling the appropriate letter(s) or letter combination(s). Hint: Redraw the Fischer projections having the longest carbon chain in the vertical direction and having similar atoms in the top and bottom portion. Classify all chiral centers in the first structure as R or S absolute configuration.

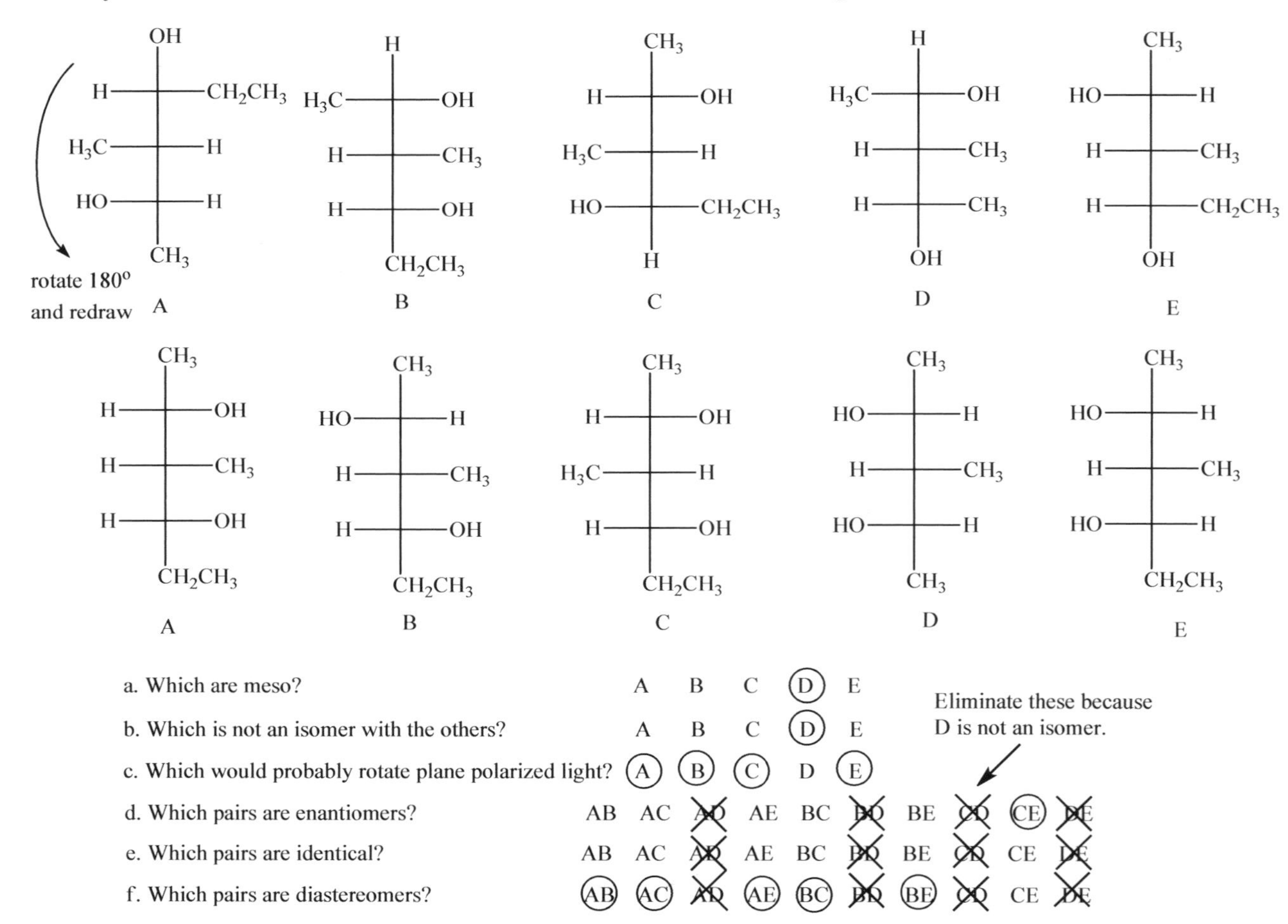

a. Which are meso? A B C (D) E

b. Which is not an isomer with the others? A B C (D) E

c. Which would probably rotate plane polarized light? (A) (B) (C) D (E)

d. Which pairs are enantiomers? AB AC ~~AD~~ AE BC ~~BD~~ BE ~~CD~~ (CE) ~~DE~~

e. Which pairs are identical? AB AC ~~AD~~ AE BC ~~BD~~ BE ~~CD~~ CE ~~DE~~

f. Which pairs are diastereomers? (AB) (AC) ~~AD~~ (AE) (BC) ~~BD~~ (BE) ~~CD~~ CE ~~DE~~

Eliminate these because D is not an isomer.

g. Draw Fischer projections of any stereoisomers of "A" which are not shown above. If there are none, indicate this.

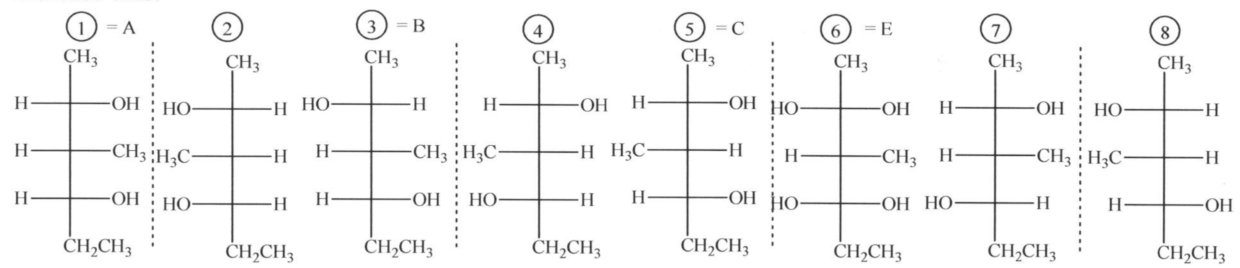

There are 8 stereoisomers. 2, 4, 7 and 8 are not shown above.

h. Would anything change if the ethyl branch was changed to a methyl branch?

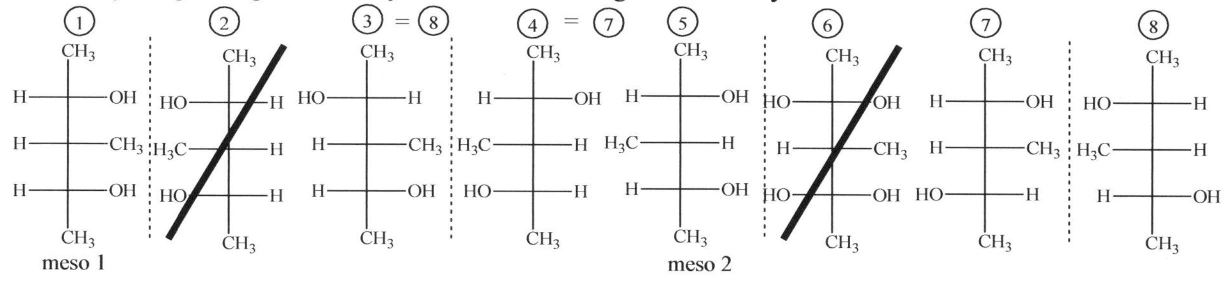

There would only be 4 stereoisomers because some would become meso and 7/8 would become identical to 3/4.

i. The structure of lobophytone A was recently determined (and the absolute configuration of all chiral centers!). It was isolated from the soft coral, *Lobophytum pauciflorum*, found in the South China Sea (Org. Lett. p.2482, 2010). Circle all chiral centers and any other stereochemical features, and calculate the maximum number of stereoisomers possible.

lobophytone A, one of seven biscembranoids found in coral living in South China Sea

E/Z possible

2^{11} possible from chiral centers = 2048

2^{12} including E/Z center = 4096

Problem 6 (p 210) - Place glyceraldehyde (2,3-dihydroxypropanal) in the proper orientation to generate a Fischer projection. Can you find any stereogenic centers? Is the molecule as a whole chiral? If so, draw each enantiomer as a 3D representation and classify all stereogenic centers as R or S.

2S 2R 2S 2R

There is only 1 chiral center, so the molecules are chiral and enantiomers.

Problem 7 (p 210) - Draw a 3D Newman projection and a sawhorse representation for each the following Fischer projections. Redraw each structure in a sawhorse projection of a stable conformation. Identify stereogenic atoms as R or S.

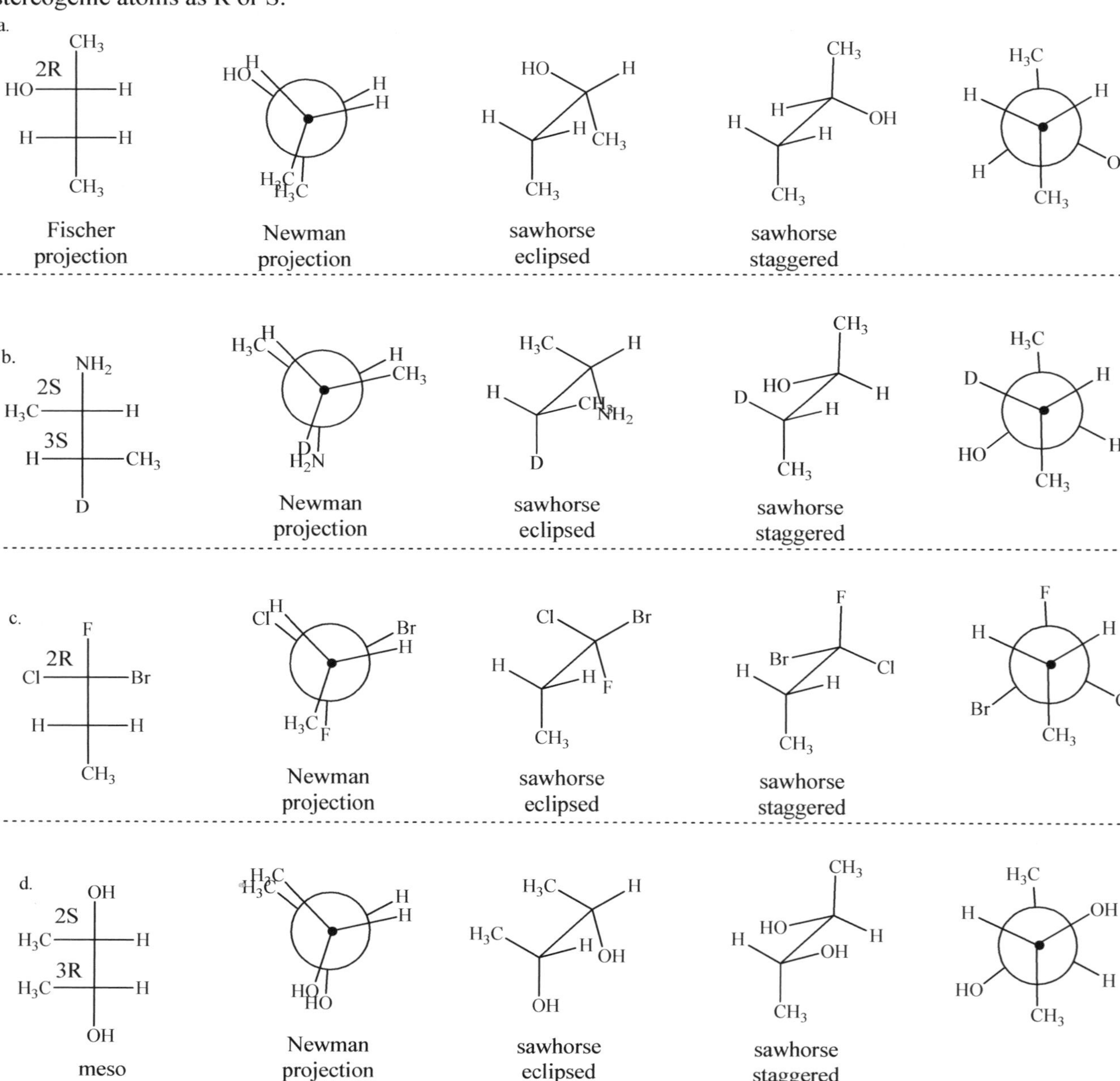

Problem 8 (p 210) - Rearrange the Fischer projections below to their most acceptable form. You may have to rotate the top and/or bottom atom(s). You also may have to twist a molecule around 180° in the plane of the paper. Assign the absolute configurations of all stereogenic centers. Using your arm and fingers is helpful here. Write the name of the first structure. How many total stereoisomers are possible for each example?

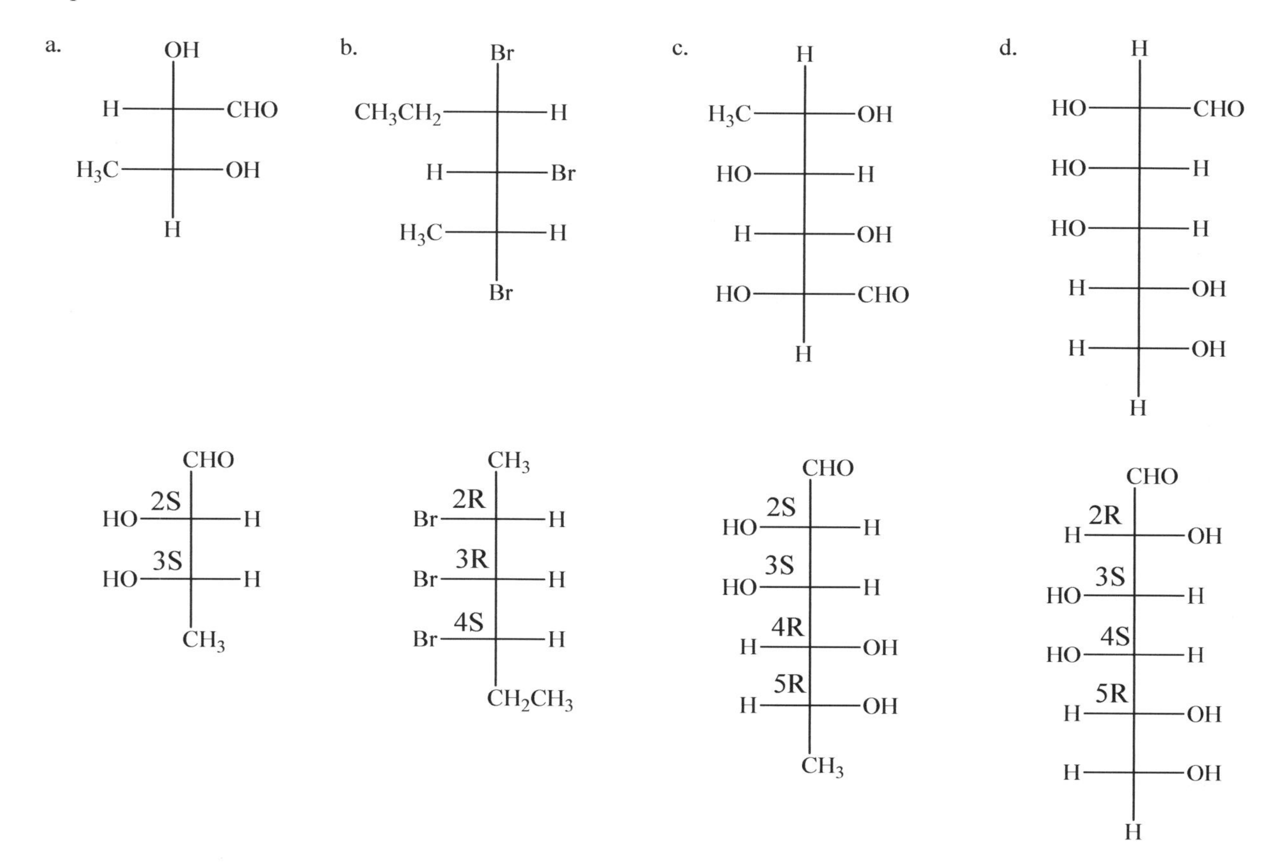

Problem 9 (p 212) – What would happen to the number of stereoisomers in each case above if the top aldehyde functionality were reduced to an alcohol functionality (a whole other set of carbohydrates!)? A generic structure is provided below to show the transformation. Nature makes some of these too. How many chiral centers does each example have? How many stereoisomers are possible for each length of carbon? What are the absolute configurations of any chiral centers? Specify the first stereoisomer as A (then B, C, etc.) and state what each relationship is to the others. Are there any meso structures.

O=C(H)–⌇–CH_2OH → Reduce aldehyde to alcohol → CH_2OH–⌇–CH_2OH

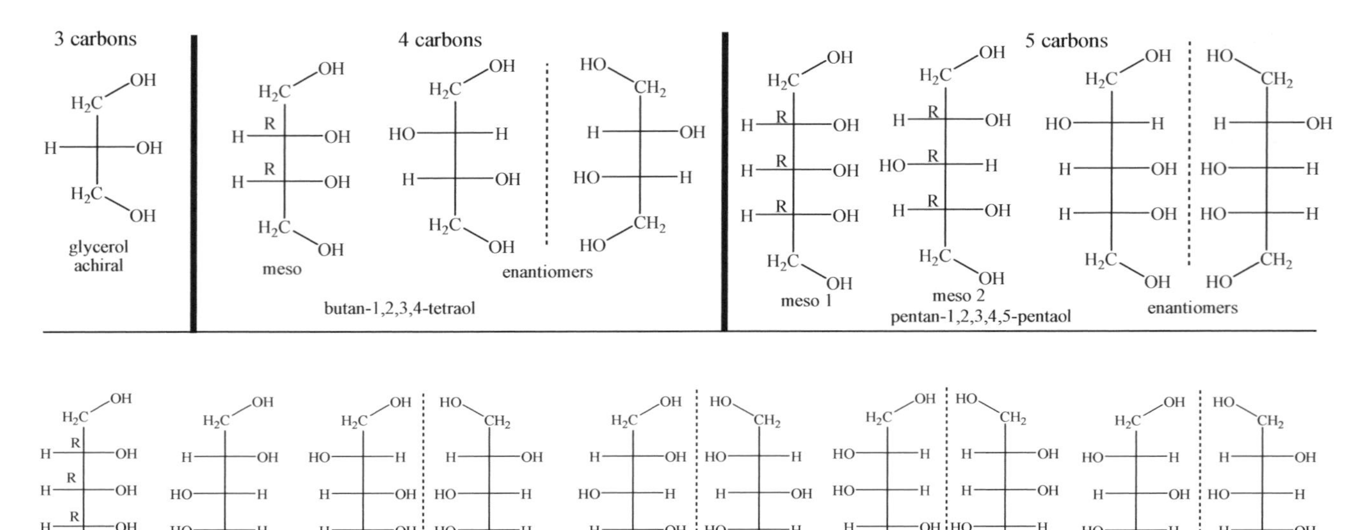

Three carbon tri-ol carbohydrate = becomes only one structure which is achiral

Four carbon tetra-ol carbohydrates = becomes three stereoisomers (one meso pattern)

Five carbon penta-ol carbohydrates = becomes four stereoisomers (two meso patterns, one duplication)

Six carbon hexa-ol carbohydrates = becomes ten stereoisomers (two meso patterns, two duplications)

Problem 10 (p 212) – How many chiral centers are found in cholesterol? How many potential stereoisomers are possible? Nature only makes one of them!

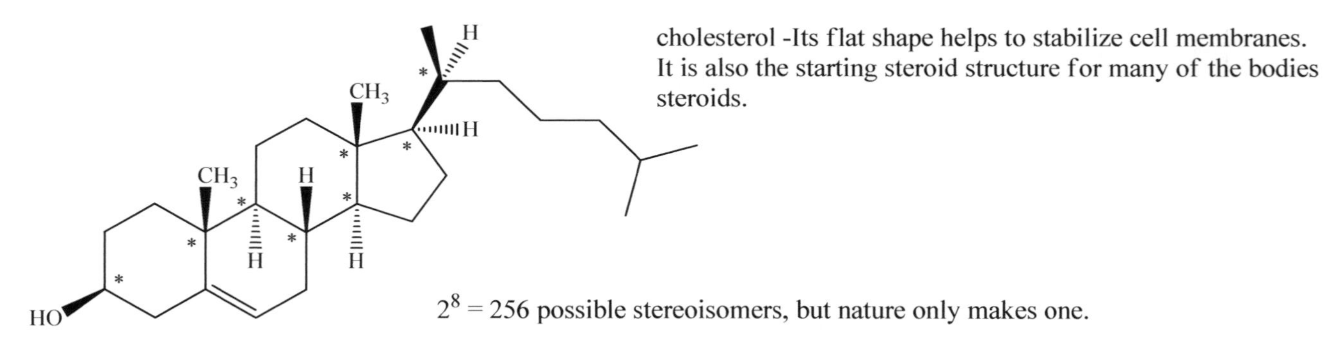

cholesterol -Its flat shape helps to stabilize cell membranes. It is also the starting steroid structure for many of the bodies steroids.

$2^8 = 256$ possible stereoisomers, but nature only makes one.

Chapter 8

Problem 1 (p 216) – Write an equation showing each of the following structures reacting as a Bronsted base, using H-A as the acid. If structures do not have any lone pairs of electrons, you will have to use the pi electrons as the base. One carbon atom of the pi bond will lose its share of the electron pair when the pi bond is broken and a new sigma bond is made using the other carbon atom and the proton. What will be the formal charge on the carbon atom losing the electrons? Hint: That carbon atom will become a carbocation. How are 'c' and 'd' related?

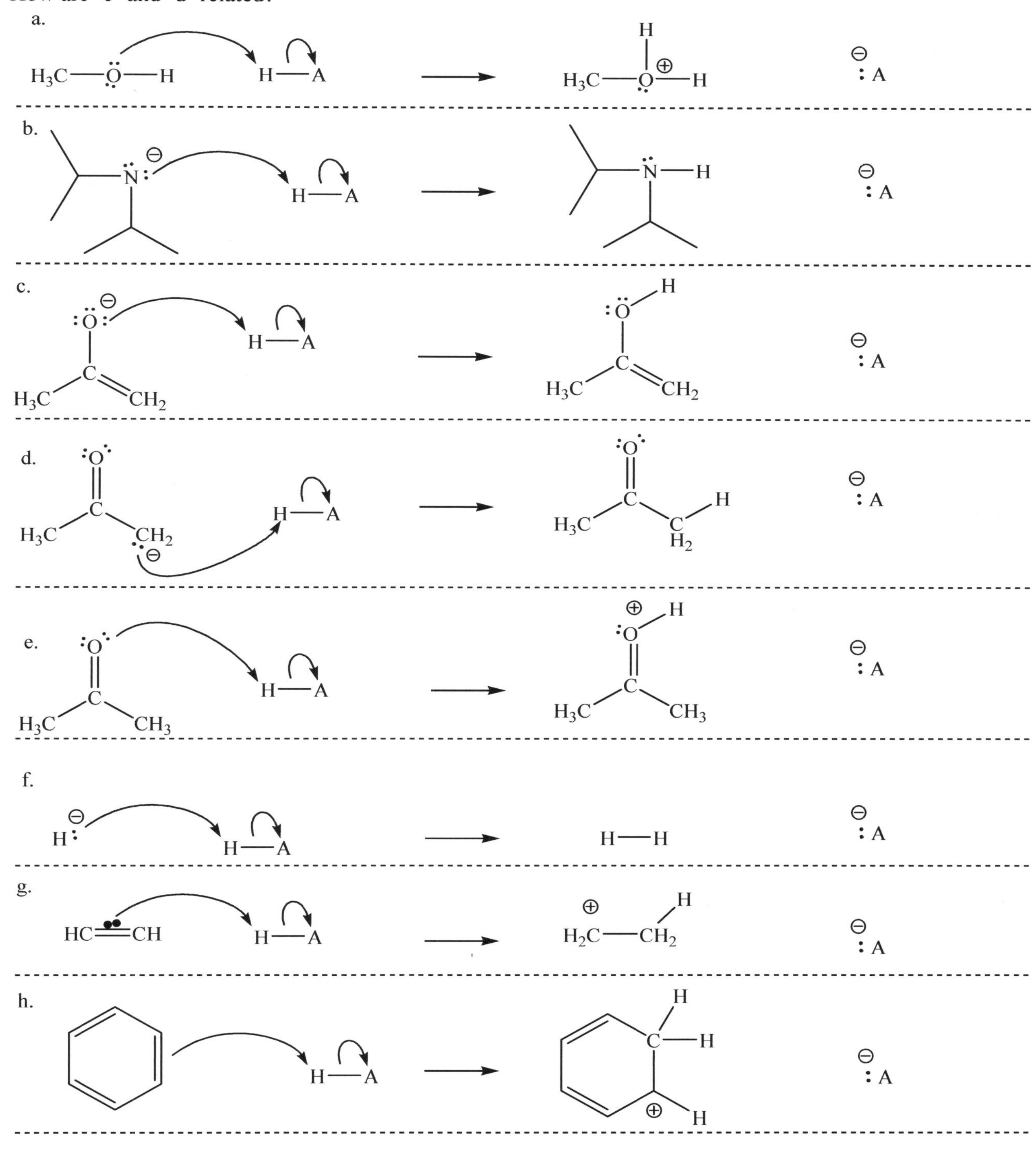

Problem 2 (p 217) – Write an equation showing each of the following reacting as a Bronsted acid, using B: as the base. You are going to have to leave two electrons behind and pay attention to formal charge.

a.

b.

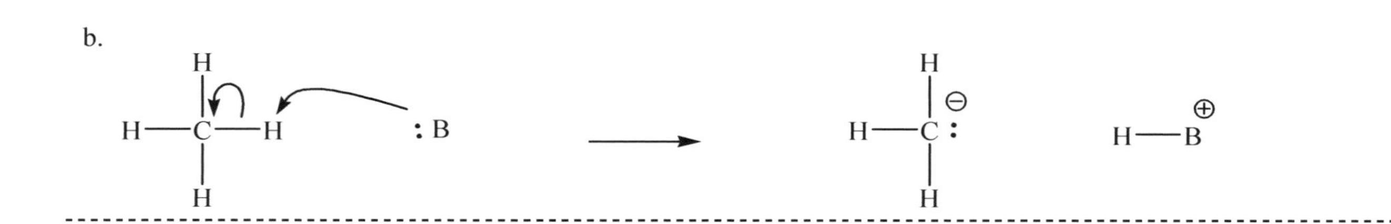

c.

H—C≡C—H   : B ⟶ H—C≡C : ⊖   H—B ⊕

d.

H_3C–C(=O)–CH_2–H   : B ⟶ H_3C–C(=O)–CH_2 ⊖   H—B ⊕

e.

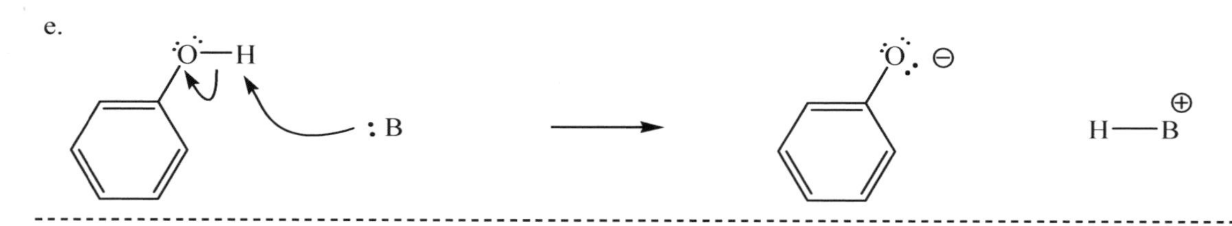

f.

H_3C—O(H)—H ⊕   : B ⟶ H_3C—O—H   H—B ⊕

g.

B :   H—O : ⊖ ⟶ : O :$^{-2}$ H—B ⊕

Problem 3 (p 217) –There are two reasonable choices for a water molecule reacting with a t-butyl carbocation. One answer involves water acting as a Bronsted base and the other involves water acting as a Lewis base (which equation goes with each term?). Write in the necessary lone pairs of electrons, curved arrows and formal charge to show each of these possibilities. What are other terms that could be used to describe the water and carbocation reactants? Match those terms with the appropriate reactant structures. This example illustrates a common competition for electron pair donors in organic chemistry: react with a hydrogen atom or react at a carbon atom.

t-butyl carbocation + H_2O → (partial yield (in one step)) → product a ($CH_2=C(CH_3)_2$) + H_3O^+

t-butyl carbocation + H_2O → (partial yield (in two steps)) → $(CH_3)_3C-OH_2^+$ + H_2O → product b ($(CH_3)_3C-OH$) + H_3O^+

Problem 4 (p 219) - Indicate whether the following compounds are stronger or weaker acids than water ($K_a = 10^{-16}$). Write an arrow-pushing mechanisms with generic base, B: (B = HO^-, B-H^+ = H_2O).

a. $HO-SO_2-OH$ ($K_a = 10^3$) + :B ⇌ $HO-SO_2-O^-$ + $H-B^+$

$$\frac{K_a\,(H_2SO_4) = 10^3}{K_a\,(H_2O) = 10^{-16}} = 10^{19}$$

Much stronger than water.

b. $H-C\equiv C-H$ ($K_a = 10^{-25}$) + :B ⇌ $H-C\equiv CH^-$ + $H-B^+$

$$\frac{K_a\,(HCCH) = 10^{-25}}{K_a\,(H_2O) = 10^{-16}} = 10^{-9}$$

Weaker than water.

c. $H_3C-C(=O)-OH$ ($K_a = 10^{-5}$) + :B ⇌ $H_3C-C(=O)-O^-$ + $H-B^+$

$$\frac{K_a\,(HCCH) = 10^{-5}}{K_a\,(H_2O) = 10^{-16}} = 10^{11}$$

Stronger than water.

d. H_3C-CH_3 ($K_a = 10^{-50}$) + :B ⇌ $H_3C-CH_2^-$ + $H-B^+$

$$\frac{K_a\,(HCCH) = 10^{-50}}{K_a\,(H_2O) = 10^{-16}} = 10^{-34}$$

Stronger than water.

Problem 5 (p 219) - Indicate whether the following bases are stronger or weaker bases than hydroxide. The K_a values given are for the conjugate acids. You need to add a proton to evaluate each base from the point of view of its conjugate acid and then invert your conclusion to decide "base" ability (stronger base = weaker acid). That goes for hydroxide too (K_a of its conjugate acid = 10^{-16}. What is its conjugate acid?). Write arrow-pushing mechanisms with generic acid, H-A.

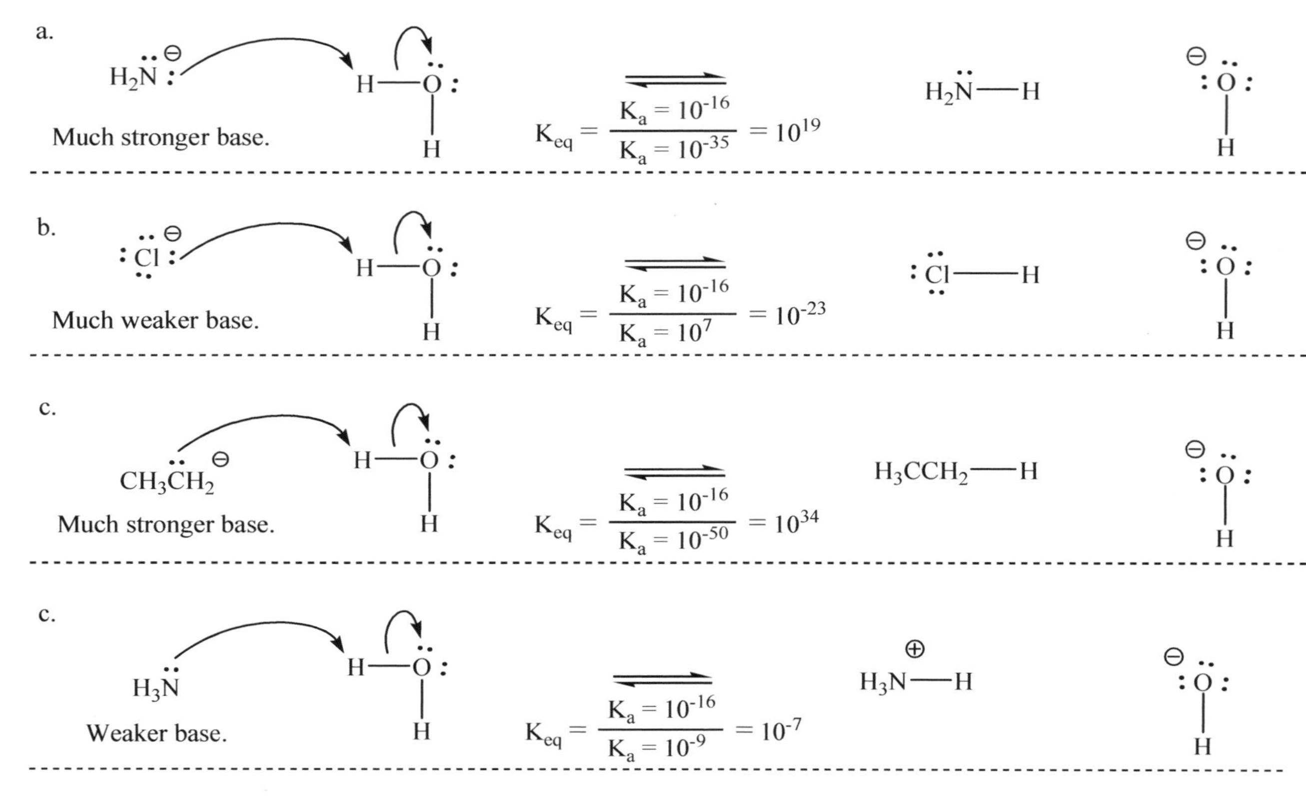

Problem 6 (p 220) – Order the acids from strongest acid (= 1) to weakest acid in each series? What is the basis for your answers? What does that tell you about the stability of the conjugate bases? What about their basicities?

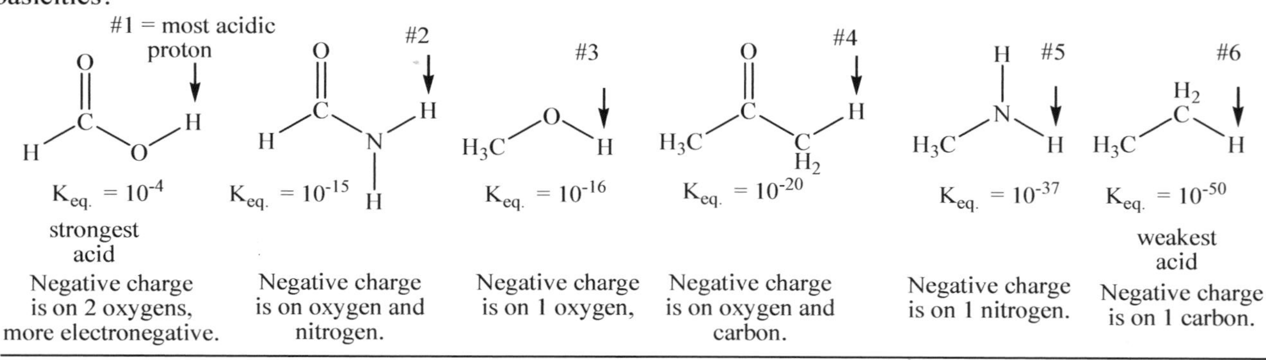

Problem 7 (p 220) - Are the bases below stronger or weaker than hydroxide (K_a of conjugate acid = 10^{-16})? The acidity constant provided (K_a) is that of each base's conjugate acid. Write out each conjugate acid. (Hint: Figure out the relative strengths of the conjugate acids before you predict the relative strengths of the bases.)

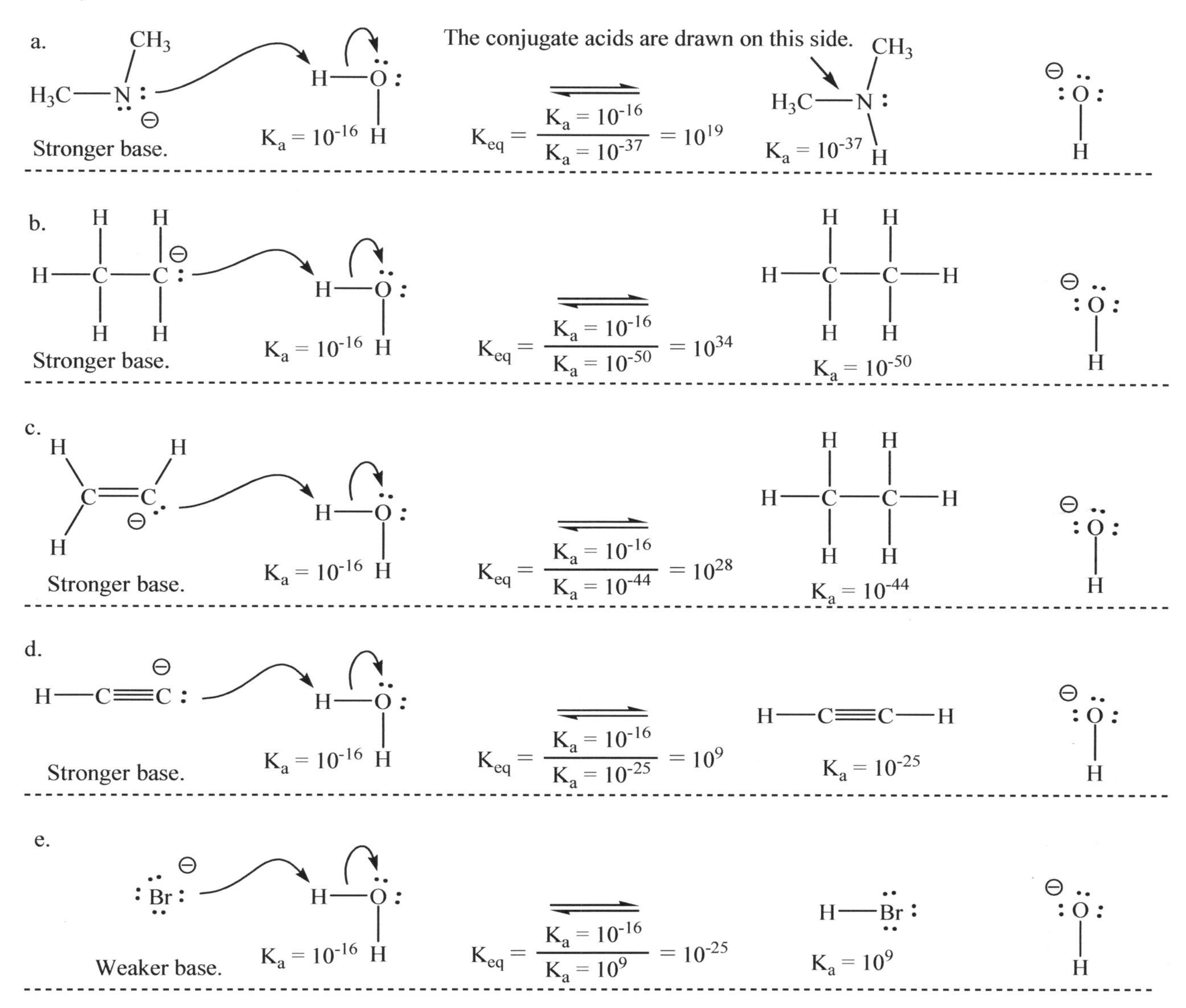

Problem 8 (p 220) - Predict whether the products or the reactants are favored in each of the following equilibria. An equation is really a combination of two acid/base reactions (K_a), one written in the forward direction and one written in the reverse direction. Put the K_a of the forward reacting acid in the numerator (normal K_a expression) and the K_a of the reverse reacting acid in the denominator (inverted). The ratio shows the balance between the two acid/base reactions (K_{eq}). Add in curved arrows to show electron movement.

Example

$$B: \;+\; H—A \;\rightleftharpoons\; B^{\oplus}—H \;+\; :A^{\ominus}$$

$K_a(HA)$ $\qquad$ $K_a(HB)$

$$K_{eq.} = \frac{K_a(HA)}{K_a(HB)}$$

$K_a(HA)$ ← normal K_a expression

$K_a(HB)$ ← inverse K_a expression

a. NH_2^{-} + $H-C\equiv C-H$ ($K_a = 10^{-25}$) $\overset{K_{eq} = ?}{\rightleftharpoons}$ NH_3 ($K_a = 10^{-37}$) + $H-C\equiv C:^{-}$

$$K_{eq} = \frac{K_a = 10^{-25}}{K_a = 10^{-37}} = 10^{12}$$

b. $H_3C-C(=O)-O-H$ ($K_a = 10^{-5}$) + ethoxide $\overset{K_{eq} = ?}{\rightleftharpoons}$ $H_3C-C(=O)-O^{-}$ + H–O–ethyl ($K_a = 10^{-16}$)

$$K_{eq} = \frac{K_a = 10^{-5}}{K_a = 10^{-16}} = 10^{11}$$

c. pyridinium N–H ($K_a = 10^{-5}$) + H_2O $\overset{K_{eq} = ?}{\rightleftharpoons}$ pyridine + H_3O^{+} ($K_a = 10^{+2}$)

$$K_{eq} = \frac{K_a = 10^{-5}}{K_a = 10^{2}} = 10^{-7}$$

Problem 9 (p 221) - Estimate whether the equilibrium in each equation below would lie to the left (reactants favored) or the right (products favored). The acid is written first in each pair and the base is written second. The K_a of each conjugate acid is written in parentheses. (For bases, you have to add a proton to see what the acid looks like.) Use curved arrows to write a mechanism for the proton transfer in each part.

a. $H-C\equiv C-H$ ($K_a = 10^{-25}$) + $^{-}:CH_3$ $\overset{K_{eq} = ?}{\rightleftharpoons}$ $H-C\equiv C:^{-}$ + CH_4 ($K_a = 10^{-50}$)

$$K_{eq} = \frac{K_a = 10^{-25}}{K_a = 10^{-50}} = 10^{25}$$

far to the right

b. CH_3-NH-H ($K_a = 10^{-37}$) + $CH_3CO_2^{-}$ $\overset{K_{eq} = ?}{\rightleftharpoons}$ CH_3-NH^{-} + CH_3CO_2H ($K_a = 10^{-5}$)

far to the left

$$K_{eq} = \frac{K_a = 10^{-37}}{K_a = 10^{-5}} = 10^{-32}$$

c. $H_3C-C(=O)-CH_2-H$ ($K_a = 10^{-20}$) + $^{-}O-CH_2-CH_3$ $\overset{K_{eq} = ?}{\rightleftharpoons}$ $H_3C-C(=O)-CH_2^{-}$ + $H-O-CH_2-CH_3$ ($K_a = 10^{-16}$)

lies to the left

$$K_{eq} = \frac{K_a = 10^{-20}}{K_a = 10^{-16}} = 10^{-4}$$

d. phenol ($K_a = 10^{-10}$) + carbonate (CO_3^{2-}) $\overset{K_{eq} = ?}{\rightleftharpoons}$ phenoxide + HCO_3^{-} ($K_a = 10^{-10}$)

evenly balanced

$$K_{eq} = \frac{K_a = 10^{-10}}{K_a = 10^{-10}} = 1$$

Problem 10 (p 222) - What is the approximate ΔG for each reaction (in water)? What is the order of acidity in the following group (strongest acid = 1)? Write each reaction with water as the base (curved arrows, etc.).

a. $H_2SO_4 + H_2O \rightleftharpoons HSO_4^- + H_3O^+$

$K_a = 10^{+10}$

a. $\Delta G = 1.4(pK_a) = 1.4(-5) = -7$ kcal/mole (very exothermic)

b. $H_2O + H_2O \rightleftharpoons HO^- + H_3O^+$

$K_{eq.} = 10^{-16}$

b. $\Delta G = 1.4(pK_a) = 1.4(16) = +22$ kcal/mole (very endothermic)

c. $R{-}NH_2 + H_2O \rightleftharpoons R{-}NH^- + H_3O^+$

$K_a = 10^{-37}$

c. $\Delta G = 1.4(pK_a) = 1.4(37) = +52$ kcal/mole (very endothermic)

d. $H_3C{-}OH_2^+ + H_2O \rightleftharpoons H_3C{-}OH + H_3O^+$

$K_{eq.} = 10^{+3}$

d. $\Delta G = 1.4(pK_a) = 1.4(-3) = -4$ kcal/mole (exothermic)

e. $H{-}C{\equiv}C{-}H + H_2O \rightleftharpoons H{-}C{\equiv}C:^- + H_3O^+$

$K_a = 10^{-25}$

e. $\Delta G = 1.4(pK_a) = 1.4(25) = +35$ kcal/mole (very endothermic)

f. $CH_3CH_3 + H_2O \rightleftharpoons CH_3CH_2^- + H_3O^+$

$K_a = 10^{-50}$

f. $\Delta G = 1.4(pK_a) = 1.4(50) = +70$ kcal/mole (very endothermic)

Problem 11 (p 223) - Draw a PE vs. POR diagram for “d” and “e” above. Show an arrow pushing mechanism using general base, B: ⁻, and the conjugate acid and base for each reaction. Assume a relative starting energy value of zero as a reference point.

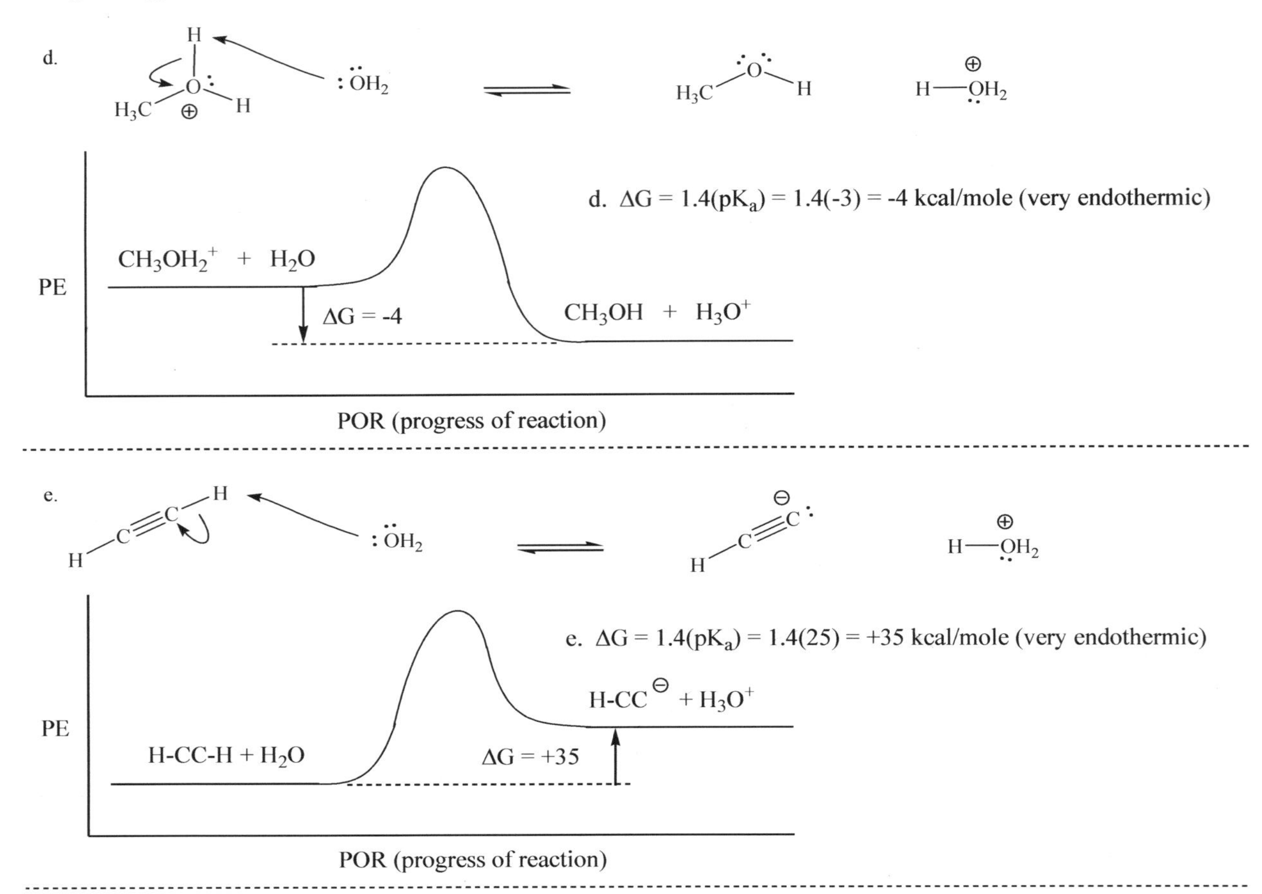

Problem 12 (p 228) - Order the following bases in increasing strength below (1=strongest). Are these differences big or small? Provide an explanation for your choice.

	$:CH_3^{\ominus}$	$:NH_2^{\ominus}$	$:OH^{\ominus}$	$:F:^{\ominus}$
conjugate acid's pK_a =	50	37	16	3

A higher Z_{eff} (and more electronegative) atom can stabilize negative charge better, when in the same row and approximately the same size (volumn). This means fluoride is the most stable negative charge and least basic,, while carbon is the least stable and most basic.

Problem 13 (p 228) - Order the substituents below, from most (=1) to least electron withdrawing, according to their apparent inductive effect based on the pK_a for the given acids (of the O-H bond). Provide a possible explanation for the relative order of acidities. Similar to above but on a much reduced scale.

Acid	F–O–H	Cl–O–H	Br–O–H	I–O–H	H–O–H (reference atom)
pK_a	does not exist	$pK_a = 7.5$	$pK_a = 8.7$	$pK_a = 11$	$pK_a = 16$

In all of the examples an oxygen carries the negative charge in the conjugate base. Water is the reference compound and it is clear that as the atom attached to the oxygen gets more electronegative the negative charge on oxygen gets more stable (because the pK_a is lower). The greater inductive withdrawal by Cl > Br > I > H leads to stronger acids (more stable conjugate bases). If HOF did exist we would expect that it would be the strongest acid.

Problem 14 (p 229) –
a. Using the data below, decide if the given substituents on the CH_2's in each carboxylic acid are electron withdrawing (stabilizing the carboxylate anion) or electron donating (destabilizing the carboxylate anion) *relative to a hydrogen atom*. Order the substituents from most inductively electron withdrawing (= 1) to least inductively electron withdrawing (…or actually electron donating).

	substituent on H_2C	pK_a
a	CO_2^-	5.70
b	CH_3	4.87
c	H (reference atom)	4.74
d	phenyl	4.31
e	OH	3.83
f	OCH_3	3.57
g	I	3.18
h	Br	2.90
i	CO_2H	2.85
j	Cl	2.85
k	F	2.59
l	CN	2.47
m	$\oplus NH_3$	2.35
n	NO_2	1.68

Electron withdrawing substituents help to stabilize the negative charge better in the conjugate base, so the acid is stronger (lower pK_a). If a substituent is electron donating then the negative charge will be less stable in the conjugate base (higher pK_a). The only 2 substituents that are electron donating are a (a carboxylate group, because it has a negative charge already present) and b (because R groups are inductively donating). All of the other groups are inductively withdrawing relative to hydrogen.

b. Are inductive withdrawing effects always stabilizing? What kind of inductive effect (stabilizing or destabilizing) would be expected with fluorine (F-) versus methyl (CH_3-), above, for a positive reaction center such as the carbocation below? (Reference X = "H")

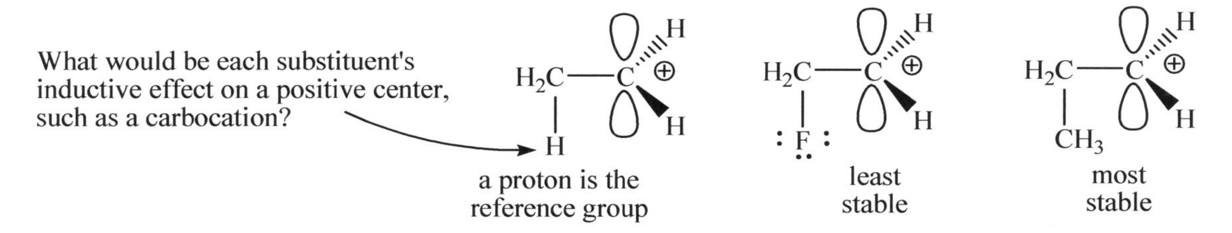

What would be each substituent's inductive effect on a positive center, such as a carbocation?
a proton is the reference group
least stable
most stable

From above we can see that R groups (alkyl groups) are inductive electron donating and that should help an electron deficient carbocation. On the other hand a fluorine (the most electronegative element) would pull electron density away from the very electron poor carbocation, making it even less stable. Both of these extremes are judged relative to a hydrogen atom in the same position.

Problem 15 (p 229) – a. Order the substituents below, from most (=1) to least electron withdrawing, according to their apparent inductive effect based on the pK_a for the given acids (of the O-H bond). Are any of the substituent groups electron donating relative to the hydrogen atom (the reference atom)? Provide a possible explanation for the relative order of acidities.

H_3C—O—H $pK_a = 16$ H_2N—O—H $pK_a = 14$ HO—O—H $pK_a = 12$ $(H_3C)_3C$—O—H $pK_a = 19$ H—O—H (reference atom) $pK_a = 16$

A higher pK_a (than 16, the reference value) means that the conjugate base is harder to form and less stable. We would argue here that the t-butyl group as 3 additional inductively donating R groups (the methyls) that would destabilize negative charge. In all likely hood there is also a solvation effect that we are ignoring. The other examples are as discussed above, with the most electronegative atom (Cl) stabilizing the negative charge the best (lowest pK_a), then Br, than I.

b. What atom loses the proton in each molecule? Why?

H_2N—O—H $pK_a = 14$

Both atoms are neutral so the negative charge is more stable on the more electronegative atom, oxygen. There is no delocalization here.

$H_3N^{\oplus}$—O—H $pK_a = 8$

In this example, there is a positive charge on the nitrogen atom which becomes neutral when the acidic proton is lost. Quenching the positive charge is better than making a negative charge separated from the positive charge, so the nitrogen atom loses the proton over oxygen here.

Problem 16 (p 230) - Are the indicated pK_a's consistent with what you would expect? Explain why.

a

butanoic acid $pK_a = 4.8$ 4-chlorobutanoic acid $pK_a = 4.5$ 3-chlorobutanoic acid $pK_a = 4.0$ 2-chlorobutanoic acid $pK_a = 2.8$

The reference acid is butanoic acid (the first one). Each of the other acids should be stronger because if the inductively withdrawing effect of chlorine. The chlorine that is closest to where the negative charge is on the conjugate base (the oxygen atoms) should produce the most stable conjugate base (lowest pK_a), which it does. As the chlorine gets farther away from the anion site its inductive withdrawal gets weaker and the pK_a gets larger.

b

CH_3COOH $pK_a = 4.7$ $ClCH_2COOH$ $pK_a = 2.9$ $Cl_2CHCOOH$ $pK_a = 1.3$ Cl_3CCOOH $pK_a = 0.7$ F_3CCOOH $pK_a = 0.2$

The reference acid is ethanoic acid (the first one). Each of the other acids should be stronger because if the inductively withdrawing effect of chlorine. All of the chlorine (and fluorine) atoms are the same distance from negative charge on the conjugate base (the oxygen atoms). More chlorine atoms should help stabilize the negative charge better and produce the most stable conjugate base (lowest pK_a), which it does. Switching in fluorine for chlorine helps the negative charge even more because of its stronger electron withdrawing effect, and the pK_a gets even lower.

Problem 17 (p 230) - a. Order the carbanions below from most (=1) to least stable. (R represents an alkyl substituent.). Explain your order.

least stable — $R_3C:^-$, $R_2CH:^-$, $RCH_2:^-$, $H_3C:^-$ — most stable

R groups (alkyl groups) are inductively donating which should destabilize negative charge. The tertiary carbanion is least stable and the methyl carbanion is most stable.

b. Order the carbocations below from most (=1) to least stable. (R represents an alkyl substituent.). Explain your order.

most stable — R_3C^+ , R_2CH^+ , RCH_2^+ , H_3C^+ — least stable

R groups (alkyl groups) are inductively donating which should stabilize positive charge. The tertiary carbocation is most stable and the methyl carbocation is least stable.

Problem 18 (p 230) - Rationalize the following series of acidities. Using the strongest acid, write out an acid/base equation with curved arrows to show electron flow, lone pairs of electrons and formal charge using B:$^-$ as the base.

a.

	Cl_3C-CH_2-O-H	Cl_2CH-CH_2-O-H	$ClCH_2-CH_2-O-H$	$CH_3O-CH_2-CH_2-O-H$
pK_a =	12	13	14.3	14.8

reference acid: $H-CH_2-CH_2-O-H$ 15.9

All of these are stronger than the reference acid (ethanol, the last one) because of the inductive withdrawal of the electronegative atoms. Just as above, the more chlorine atoms the better to help stabilize the negative charge. A lower pK_a means a lower energy to form the conjugate base, meaning a stronger acid.

b.

reference subsituent → (H)

	$H-CH_2-O-H$	H_3C-CH_2-O-H	$(H_3C)CH(CH_3)-O-H$	$(H_3C)_2C(CH_3)-O-H$
pK_a =	15.5	15.9	17.1	19.2

Methanol (the first one) is the strongest acid in this series because each time an R group is addes, the inductive donation destabilizes the negative charge in the conjugate base. Less stable means a higher pK_a which means a weaker acid.

Problem 19 (p 232) – Rationalize the following differences in pK_a values.

a.

$H_3C{-}C{\equiv}N^{\oplus}{-}H$ — pK_a = -10

$(H_3C)_2C{=}N^{\oplus}H_2$ — pK_a = +5

$H_3C{-}CH_2{-}N^{\oplus}H_3$ — pK_a = +9

	%s	%p
sp	50	50
sp^2	33	67
sp^3	25	75

The more s character in the hybrid orbital the more electronegative it is and the better able it is to hold a negative charge like formed in the conjugate base. The sp nitrogen is clearly the strongest acid (most stable conjugate base) by 10^{15}! The sp^2 nitrogen is better able to carry the negative charge than sp^3 by 10^4.

b.

$H{-}C{\equiv}C{-}H$ — pK_a = +25

$H{-}C{\equiv}N{:}$ — pK_a = +9

$H{-}C{\equiv}O^{\oplus}{:}$ — pK_a = ? (you guess, most likely a very negative number)

All carbon atoms in these compounds are sp hybridized. The difference here is the electronegativity of the atom they are attached to. Oxygen is more electronegative than nitrogen which is more electronegative than carbon. Also, there is a positive charge on the oxygen example which is very high energy (meaning the proton is lost very easily). There is difference in acidity of 10^{16} between ethyne and hydrogen cyanide (just because of the electronegativity between carbon and nitrogen. If a similar difference was present between nitrogen and oxygen we could estimate the pK_a of protonated carbon monoxide at -7 (or $K_a = 10^7$, a very strong acid).

Problem 20 (p 233) – Propose an explanation for the following differences in acidities.

O, OH — pK_a = 4.8

O, OH — pK_a = 4.4

O, OH — pK_a = 2.6

ΔpK_a = 0.4 = a factor of 2.5

ΔpK_a = 1.8 = a factor of 63

This series shows that the inductive withdrawal of more electronegative hybrid orbitals is similar to more electronegative atoms.

O, OH — pK_a = 4.8

O, OH — pK_a = 4.4

O, OH — pK_a = 2.6

ΔpK_a = 0.4 = a factor of 2.5

ΔpK_a = 1.8 = a factor of 63

This series shows that the inductive withdrawal of more electronegative hybrid orbitals is similar to more electronegative atoms.

Problem 21 (p 233) - Predict and explain the order of basicities for the following compounds (1 = most **basic).**

a. most basic $CH_3CH_2^{\ominus}:$ $H_2C{=}CH^{\ominus}:$ $H{-}C{\equiv}C^{\ominus}:$ least basic

This is just the reverse of the acid arguments above. The orbitals best able to donate electrons are the least electronegative, so sp^3 is more basic than sp^2 is more basic than sp (where electrons are held the tightest).

b.

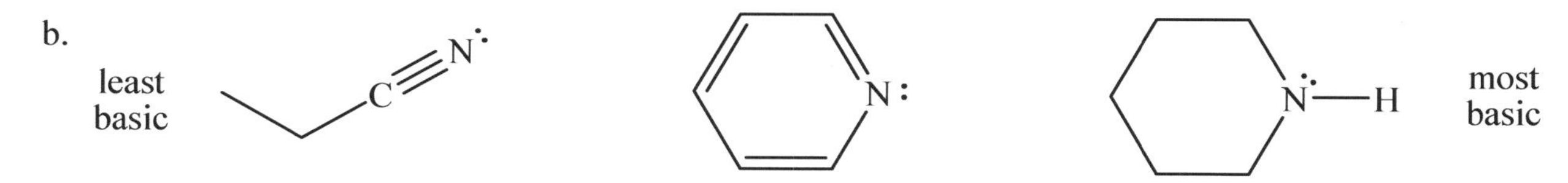

Similar argument to above. The orbitals best able to donate electrons are the least electronegative, so sp^3 is more basic than sp^2 is more basic than sp (where electrons are held the tightest).

Problem 22 (p 235) - Propose an explanation for the following pK_a's.

a weaker acid H–O–H $pK_a = 16$ H–S–H $pK_a = 7$ H–Se–H $pK_a = 4$ H–Te–H $pK_a = 3$ stronger acid

The negative charge is on an increasingly larger atom with the same Z_{eff} (= +6). More delocalized electron density is more stable (easier to form) so the conjugate acid is stronger (lower pK_a).

---

b H_3C–O–H $pK_a = 16$ H_3C–S–H $pK_a = 10$ Same argument as part a.

Problem 23 (p 235) - Which is the stronger acid in each part? Explain your choice.

a. HCl (pK_a = -7) is a stronger acid than H_2S (pK_a = 7) because they are in the same row and chlorine has a greater Z_{eff} = +7) than sulfur (Z_{eff} = +6), making chlorine more electronegative than sulfur.

---

b. PH_4^+ (pK_a = 0) is a stronger acid than NH_4^+ (pK_a = 9) because they are in the same column (same Z_{eff} = +5) and phosphorous is larger so the electrons are more delocalized in the conjugate base and more stable.

---

c. The same argument as b. SiH_4 (pK_a < 50?) is a stronger acid than CH_4 (pK_a = 50) because they are in the same column (same Z_{eff} = +4) and silicon is larger so the electrons are more delocalized in the conjugate base and more stable.

Problem 24 (p 235) - Which is the stronger base in each part (better electron pair donor)? Explain your choice.

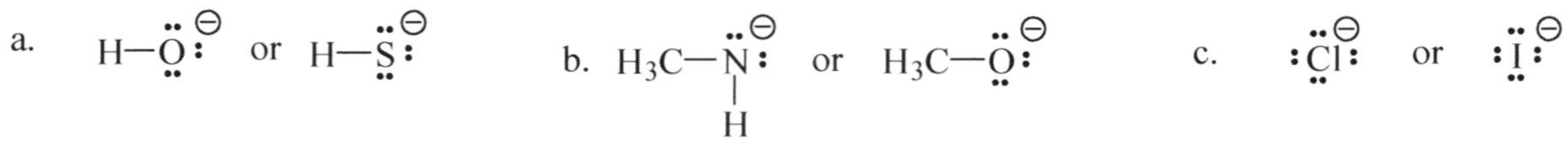

a. $H-\ddot{O}:^{\ominus}$ or $H-\ddot{S}:^{\ominus}$ b. $H_3C-\ddot{N}(H):^{\ominus}$ or $H_3C-\ddot{O}:^{\ominus}$ c. $:\ddot{Cl}:^{\ominus}$ or $:\ddot{I}:^{\ominus}$

a. Both oxygen and sulfur are in the same column (same $Z_{eff} = +5$) and sulfur is larger so the electrons are more delocalized in the anion and more stable. Therefore hydroxide is more basic and has a weaker acid (water).

---

b. The nitrogen anion ($Z_{eff} = +5$) is more basic than the oxygen anion ($Z_{eff} = +6$) because they are in the same row, making the oxygen anion more stable and less reactive.

---

c. The same argument as a. Both iodide and chloride are in the same column (same $Z_{eff} = +7$). Because iodide is larger the electrons are more delocalized in the conjugate base and more stable, so less reactive (less basic).

Problem 25 (p 236) – Explain the relative pK_a's in each part. Show all relevant structures in your explanation. Write an equation for each example using a general base, B: Include curved arrows to show electron movement, include formal charge and all lone pairs. In part c, can you tell which conjugate acid is more stable from their pK_a's?

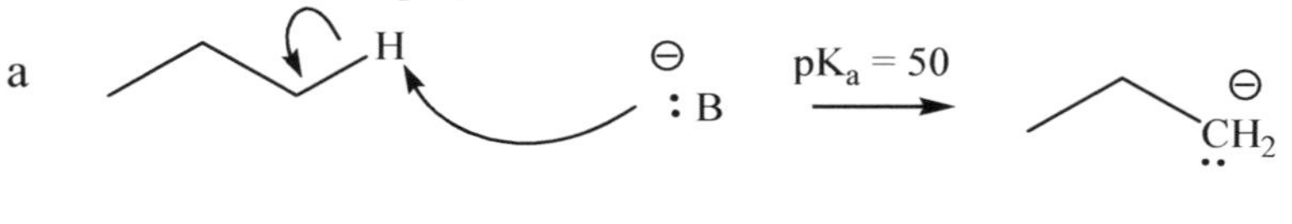

a

$:B^{\ominus}$ $pK_a = 50$ → $\ddot{C}H_2^{\ominus}$

The lone pair electron in the carbanion are completely localized on carbon, one of the least stable anions.

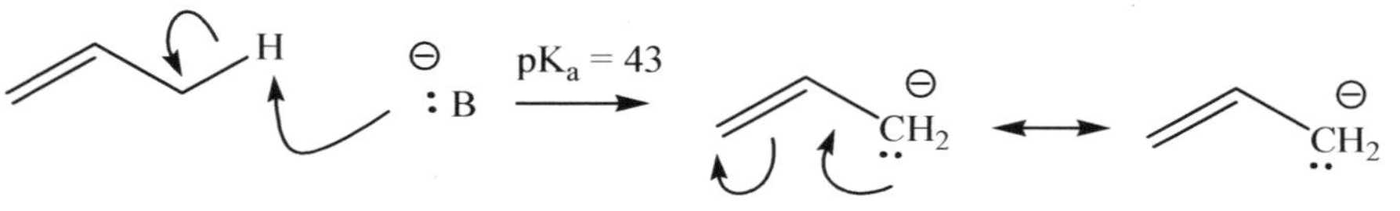

$:B^{\ominus}$ $pK_a = 43$ → $\ddot{C}H_2^{\ominus}$ ⟷ $\ddot{C}H_2^{\ominus}$

The lone pair electron in the delocalized,making the carbanion more stable by a factor of 10^7.

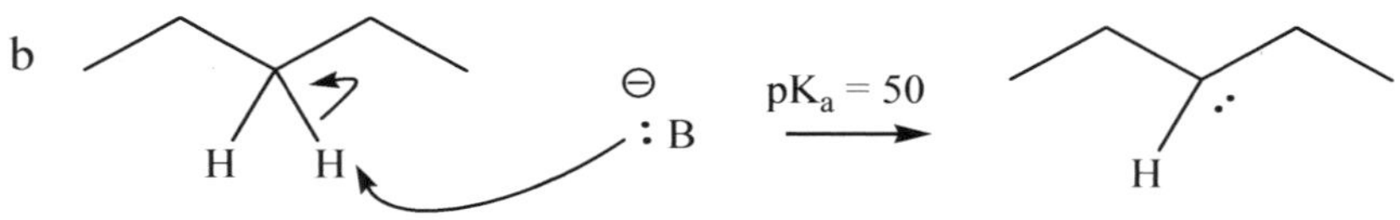

---

b

H H $:B^{\ominus}$ $pK_a = 50$ → H

The lone pair electron in the carbanion are completely localized on carbon, one of the least stable anions.

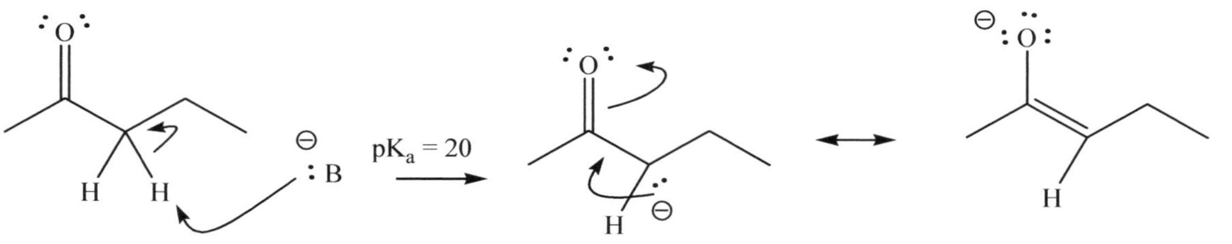

H H $:B^{\ominus}$ $pK_a = 20$ → H ⟷ H

The lone pair electron in the delocalized onto the oxygen atom making the carbanion more stable by a factor of 10^{30}.

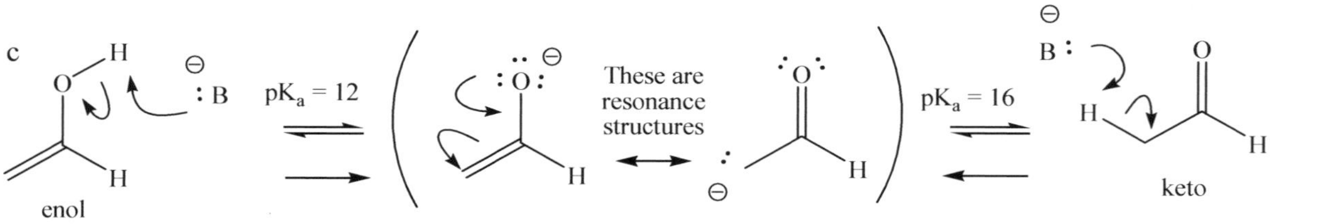

---

c

enol: O H H $:B^{\ominus}$ $pK_a = 12$ ⇌ (H ⟷ H) These are resonance structures ⇌ $pK_a = 16$ $B:^{\ominus}$ H H keto

It's easier to lose the proton from the oxygen than the carbon and that's why the pKa is lower on the enol than the keto tautomer.

Combination Effects - Inductive and Resonance Together

Analyze each example in the following series of organic acids and explain the relative differences in acidity.

$pK_a = 50$ $pK_a = 40$ $pK_a = 20$

$pK_a = 37$ $pK_a = 28$ $pK_a = 15$

$pK_a = 16$ $pK_a = 10$ $pK_a = 5$

$pK_a = 8$ $pK_a = 6.5$ $pK_a = 3.1$

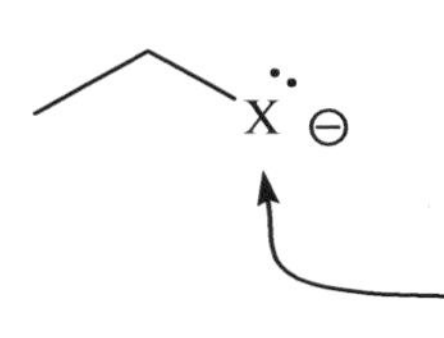

In the first column the negative charge of the conjugate base is fully localized on the atom losing the proton (first row = C, second row = N, third row = O and last row = S). In the first 3 rows a more electronegative atom carries the negative charge better (as they are all similar in size). The fourth row has sulfur, similar to oxygen in Zeff, but larger in size. The electrons are more delocalized so the conjugate acid is stronger. This argument is true for all 3 columns. What is different in each column is additional resonance in present in the second column on 3 additional carbon atoms in the aromatic ring. In the third column there are only 2 resonance structures, but both have the negative charge on an additional oxygen which is is better than 3 additional resonance structures on carbon.

aromatic resonance structures

carbonyl structures - all show that the negative charge is partially localized on oxygen and one other atom.

H_3C—C ⟷ H_3C—C

Problem 26 (p 237) – Which is the weaker base in each pair below. Explain your answers. Match each pK_a with its acid.

a

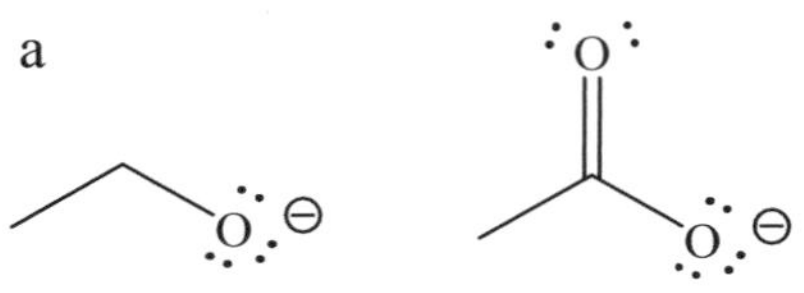

pK_a's = 5, 16

The weaker base is the more stable anion. The alkoxide, with the full negative charge on one oxygen is less stable than carboxylate which has resonance delocalization of the negative charge on two oxygens. The alkoxide is less stable by 10^{11} and much more basic than the carboxylate.

b

pK_a's = 9, 25

Both anions have negative charge localized in an sp orbital, which is relatively good because of the 50% s character of the orbital. The s orbital is more electronegative than a p orbital because the electrons are closer to the nucleus and held more tightly. Cyanide has an additional advantage due to the greater electronegativity of nitrogen over carbon. This makes cyanide more stable by a factor of 1016 judging from the pK_a values. More stable is less basic.

c

pK_a's = 43, 50

Both anions have negative charge on carbon which is not a very stable place for negative charge. However, the allylic anion (first example) is resonance delocalized, making it more stable than the sp3 carbanion by 10^7. More stable is less basic.

Problem 27 (p 237) – Identify the most acidic hydrogen in each of the following compounds. Explain your choices. (Consult your pK_a table, if necessary.)

a.

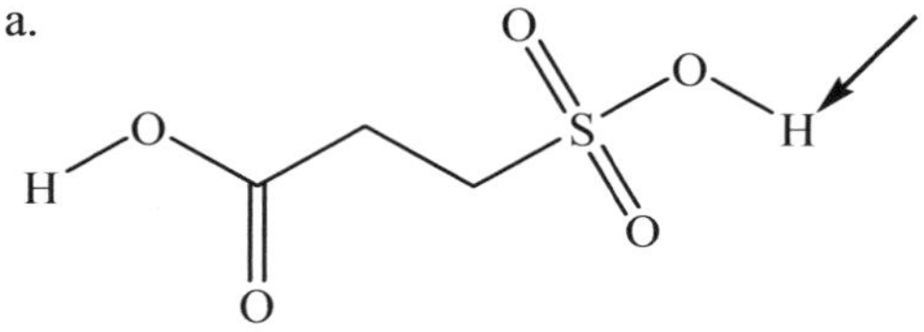

The H of the sulfonic acid is more acidic by far (pK_a = -3) since the negative charge can be delocalized onto 3 different oxygen atoms and has an additional electronegative inductive effect from the sulfur. The pK_a of the carboxylic acid is around 5, or 10^8 times less acidic.

---

b.

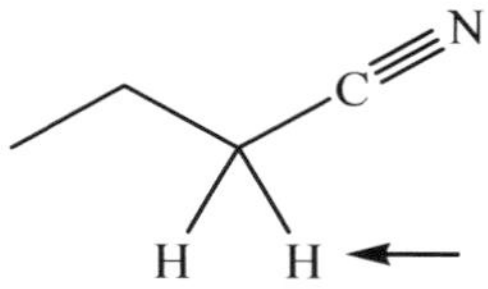

The protons next to the nitrile group are most acidic since the anion of the conjugate base can be delocalized onto the sp nitrogen atom (pK_a = 25) versus pK_a of the normal sp^3 CH bonds of about 50.

---

c.

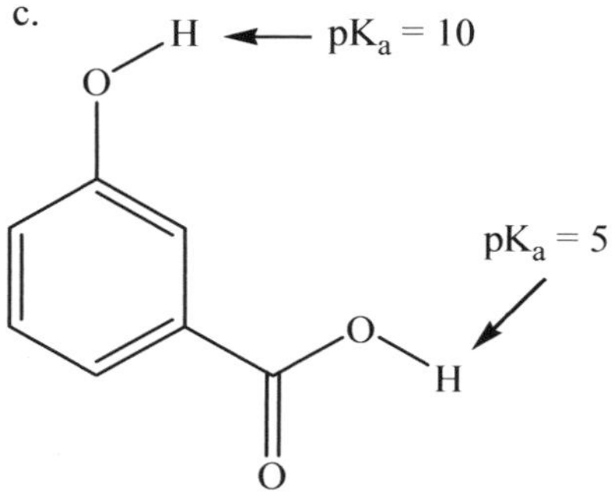

pK_a = 10

pK_a = 5

The carboxylic acid protons are 10^5 times more acidic because the negative charge in the conjugate base is delocalized over 2 oxygen atoms. The phenolic proton is more acidic than an alcohol (ROH, pKa = 16) because of resonance, but only one resonance structure has the negative charge on oxygen, with 3 resonance structures having negative charge on carbon. One extra oxygen easily beats 3 extra carbons.

---

d. pKa ≈ 20 (a), pKa ≈ 11 (b), pKa ≈ 50 (c), pKa ≈ 50 (d)

Protons b are most acidic because the negative charge can be delocalized over 2 oxygens and 1 carbon. Protons a are next because negative charge can be delocalized over 1 carbon and 1 oxygen. Protons c and d are least acidic because the negative charge of the conjugate base is fully localized on carbon. Protons c are probably a little more acidic that protons d because they are closer to the oxygen atom and would have a stronger electron withdrawing effect.

Problem 28 (p 238) – Supply the necessary curved arrows, lone pairs of electrons and/or formal charge to show how the first step each reaction proceeds. All of these reactions are all simple proton transfer reactions generating an anion (reactions e-j generate carbanions). Generally, there is some stabilizing feature that allows a carbanion to form via acid/base chemistry, such as inductive and/or resonance effects. In working the problem below, show any important resonance structures or identify the inductive effect that makes the reaction possible.

a.

$pK_a = 16$ — sodium hydride (given, very strong base) — $K_{eq} = \frac{10^{-16}}{10^{-37}} = 10^{+21}$ — sodium alkoxide (strong base/nucleophile) — H—H, $pK_a = 37$, hydrogen gas bubbles away

KH is also used with t-butyl alcohol (2-methylpropan-2-ol) to make potassium t-butoxide, but not in our course.

---

b.

$pK_a = 5$ — sodium hydroxide (given) — $K_{eq} = \frac{10^{-5}}{10^{-16}} = 10^{+11}$ — sodium carboxylate (strong base/nucleophile) — $pK_a = 16$, water

---

c.

$pK_a = 8$ — sodium hydroxide (given) — $K_{eq} = \frac{10^{-8}}{10^{-16}} = 10^{+8}$ — sodium thiolate (strong base/nucleophile) — $pK_a = 16$, water

---

d.

$pK_a = 37$ — $:CH_2CH_2CH_2CH_3$, n-butyl lithium is commercially available (given) — $K_{eq} = \frac{10^{-37}}{10^{-50}} = 10^{+13}$ — = LDA

Formation of lithium diisopropyl amide (LDA) using butyl lithium.

$H—CH_2CH_2CH_2CH_3$, $pK_a = 50$, butane

---

e.

R₂N⁻ Li⁺ (LDA (make)) + ketone ($pK_a = 20$) →(-78 °C) enolate (resonance) + $R_2N\text{-}H$ ($pK_a = 37$)

$K_{eq} = \frac{10^{-20}}{10^{-37}}$

$K_{eq} = 10^{+17}$

---

f.

LDA (make) + ester ($pK_a = 25$) →(-78 °C) enolate (resonance) + $R_2N\text{-}H$ ($pK_a = 37$)

$K_{eq} = \frac{10^{-25}}{10^{-37}}$

$K_{eq} = 10^{+12}$

---

g.

LDA (make) + nitrile ($pK_a = 30$) →(-78 °C) enolate-like (resonance) + $R_2N\text{-}H$ ($pK_a = 37$)

$K_{eq} = \frac{10^{-30}}{10^{-37}}$

$K_{eq} = 10^{+7}$

---

h.

LDA (make) (Na⁺) + terminal alkyne H—C≡C—R ($pK_a = 25$) →(-33 °C) terminal acetylide (sp carbanion) :C≡C—R + $H_2N\text{-}H$ ($pK_a = 37$, ammonia)

$K_{eq} = \frac{10^{-25}}{10^{-37}}$

$K_{eq} = 10^{+12}$

Problem 29 (p 240) –

a. Is protonation of an amine more likely or protonation of an alcohol? Explain your reasoning.

$H_3C\text{-}CH_2\text{-}NH_2$ + H—A →(?) $H_3C\text{-}CH_2\text{-}NH_3^+$ ($pK_a = 10$)

$H_3C\text{-}CH_2\text{-}OH$ + H—A →(?) $H_3C\text{-}CH_2\text{-}OH_2^+$ ($pK_a = -2$)

Nitrogen is MUCH better at sharing its electrons because it is LESS electronegative (has a smaller $Z_{eff} = +5$ than oxygen $Z_{eff} = +6$).

Problem 30 (p 241) –

a. Is protonation of an amide more likely on the oxygen atom or the nitrogen atom. Show the reaction for both and examine any possible resonance structures for clues that explain the difference.

charge is delocalized over 3 resonance structure, 2 of which have an extra bond and full octets

charge if fully localized on the nitrogen atom, not as good as above

b. Predict the more basic oxygen in a carboxylic acid. Show both oxygen atoms acting as a base and explain your choice (just like the amide in part a).

charge is delocalized over 3 resonance structure, 2 of which have an extra bond and full octets

charge if fully localized on one oxygen atom, not as good as above

same argument as above

c. Would you expect the amide C=O or the acid C=O to be more basic? Explain.

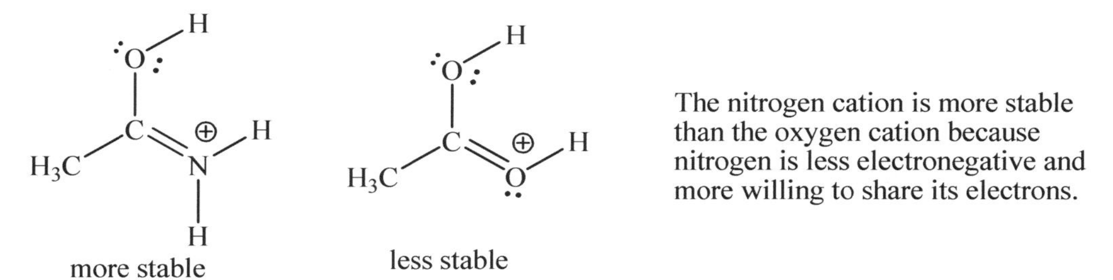

The nitrogen cation is more stable than the oxygen cation because nitrogen is less electronegative and more willing to share its electrons.

Whenever an atom with a lone pair is next to a pi bond (of any kind) it will share its electrons with the pi bond, making it more electron rich and a better electron pair donor.

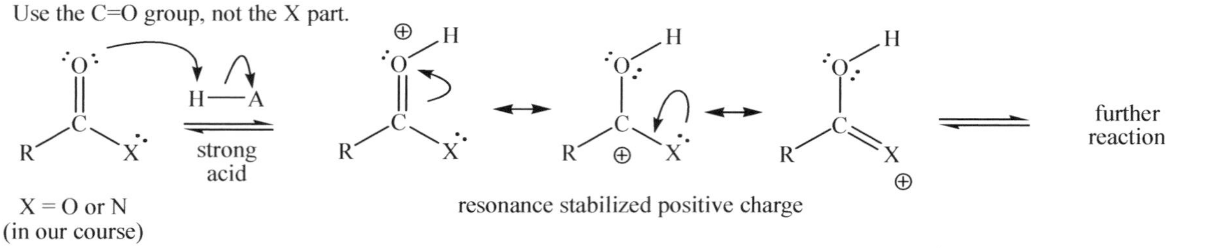

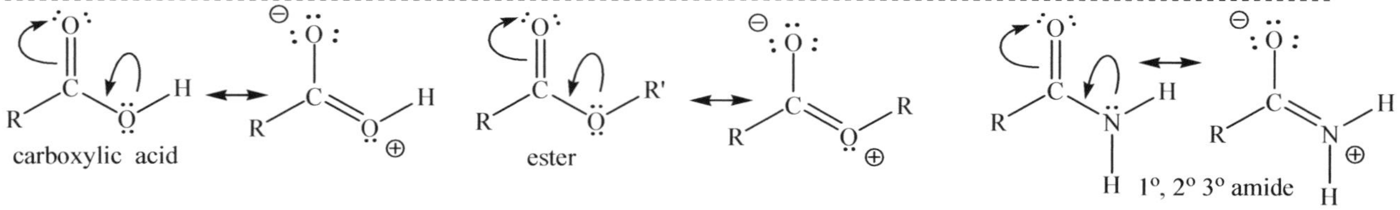

Use the C=O group, not the X part.

H—A strong acid

further reaction

X = O or N (in our course)

resonance stabilized positive charge

carboxylic acid

ester

H 1°, 2° 3° amide

This is even true when "X" is next to a C=C pi bond. If you protonate X, the positive charge is only on X, but if you protonate the pi bond (at the atom not bonded to X) you get resonance structures with positive charge on carbon and X. This is usually the better choice.

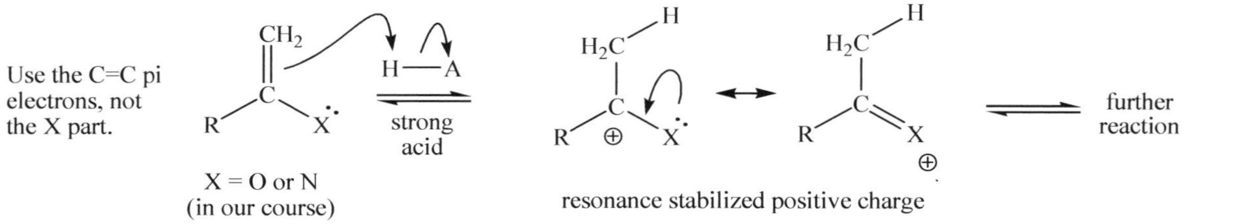

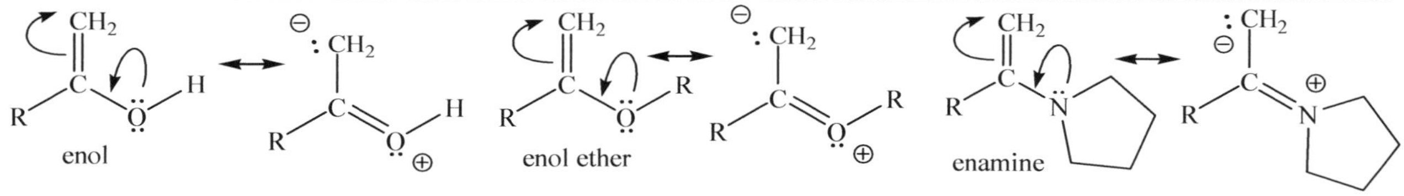

Use the C=C pi electrons, not the X part.

H—A strong acid

further reaction

X = O or N (in our course)

resonance stabilized positive charge

enol

enol ether

enamine

Problem 31 (p 241)

(The power of a proton in organic chemistry.) – Many functional groups we study are shown below. They all can react with strong acid in a similar first step (H-A, below). Supply curved arrows to show the proton transfers in the acid/base reactions. The transfer of a proton from a strong acid generally increases the potential energy from a unreactive neutral molecule to a highly reactive, protonated cation. Even though a positive formal charge is written on the heteroatom, carbon also carries much of the positive charge. Often this can be shown with a resonance structure. Placing positive charge on carbon will almost always lead to one of the three possible outcomes just mentioned above: 1. add nucleophile, 2. lose a beta hydrogen atom or 3. rearrange. Note the common themes that repeat themselves again and again. Possible outcomes from these initial steps are listed to the right. These are some of the reactions we will study in greater detail in the topics that follow.

a.

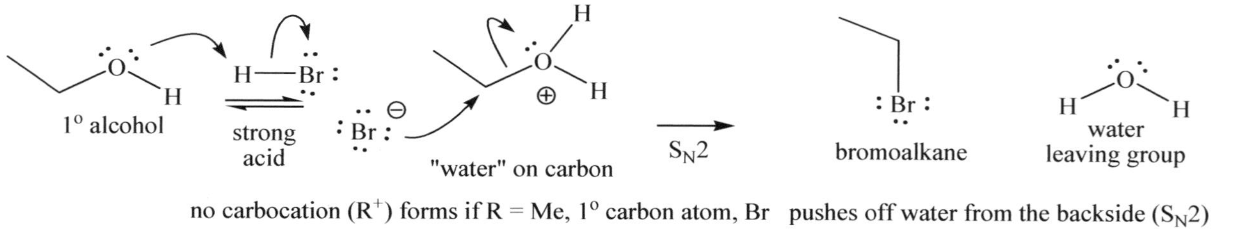

1° alcohol

H—Br: strong acid

"water" on carbon

S_N2

bromoalkane

water leaving group

no carbocation (R^+) forms if R = Me, 1° carbon atom, Br pushes off water from the backside (S_N2)

b.

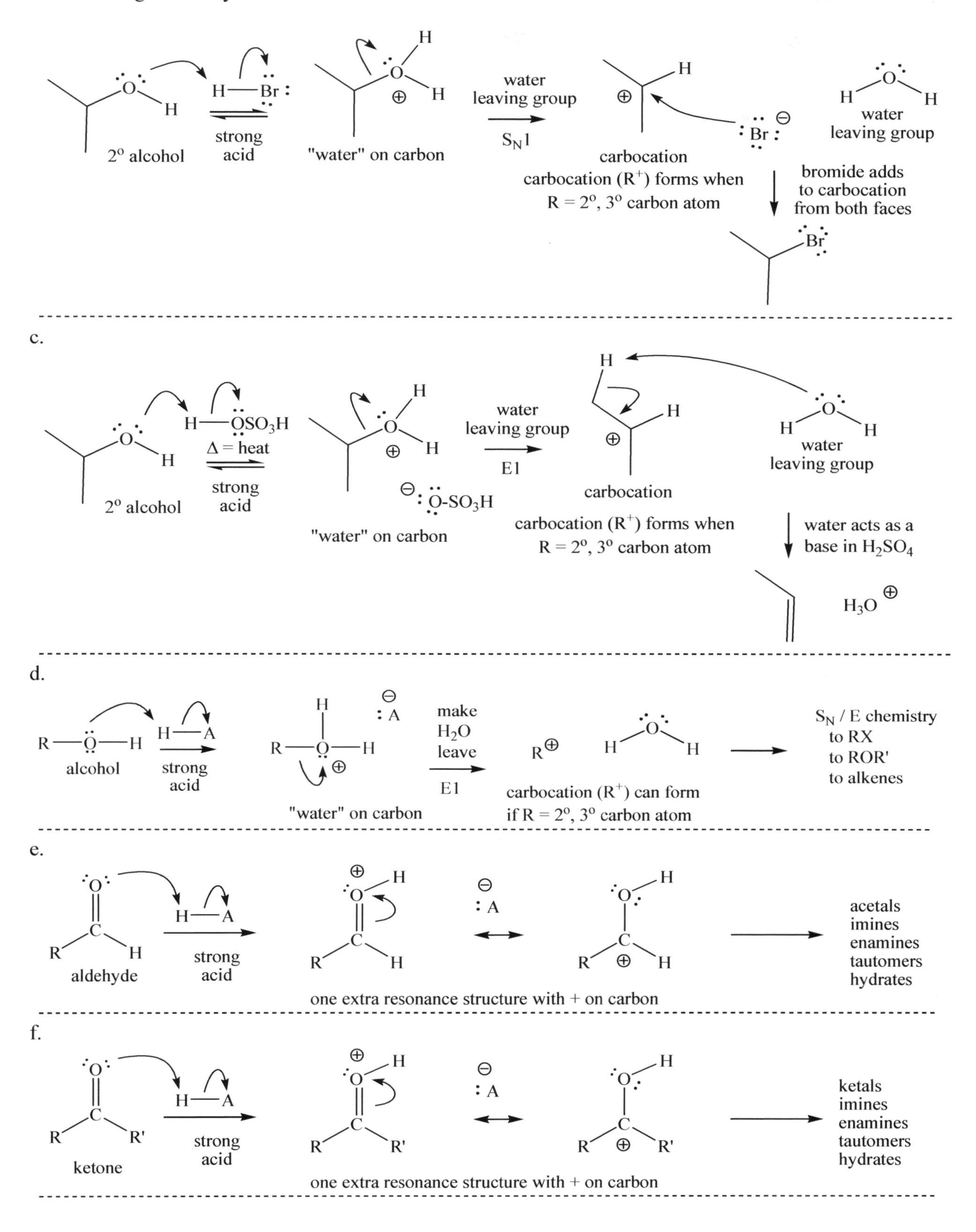

2° alcohol
strong acid
"water" on carbon
water leaving group
S_N1
carbocation
carbocation (R⁺) forms when R = 2°, 3° carbon atom
water leaving group
bromide adds to carbocation from both faces
c.
2° alcohol
Δ = heat
strong acid
"water" on carbon
water leaving group
E1
carbocation
carbocation (R⁺) forms when R = 2°, 3° carbon atom
water leaving group
water acts as a base in H_2SO_4
H_3O
d.
alcohol
strong acid
"water" on carbon
make H_2O leave
E1
carbocation (R⁺) can form if R = 2°, 3° carbon atom
S_N / E chemistry
to RX
to ROR'
to alkenes
e.
aldehyde
strong acid
one extra resonance structure with + on carbon
acetals
imines
enamines
tautomers
hydrates
f.
ketone
strong acid
one extra resonance structure with + on carbon
ketals
imines
enamines
tautomers
hydrates

g.

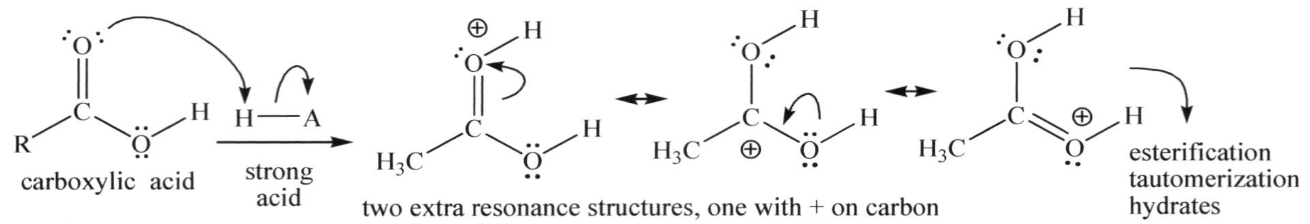

carboxylic acid

strong acid

two extra resonance structures, one with + on carbon

esterification
tautomerization
hydrates

---

h.

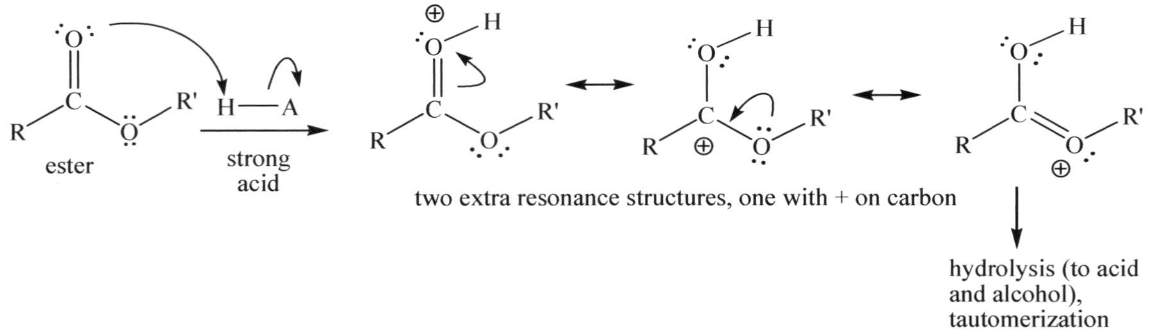

ester

strong acid

two extra resonance structures, one with + on carbon

hydrolysis (to acid and alcohol), tautomerization

---

i.

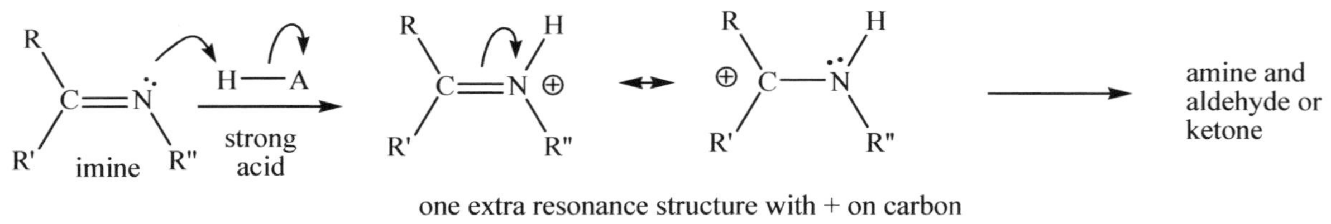

imine

strong acid

amine and aldehyde or ketone

one extra resonance structure with + on carbon

---

j.

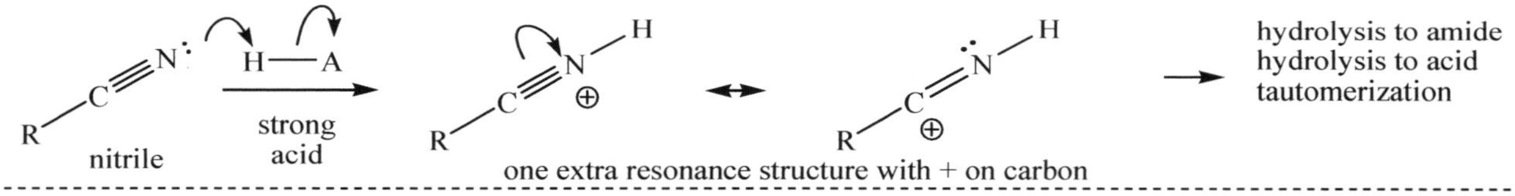

nitrile

strong acid

hydrolysis to amide
hydrolysis to acid
tautomerization

one extra resonance structure with + on carbon

---

k.

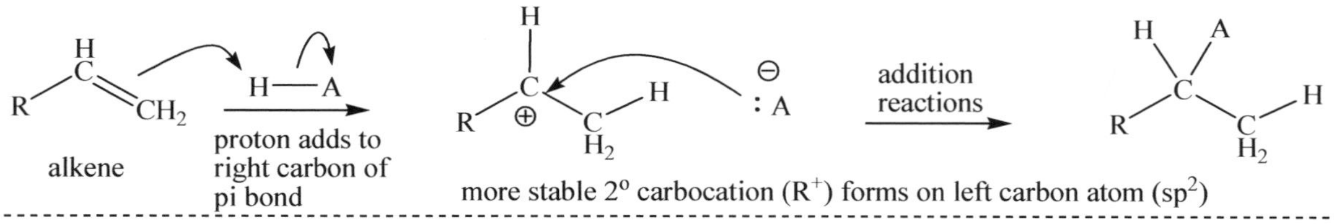

alkene

proton adds to right carbon of pi bond

addition reactions

more stable 2° carbocation (R^+) forms on left carbon atom (sp^2)

---

l.

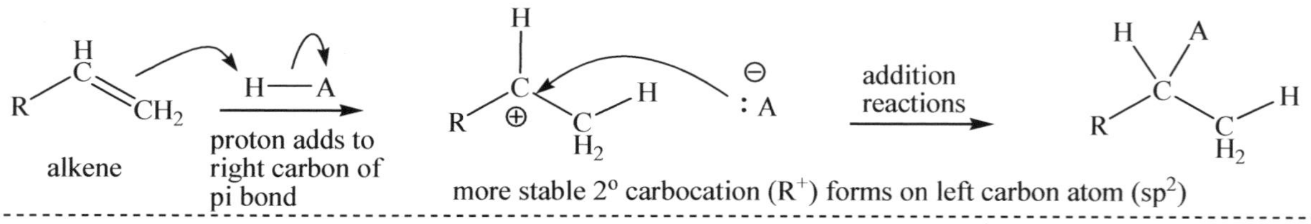

alkene

proton adds to right carbon of pi bond

addition reactions

more stable 2° carbocation (R^+) forms on left carbon atom (sp^2)

---

m.

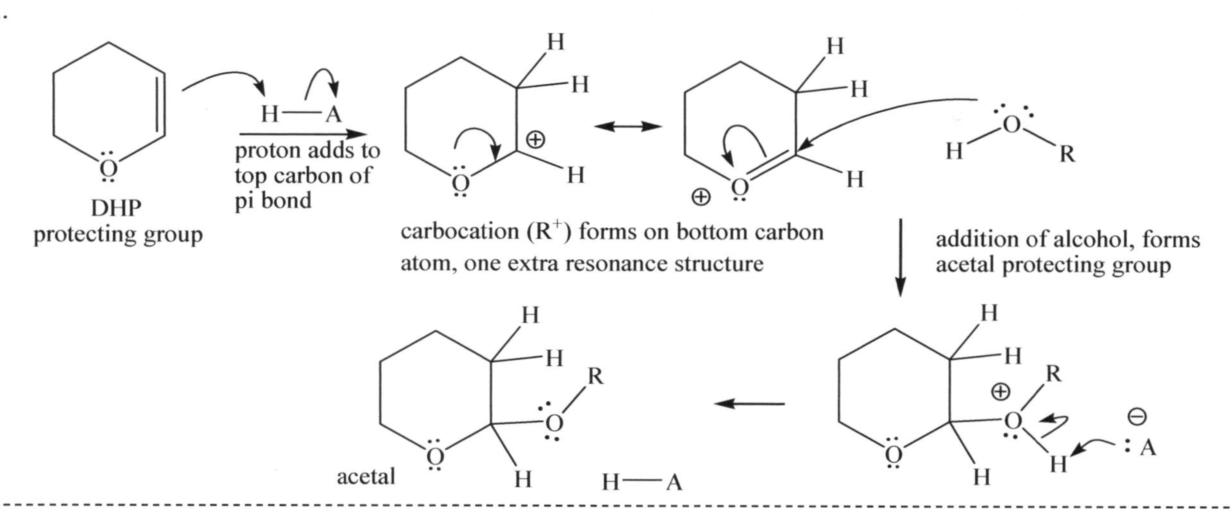
H—A
proton adds to top carbon of pi bond
DHP protecting group
carbocation (R^+) forms on bottom carbon atom, one extra resonance structure
addition of alcohol, forms acetal protecting group
acetal
H—A

n.

H—A

proton adds to right carbon of pi bond

enamine

carbocation (R^+) forms on left carbon atom, one extra resonance structure

hydrolysis to amine and ketone

etc.

o.

D—A

deuterium adds to any carbon of a pi bond

aromatic

carbocation (R^+) forms on opposite carbon of pi bond, two extra resonance structures

substitution reactions

H—A

Problem 32 (p 210) - Which nitrogen atom is most basic? (1, 2 or 3). The pK_a values of the conjugate acids of the two compounds are 5 and 10. Match the pK_a values with their appropriate acids and explain the very large difference (100,000 / 1). You have to draw the conjugate acids to show this.

1 — $pK_a = 5$

2 — $pK_a = 10$

The conjugate acid of 2 is much more stable due to the very good resonance structure with 2 nitrogen atoms carrying the positive charge (full octets).

Chapter 9

Problem 1 (p 254) - How can you tell whether the S_N2 reaction occurs with front side attack (retention), backside attack (inversion) or front and backside attack (racemization)? Use the two molecules to explain your answer. Follow the curved arrow formalism to show electron movement.

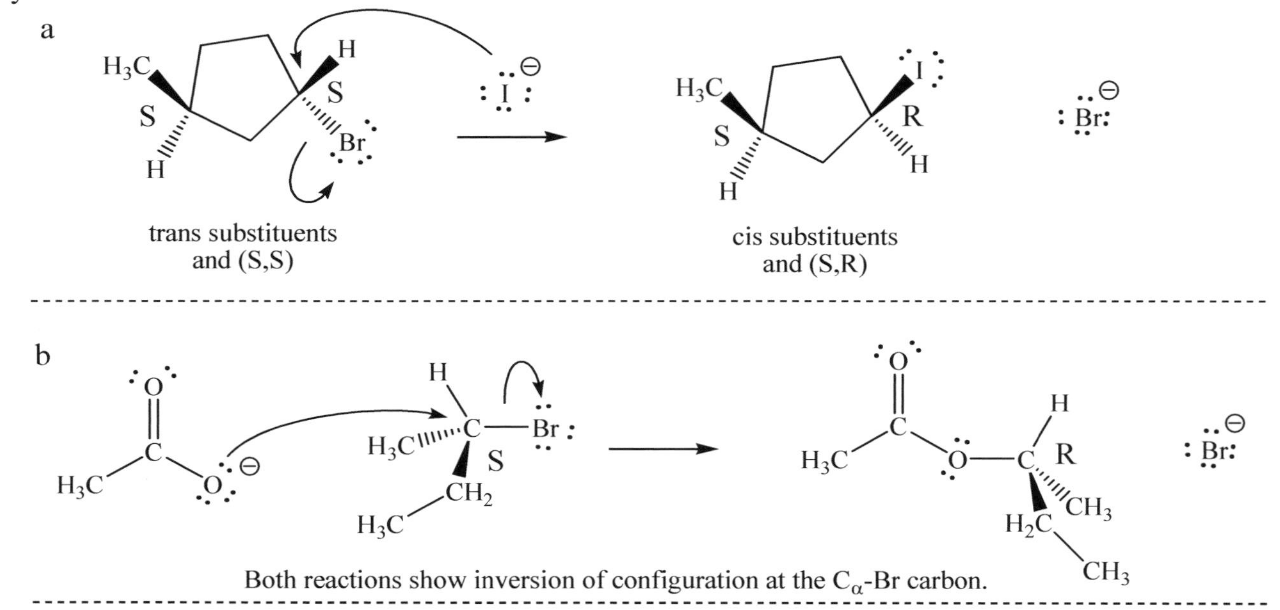

Both reactions show inversion of configuration at the C_α-Br carbon.

Problem 2 (p 254) - In each of the following pairs of nucleophiles one is a much better nucleophile than its closely related partner. Propose a possible explanation.

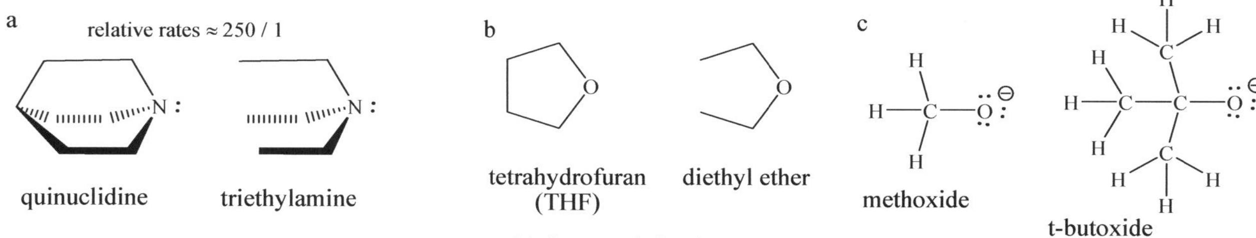

The side chains in quinuclidine are pinned back and do not inhibit attack of the nitrogen atom, whereas the side chains of triethylamine are free to rotate 360° and sometimes will be in the way of attack by the nitrogen which will slow down attack on electrophilic centers.

This is essentially the same argument as in a. The ring in THF pins the carbons back. Diethylether side chains can rotate 360° and will sometimes get in the way of nucleophilic attack, so it will react at a slower rate.

Methoxide is a much better nucleophile than t-butoxide, which is not even a nucleophile in our course. It is too sterically hindered to get close enough to the C_α carbon to form a new CC bond. It only acts as a base (except with bromomethane).

Problem 3 (pp 256-261) - Fill in the necessary details to show how each reaction works. Specify as only S_N2, S_N2 > E2, E2 > S_N2 or only E2.

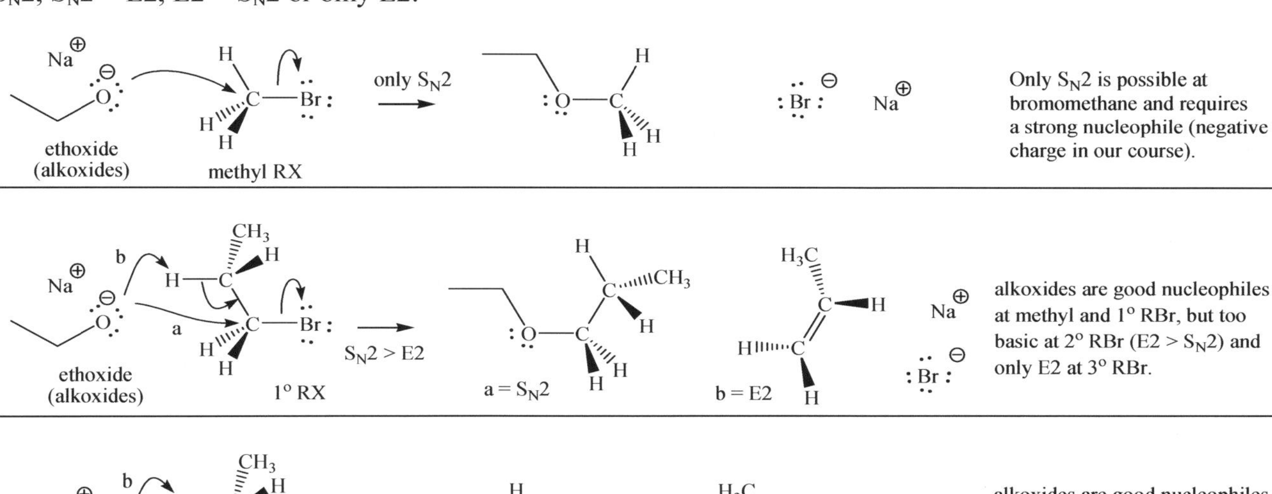

ethoxide (alkoxides) — 2° RX — E2 > S_N2 — a = S_N2 — b = E2, E and Z possible — c = E2

alkoxides are good nucleophiles at methyl and 1° RBr, but too basic at 2° RBr (E2 > S_N2) and only E2 at 3° RBr.

ethoxide (alkoxides) — 3° RX — only E2 — a = S_N2 is not possible — b = E2 — c = E2

alkoxides are good nucleophiles at methyl and 1° RBr, but too basic at 2° RBr (E2 > S_N2) and only E2 at 3° RBr.

Na⊕ ⊖S H — monohydrogen sulfide; methyl RX; only S_N2; $:Br:^{\ominus}$ Na⊕

Only S_N2 is possible at bromomethane and requires a strong nucleophile (negative charge in our course).

---

monohydrogen sulfide; 1° RX; $S_N2 > E2$; a = S_N2; b = E2; $:Br:^{\ominus}$ Na⊕

monohydrogen sulfide is a good nucleophile at methyl, 1° and 2° RBr, but only E2 at 3° RBr.

---

monohydrogen sulfide; 2° RX; $S_N2 > E2$; a = S_N2; b = E2 E and Z possible; c = E2; $:Br:^{\ominus}$ Na⊕

monohydrogen sulfide is a good nucleophile at methyl, 1° and 2° RBr, but only E2 at 3° RBr.

---

monohydrogen sulfide; 3° RX; only E2; a = S_N2 is not possible; b = E2; c = E2; $:Br:^{\ominus}$ Na⊕

monohydrogen sulfide is a good nucleophile at methyl, 1° and 2° RBr, but only E2 at 3° RBr.

---

ethyl thiolate; methyl RX; only S_N2; $:Br:^{\ominus}$ Na⊕

Only S_N2 is possible at bromomethane and requires a strong nucleophile (negative charge in our course).

---

ethyl thiolate; 1° RX; $S_N2 > E2$; a = S_N2; b = E2; $:Br:^{\ominus}$ Na⊕

sodium thiolates are good nucleophiles at methyl, 1° and 2° RBr, but only E2 at 3° RBr.

---

ethyl thiolate; 2° RX; $S_N2 > E2$; a = S_N2; b = E2 E and Z possible; c = E2; $:Br:^{\ominus}$ Na⊕

sodium thiolates are good nucleophiles at methyl, 1° and 2° RBr, but only E2 at 3° RBr.

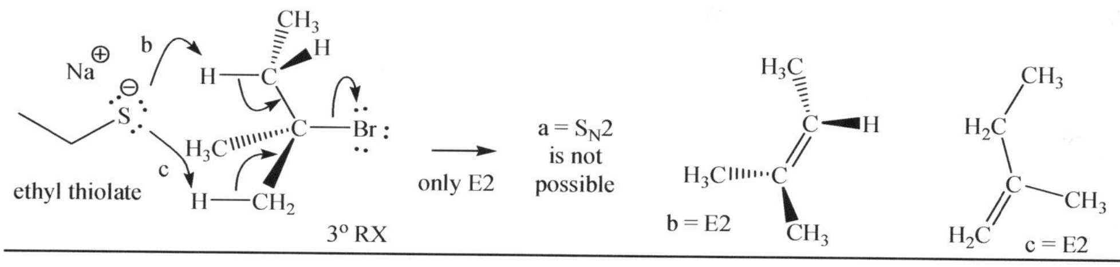

Na ⊕, ethyl thiolate, 3° RX, only E2, a = S_N2 is not possible, b = E2, c = E2

sodium thiolates are good nucleophiles at methyl, 1° and 2° RBr, but only E2 at 3° RBr.

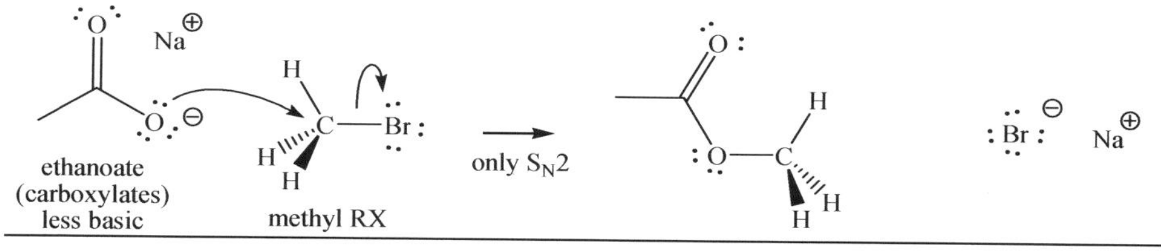

ethanoate (carboxylates) less basic, methyl RX, only S_N2

Only S_N2 is possible at bromomethane and requires a strong nucleophile (negative charge in our course).

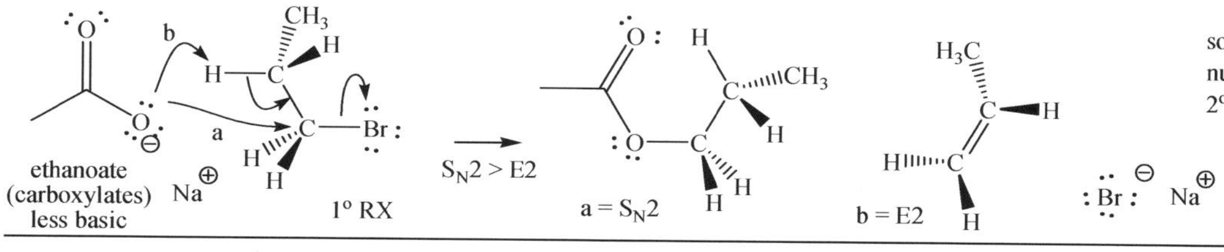

ethanoate (carboxylates) less basic, 1° RX, $S_N2 > E2$, a = S_N2, b = E2

sodium carboxylates are good nucleophiles at methyl, 1° and 2° RBr, but only E2 at 3° RBr.

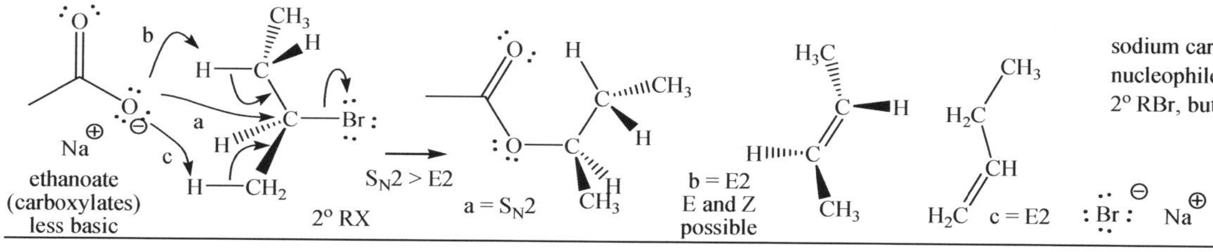

ethanoate (carboxylates) less basic, 2° RX, $S_N2 > E2$, a = S_N2, b = E2 E and Z possible, c = E2

sodium carboxylates are good nucleophiles at methyl, 1° and 2° RBr, but only E2 at 3° RBr.

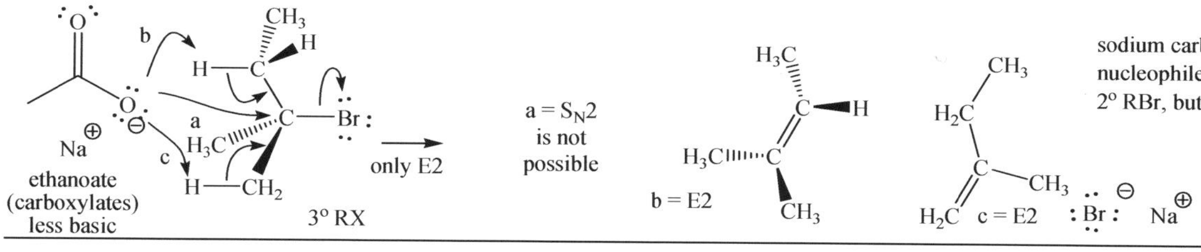

ethanoate (carboxylates) less basic, 3° RX, only E2, a = S_N2 is not possible, b = E2, c = E2

sodium carboxylates are good nucleophiles at methyl, 1° and 2° RBr, but only E2 at 3° RBr.

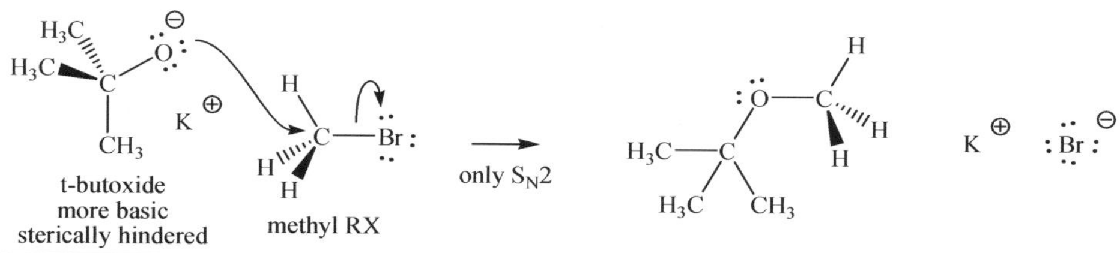

t-butoxide more basic sterically hindered, methyl RX, only S_N2

Only S_N2 is possible at bromomethane and requires a strong nucleophile (negative charge in our course).

t-butoxide more basic sterically hindered, 1° RX, $E2 > S_N2$

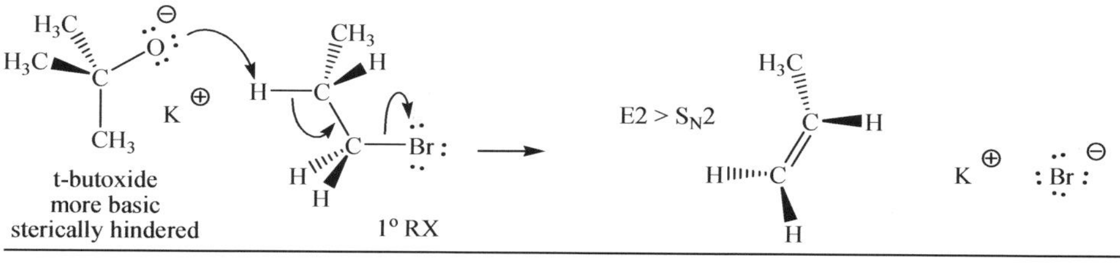

t-butoxide is too sterically large and very basic to be a good nucleophile. It acts as a base at 1°, 2° and 3° RBr (only E2 in our course, except for methyl).

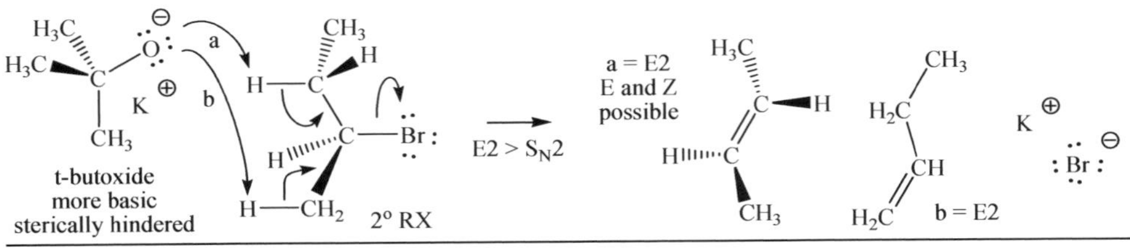

t-butoxide is too sterically large and very basic to be a good nucleophile. It acts as a base at 1°, 2° and 3° RBr (only E2 in our course, except for methyl).

H_3C H_3C C O ⊖ K ⊕ a b CH_3

t-butoxide
more basic
sterically hindered

CH_3 H H—C C—Br : H_3C H—CH_2

3° RX

only E2

H_3C C—H H_3C—C CH_3

a = E2

CH_3 H_2C C—CH_3 H_2C

b = E2

K ⊕ : Br : ⊖

t-butoxide is too sterically large and very basic to be a good nucleophile. It acts as a base at 1°, 2° and 3° RBr (only E2 in our course, except for methyl).

Na ⊕ ⊖ :N≡C:

cyanide
less basic

H H C—Br : H

methyl RX

only S_N2

:N≡C—C H H H

: Br : ⊖ Na ⊕

Only S_N2 is possible at bromomethane and requires a strong nucleophile (negative charge in our course).

Na ⊕ ⊖ :N≡C: b a

cyanide
less basic

CH_3 H H—C C—Br : H H

1° RX

S_N2 > E2

H C—CH_3 :N≡C—C H H

a = S_N2

H_3C C—H H—C H

b = E2

: Br : ⊖ Na ⊕

cyanide is a good nucleophile at methyl, 1° and 2° RBr, but only E2 at 3° RBr.

Na ⊕ ⊖ :N≡C: b a c

cyanide
less basic

CH_3 H H—C C—Br : H H—CH_2

2° RX

S_N2 > E2

H C—CH_3 :N≡C—C H CH_3 H

a = S_N2

H_3C C—H H—C CH_3

b = E2
E and Z
possible

CH_3 H_2C CH H_2C

c = E2

: Br : ⊖ Na ⊕

cyanide is a good nucleophile at methyl, 1° and 2° RBr, but only E2 at 3° RBr.

Na ⊕ ⊖ :N≡C: b c

cyanide
less basic

CH_3 H H—C C—Br : H_3C H—CH_2

3° RX

only E2

a = S_N2
is not
possible

b = E2

H_3C C—H H_3C—C CH_3

CH_3 H_2C C—CH_3 H_2C

c = E2

: Br : ⊖ Na ⊕

cyanide is a good nucleophile at methyl, 1° and 2° RBr, but only E2 at 3° RBr.

Na ⊕ ⊖ R—C≡C:

acetylides
more basic

H H C—Br : H

methyl RX

only S_N2

R—C≡C—C H H H

: Br : ⊖ Na ⊕

Only S_N2 is possible at bromomethane and requires a strong nucleophile (negative charge in our course).

acetylides more basic — 1° RX — $S_N2 > E2$ — a = S_N2 — b = E2

Acetylides are good nucleophiles at methyl and 1° RBr, but too basic at 2° RBr (E2 > S_N2) and ony E2 at 3° RBr.

acetylides more basic — 2° RX — E2 > S_N2 — a = S_N2 — b = E2, E and Z possible — c = E2

Acetylides are good nucleophiles at methyl and 1° RBr, but too basic at 2° RBr (E2 > S_N2) and ony E2 at 3° RBr.

acetylides more basic — 3° RX — only E2 — a = S_N2 is not possible — b = E2 — c = E2

Acetylides are good nucleophiles at methyl and 1° RBr, but too basic at 2° RBr (E2 > S_N2) and ony E2 at 3° RBr.

azide less basic — methyl RX — only S_N2

Only S_N2 is possible at bromomethane and requires a strong nucleophile (negative charge). Azides are reacted in a second S_N2 reaction at nitrogen using lithium aluminum hydride.

azide less basic — 1° RX — $S_N2 > E2$ — a = S_N2 — b = E2

azide is a good nucleophile at methyl, 1° and 2° RBr, but only E2 at 3° RBr. Azides are reacted in a second S_N2 reaction at nitrogen using lithium aluminum hydride.

azide less basic — 2° RX — $S_N2 > E2$ — a = S_N2 — b = E2, E and Z possible — c = E2

azide is a good nucleophile at methyl, 1° and 2° RBr, but only E2 at 3° RBr. Azides are reacted in a second S_N2 reaction at nitrogen using lithium aluminum hydride.

Example of S_N2 reaction at nitrogen in azides to form primary amines (after acidic workup)

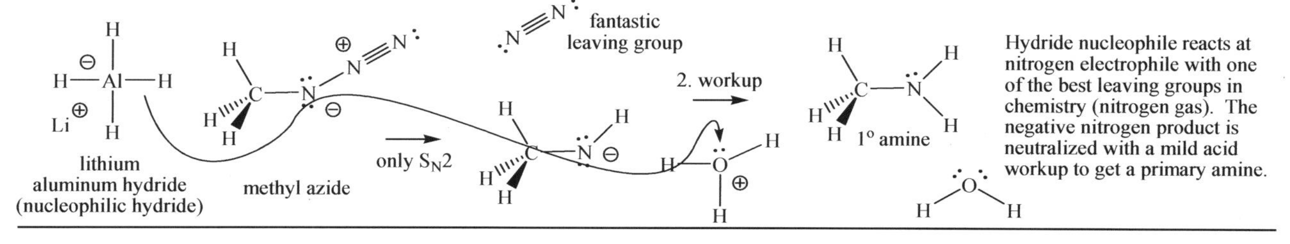

Hydride nucleophile reacts at nitrogen electrophile with one of the best leaving groups in chemistry (nitrogen gas). The negative nitrogen product is neutralized with a mild acid workup to get a primary amine.

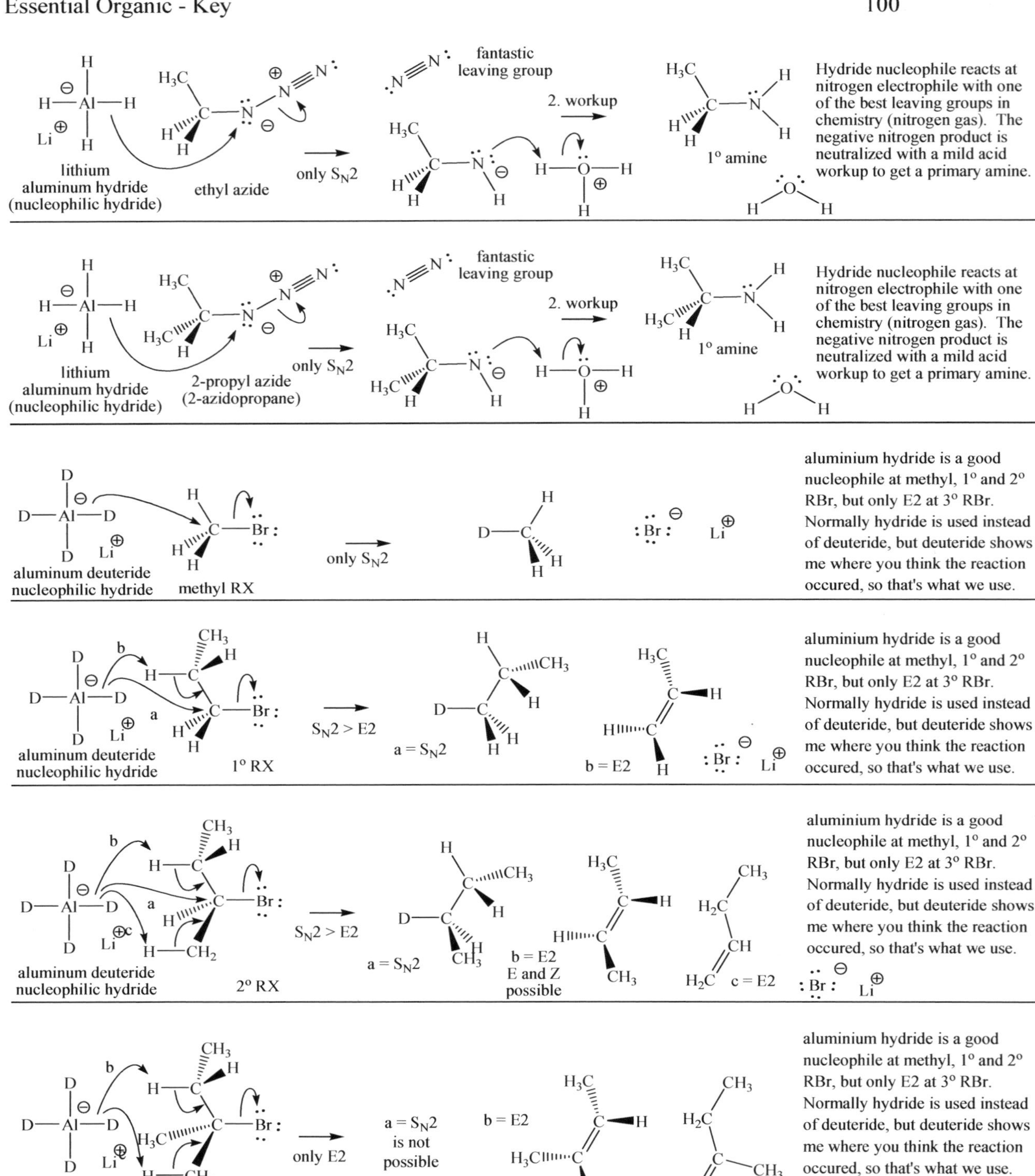
H
H—Al—H
Li
H
lithium aluminum hydride (nucleophilic hydride)
ethyl azide
only S_N2
fantastic leaving group
2. workup
1° amine
Hydride nucleophile reacts at nitrogen electrophile with one of the best leaving groups in chemistry (nitrogen gas). The negative nitrogen product is neutralized with a mild acid workup to get a primary amine.
lithium aluminum hydride (nucleophilic hydride)
2-propyl azide (2-azidopropane)
only S_N2
fantastic leaving group
2. workup
1° amine
Hydride nucleophile reacts at nitrogen electrophile with one of the best leaving groups in chemistry (nitrogen gas). The negative nitrogen product is neutralized with a mild acid workup to get a primary amine.
aluminum deuteride nucleophilic hydride
methyl RX
only S_N2
aluminium hydride is a good nucleophile at methyl, 1° and 2° RBr, but only E2 at 3° RBr. Normally hydride is used instead of deuteride, but deuteride shows me where you think the reaction occured, so that's what we use.
aluminum deuteride nucleophilic hydride
1° RX
S_N2 > E2
a = S_N2
b = E2
aluminium hydride is a good nucleophile at methyl, 1° and 2° RBr, but only E2 at 3° RBr. Normally hydride is used instead of deuteride, but deuteride shows me where you think the reaction occured, so that's what we use.
aluminum deuteride nucleophilic hydride
2° RX
S_N2 > E2
a = S_N2
b = E2
E and Z possible
c = E2
aluminium hydride is a good nucleophile at methyl, 1° and 2° RBr, but only E2 at 3° RBr. Normally hydride is used instead of deuteride, but deuteride shows me where you think the reaction occured, so that's what we use.
aluminum deuteride nucleophilic hydride
3° RX
only E2
a = S_N2 is not possible
b = E2
c = E2
aluminium hydride is a good nucleophile at methyl, 1° and 2° RBr, but only E2 at 3° RBr. Normally hydride is used instead of deuteride, but deuteride shows me where you think the reaction occured, so that's what we use.

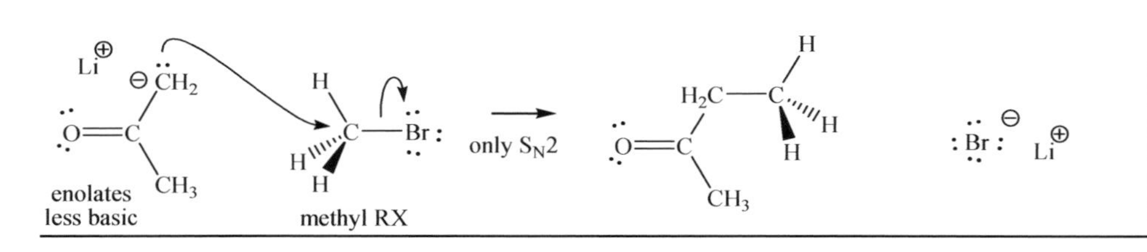
enolates less basic
methyl RX
only S_N2

Only S_N2 is possible at bromomethane and requires a strong nucleophile (negative charge). There are many variations of this enolate reaction, but this is the only one we will use.

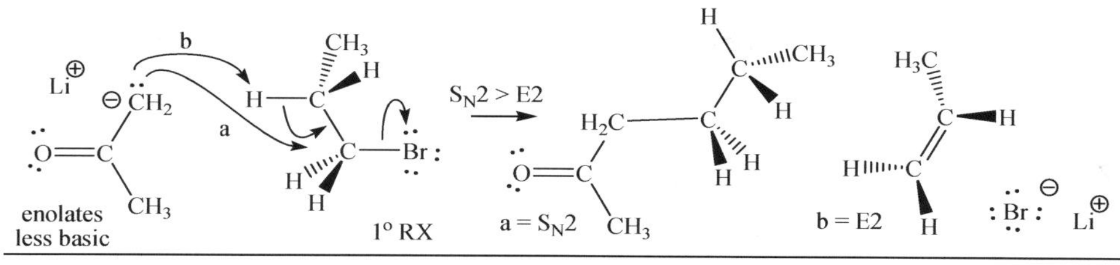

$Li^{\oplus}$ $\ominus CH_2$ b a $O=C$ CH_3 enolates less basic H—C CH_3 H C—Br : H H 1° RX

$S_N2 > E2$

H_2C—C H CH_3 H H H $O=C$ CH_3 a = S_N2

H_3C C—H H C H b = E2

: Br : $\ominus$ $Li^{\oplus}$

enolates are good nucleophiles at methyl, 1° and 2° RBr, but only E2 at 3° RBr. There are many variations of this enolate reaction, but this is the only one we will use.

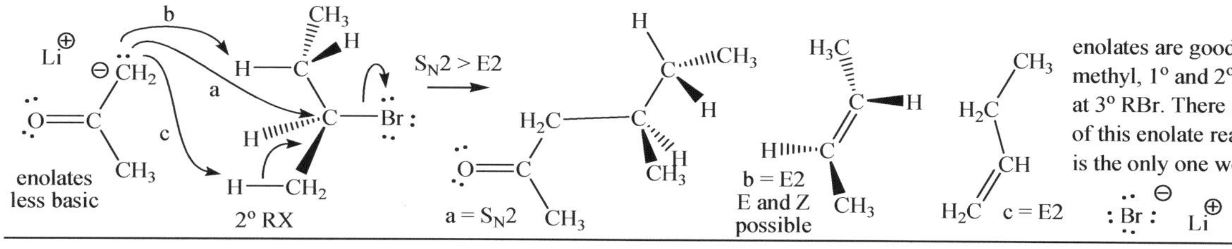

$Li^{\oplus}$ $\ominus CH_2$ b a c $O=C$ CH_3 enolates less basic H—C CH_3 H H C—Br : H—CH_2 2° RX

$S_N2 > E2$

H_2C—C H CH_3 H H CH_3 $O=C$ CH_3 a = S_N2

H_3C C—H H C CH_3 b = E2 E and Z possible

CH_3 H_2C CH H_2C c = E2

: Br : $\ominus$ $Li^{\oplus}$

enolates are good nucleophiles at methyl, 1° and 2° RBr, but only E2 at 3° RBr. There are many variations of this enolate reaction, but this is the only one we will use.

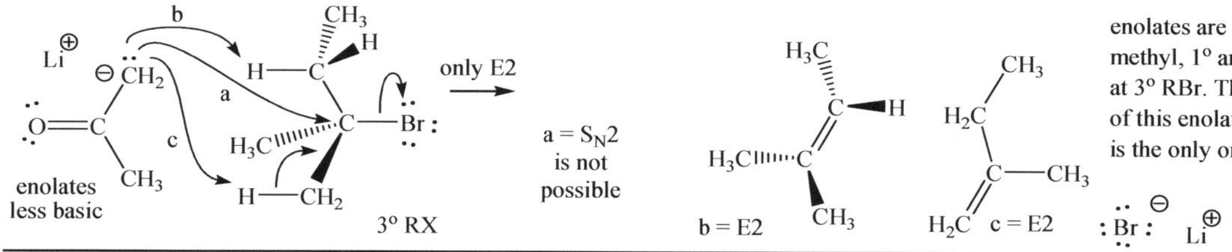

$Li^{\oplus}$ $\ominus CH_2$ b a c $O=C$ CH_3 enolates less basic H—C CH_3 H H_3C C—Br : H—CH_2 3° RX

only E2

a = S_N2 is not possible

H_3C C—H H_3C C CH_3 b = E2

CH_3 H_2C C—CH_3 H_2C c = E2

: Br : $\ominus$ $Li^{\oplus}$

enolates are good nucleophiles at methyl, 1° and 2° RBr, but only E2 at 3° RBr. There are many variations of this enolate reaction, but this is the only one we will use.

Problem 4 (p 262) – Hydrolysis of ester in aqueous base provides an alcohol synthesis and carboxylate anion. Write a complete mechanism and show the products?

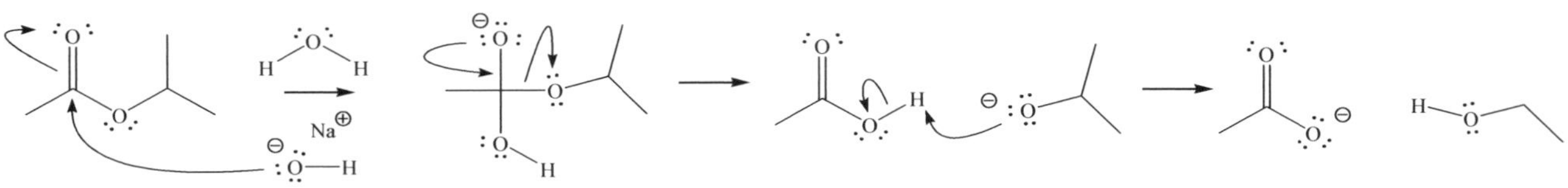

Problem 5 (p 263) – Write an arrow pushing mechanism for each of the following reactions. NaH and KH only act as a base in our course.

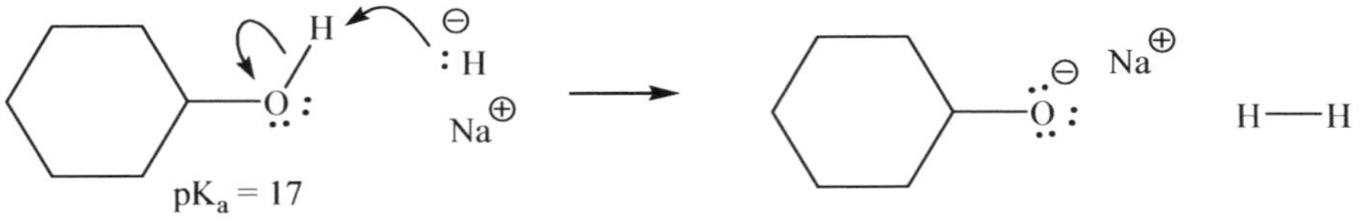

$pK_a = 17$

	pK_a
ROH	16-19
H-H	37

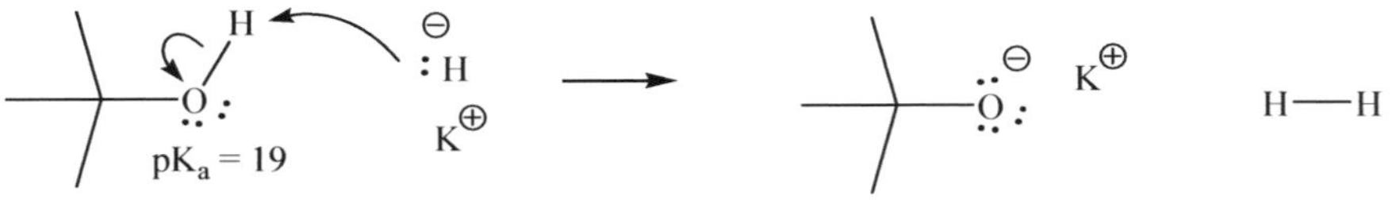

$pK_a = 19$

Problem 6 (p 263) – Write an arrow pushing mechanism for the following reactions. $LiAlH_4$ and $NaBH_4$ only react as nucleophiles in our course.

a.

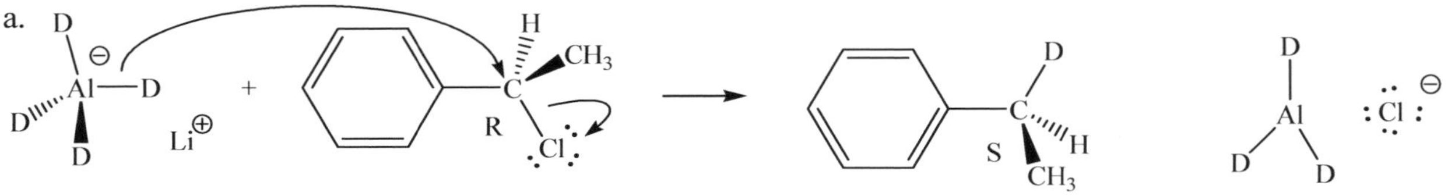

b.

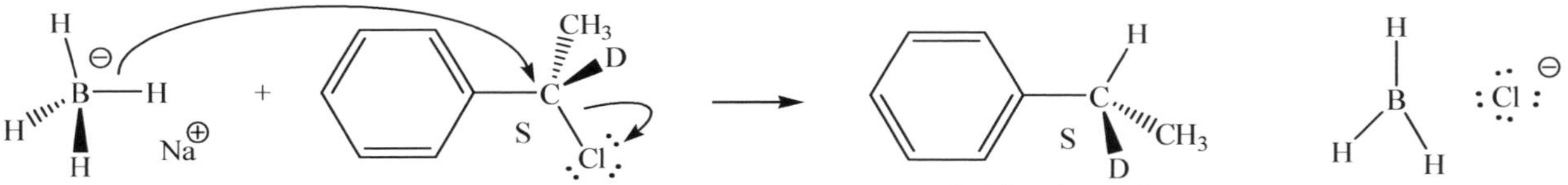

priorities changed
(D was 4, now is 3)

Problem 7 (p 264) – It is hard to tell where the hydride was introduced since there are usually so many other hydrogens in organic molecules. Where could have X have been in the reactant molecule? S_N2 reactions only occur at sp3 carbon in our course, so X would have to be in the side chain. There are no obvious clues. Which position(s) for X would likely be more reactive with the hydride reagent? Could we tell where X was if we used $LiAlD_4$?

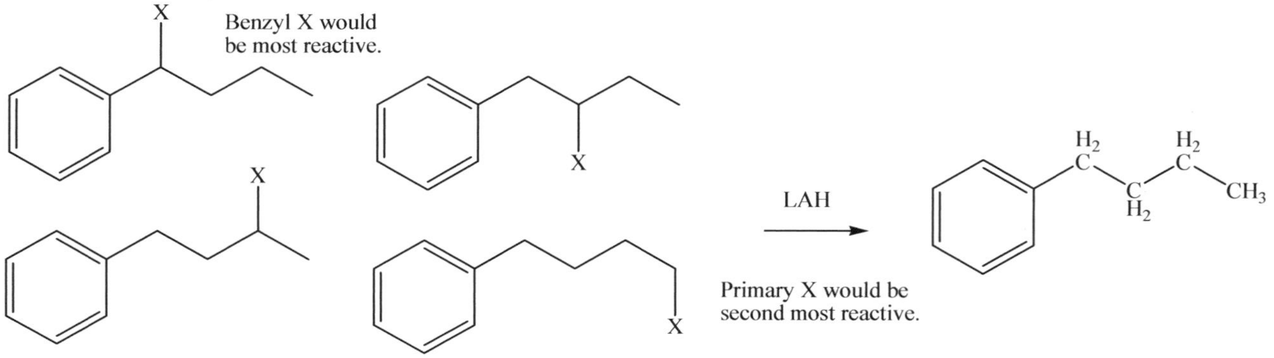

Benzyl X would be most reactive.

LAH

Primary X would be second most reactive.

X could have been on any sp^3 carbon (X = Cl, Br, I). If LiAlD4 was used, a deuterium would show us the carbon that was attacked.

Problem 8 (p 264-266) – Predict the major products (S_N2 or E2) and only show the S_N2 product, if formed. We will develop the E2 reaction next.

a.

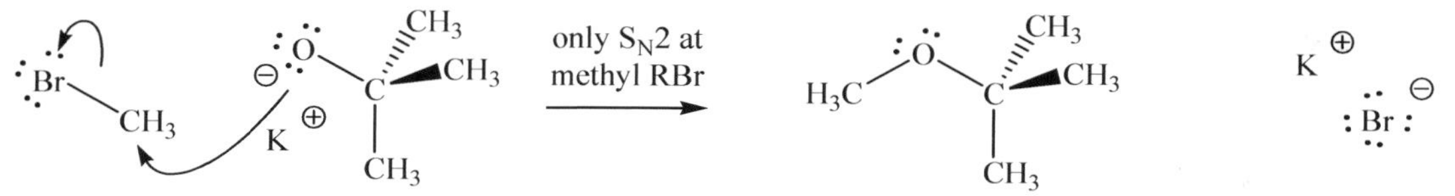

only S_N2 at methyl RBr

b.

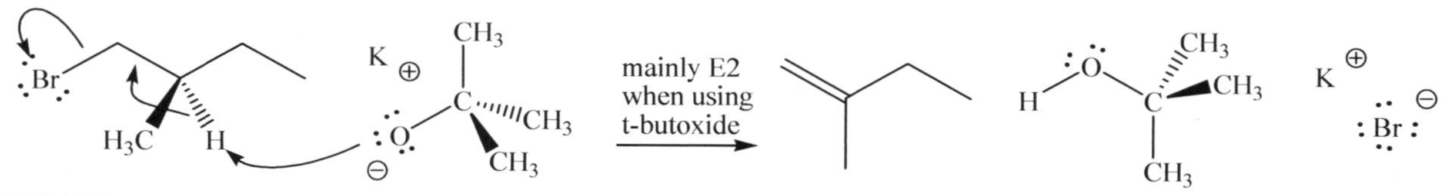

mainly E2 when using t-butoxide

c.

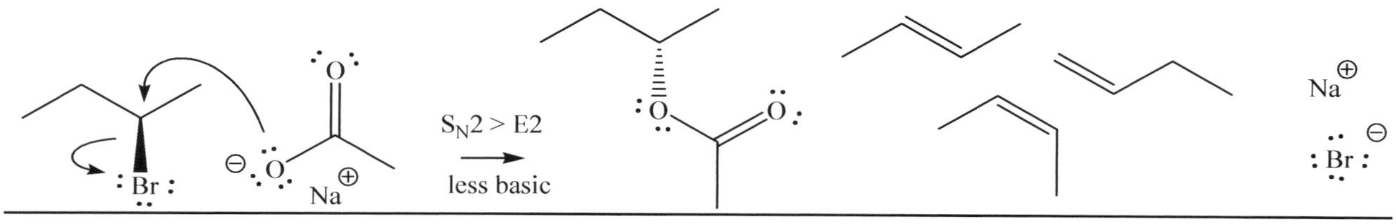

S_N2 > E2

less basic

d.

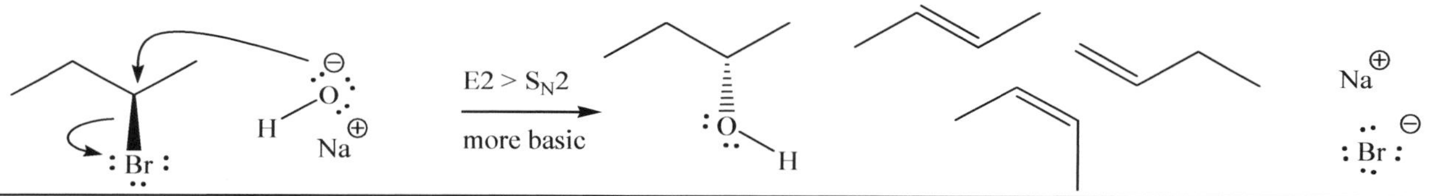

E2 > S_N2

more basic

e.

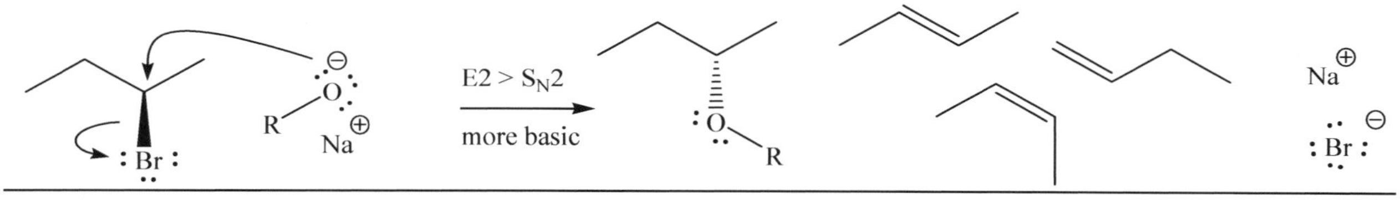

E2 > S_N2

more basic

f.

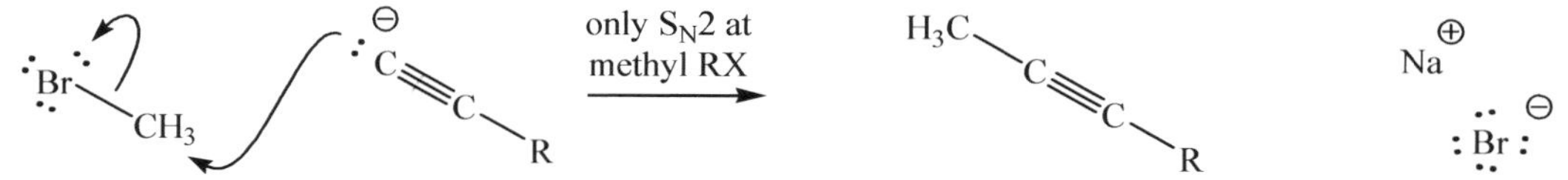

only S_N2 at methyl RX

g.

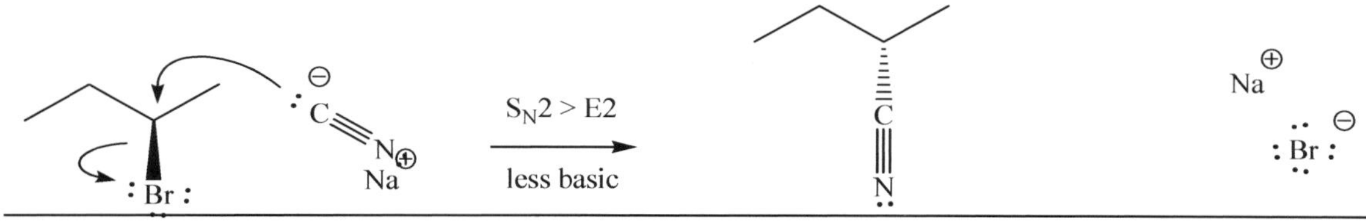

S_N2 > E2

less basic

h.

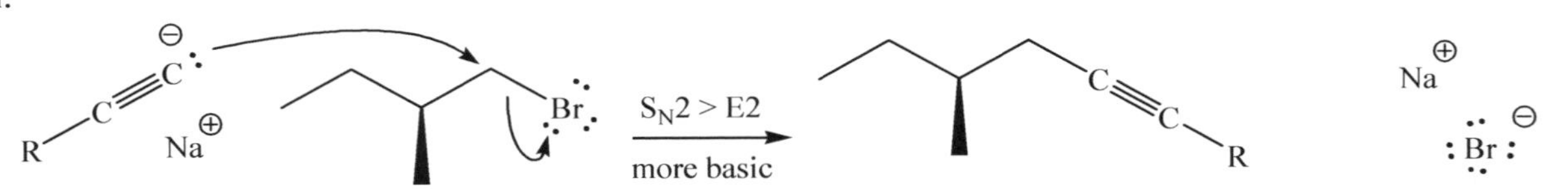

i

H
: Br :
C
C
R
Na
$E2 > S_N2$
more basic
Na
: Br :

j.

: Br :
N
N
N
Na
$S_N2 > E2$
less basic
: N
N
N
Na
: Br :

k.

: Br :
S
H
Na
$S_N2 > E2$
less basic
: N
N
N
Na
: Br :

l.

H
CH_2
H_3C
CH_3
C
: Br :
S
R
Na
only E2
less basic
Na
: Br :

m.

: Br :
D
D—B—D
D
Na
$S_N2 > E2$
less basic
D
Na
: Br :

n.

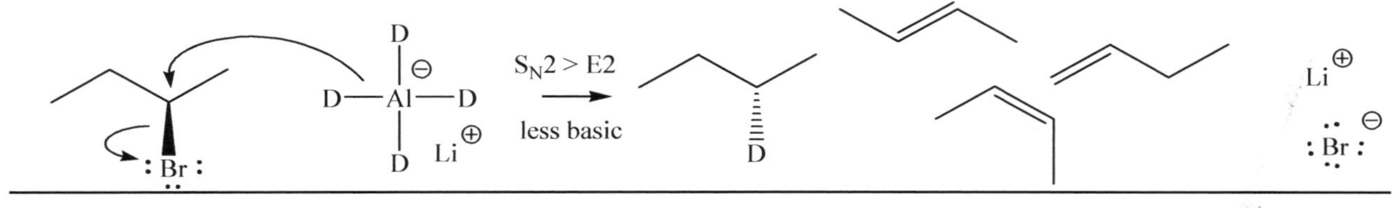

o.

only E2

less basic

p.

$S_N2 > E2$

less basic

q.

$S_N2 > E2$

less basic

Problem 9 (p 268) – How many total hydrogen atoms are on C_β carbons in the given RX compound? How many different types of hydrogen atoms are on C_β carbons (a little tricky)? How many different products are possible? Hint - Be careful of the simple CH_2. The two hydrogen atoms appear equivalent, but E/Z (cis/trans) possibilities are often present. (See below for relative expected amounts of the E2 products.) Which elimination products would you expect to be the major and minor products. How would an absolute configuration of C_α as R, compare to C_α as S? What about $C_{\beta 1}$ as R versus S?

Six total C_β-H
Four types of protons.

4 (all 3 methyl protons)

$C_{\beta 1}$ product
(3S,4R) -> Z alkene

(3S,4S)

E

$C_{\beta 2}$-H_a product

Z

$C_{\beta 2}$-H_b product

neither

$C_{\beta 2}$ product

Problem 10 (p 268) -Provide an explanation for any unexpected deviations from our general rule for alkene stabilities above. A less negative potential energy is more stable.

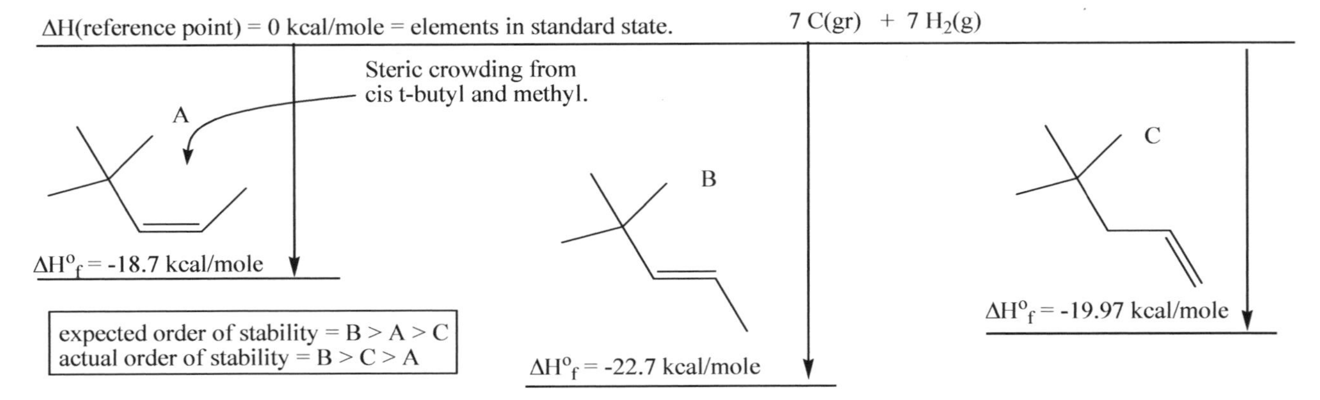

The more negative ΔH indicates a more stable energy, since all three alkenes are created from the same amount of carbon and hydrogen. The 3rd structure is expected to be the least stable (largest negative energy) because it is monosubstituted. The second structure is expected to be the most stable (smallest energy) and it is. The first structure is expected to be the middle energy, but instead it is the least negative energy (least stable). The large t-butyl group next to the methyl group destabilizes the alkene (larger negative value) so that it becomes less stable than the monosubstituted alkene.

Problem 11 (p 268) – Using the most stable alkyne as a reference point, the second most stable alkyne is 4.6 kcal/mole less stable and the least stable alkyne is 4.8 kcal/mole less stable than the second most stable alkyne. Order the alkynes in relative stability (1 = most stable) and provide a possible explanation.

3	2	1
H—C≡C—H	R—C≡C—H	R—C≡C—R

"R" = a simple alkyl group (inductively donating relative to a hydrogen atom)

The sp carbon atoms are more electronegative than sp^3 or sp^2 carbon atoms. The inductively donating R groups donate electron density to those carbon atoms, helping to stabilize their electronegatie pull.

Problem 12 (p 268) – One of the following reactions produces over 90% S_N2 product and one of them produces about 85% E2 product in contrast to our general rules (ambiguity is organic chemistry's middle name). Match these results with the correct reaction and explain why they are different.

pK_a = 16 (> 90% S_N2)

Less basic and less sterically hindered than t-butoxide. S_N2 is the main product at methyl and primary RX.

---

pK_a = 19 (≈85% E2)

More basic and more sterically hindered than simple alkoxides. E2 is the main product, even at 1° RX.

Problem 13 (p 269) - Propose an explanation for the following table of data. Write out the expected products and state by which mechanism they formed. Nu: $^{-}$ /B: $^{-}$ = $CH_3CO_2^{-}$ (a weak base, but good nucleophile).

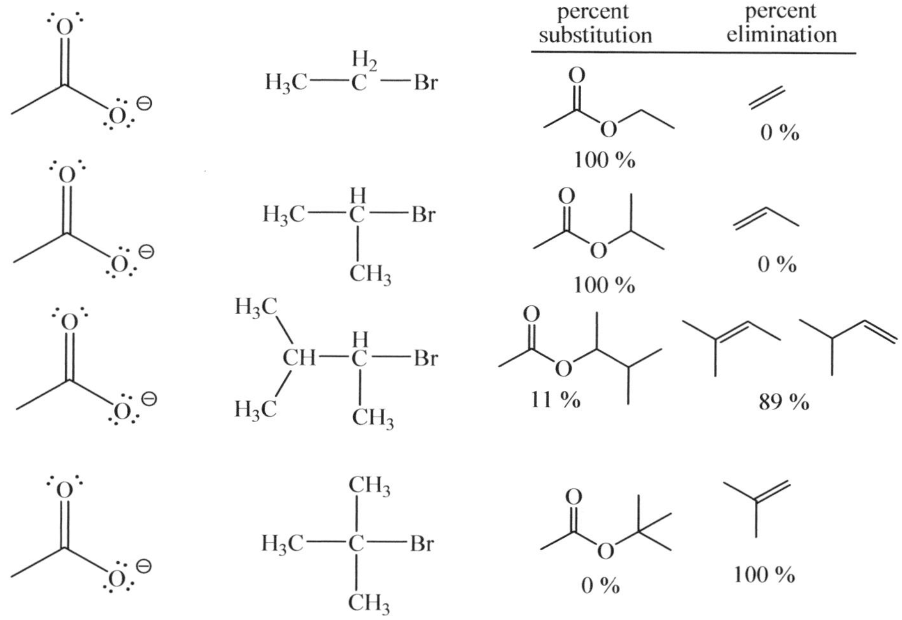

RX	percent substitution	percent elimination	
$H_3C—CH_2—Br$	100 %	0 %	Good nucleophile reacting with a 1° RX center. S_N2 is expected as the main product.
$H_3C—CH(CH_3)—Br$	100 %	0 %	Good nucleophile reacting with a 2° RX center. S_N2 is expected as the main product.
$(H_3C)_2CH—CH(CH_3)—Br$	11 %	89 %	Good nucleophile reacting with a 2° RX center. S_N2 is expected as the main product. In this example the C_α is secondary and C_β position tertiary. This slows S_N2 so that E2 becomes the main product.
$H_3C—C(CH_3)_2—Br$	0 %	100 %	Good nucleophile reacting with a 3° RX center. E2 is expected as the only product.

Problem 14 (p 269) - A stronger base (as measured by a higher pK_a of its conjugate acid) tends to produce more relative amounts of E2 compared to S_N2, relative to a second (weaker) base/nucleophile. Greater substitution (R > H) at C_α and C_β also increases the proportion of E2 product, because the greater steric hindrance slows down the competing S_N2 reaction. Use this information to make predictions about which set of conditions in each part would produce relatively more elimination product. Briefly, explain your reasoning. Write out all expected elimination products. Are there any examples below where one reaction (S_N2 or E2) would completely dominate?

a.

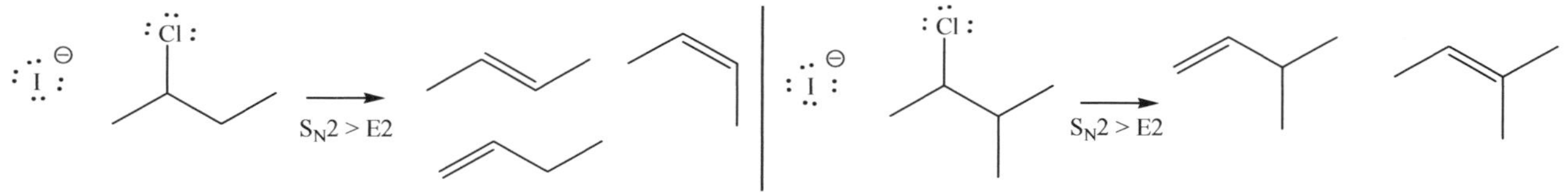

$:\ddot{I}:^{\ominus}$ + :Cl: → S_N2 > E2 | $:\ddot{I}:^{\ominus}$ + :Cl: → S_N2 > E2

More E2 product because RCl is more sterically hindered.

b.

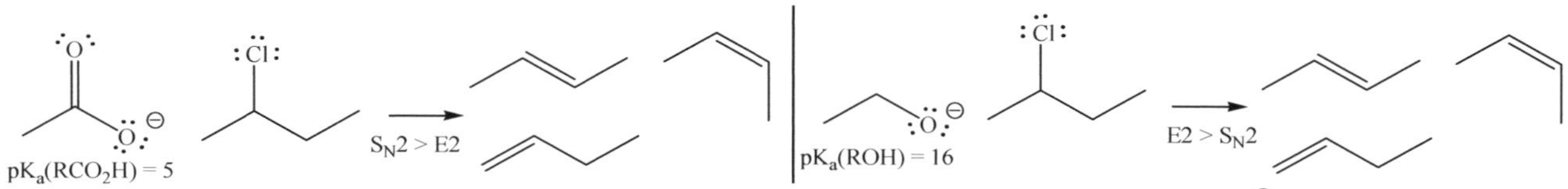

$pK_a(RCO_2H) = 5$ → S_N2 > E2 | $pK_a(ROH) = 16$ → E2 > S_N2

More E2 product because $RO^{\ominus}$ is more basic.

c.

$S_N2 > E2$

$E2 > S_N2$

More E2 product because RCl is more sterically hindered.

d.

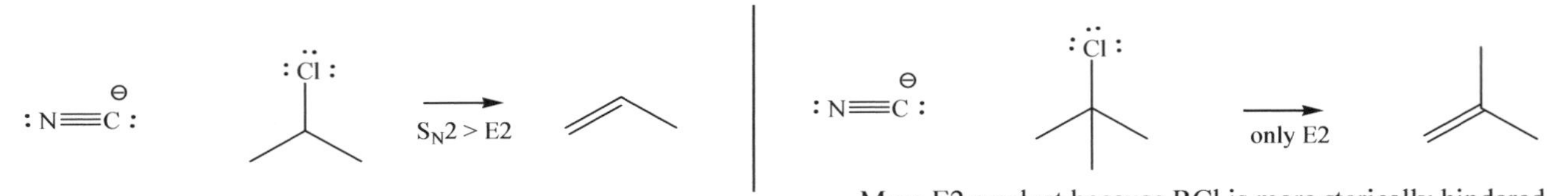

$S_N2 > E2$

only E2

More E2 product because RCl is more sterically hindered.

e.

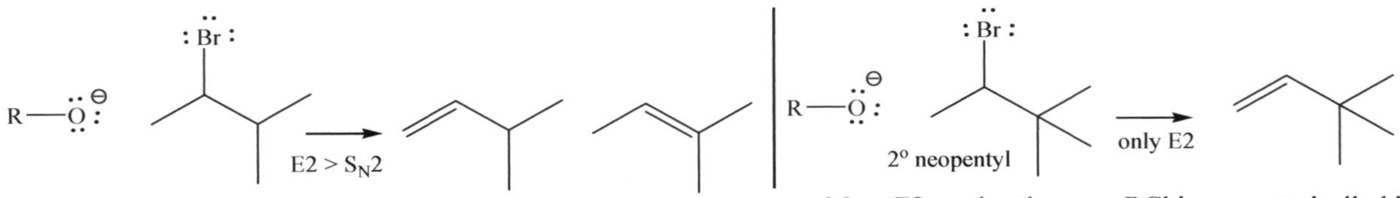

$E2 > S_N2$

2° neopentyl

only E2

More E2 product because RCl is more sterically hindered.

f.

$E2 > S_N2$

$pK_a(RCCH) = 25$

More E2 product because acetylides are more basic than cyanide.

$pK_a(NCH) = 9$

$S_N2 > E2$

Problem 15 (p 270) - (2R,3S)-2-bromo-3-deuteriobutane when reacted with potassium ethoxide produces cis-2-butene having deuterium and trans-2-butene not having deuterium. The diastereomer (2R,3R)-2-bromo-3-deuteriobutane under the same conditions produces cis-2-butene not having deuterium and trans-2-butene having deuterium present. Explain these observations by drawing the correct 3D structures, rotating to the proper conformation for elimination and showing an arrow pushing mechanism leading to the observed products. (Protium = H and deuterium = D; H and D are isotopes. Their chemistries are similar, but we can tell them apart.)

(2R,3S)-2-bromo-3-deuteriobutane

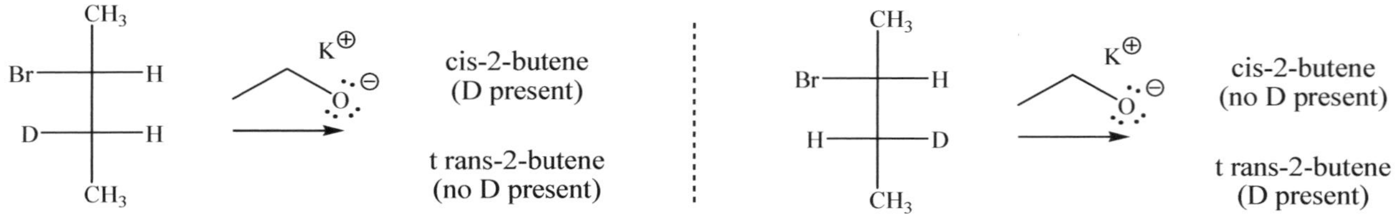

cis-2-butene (D present)

t rans-2-butene (no D present)

(2R,3S)

cis-2-butene (no D present)

t rans-2-butene (D present)

(2R,3R)

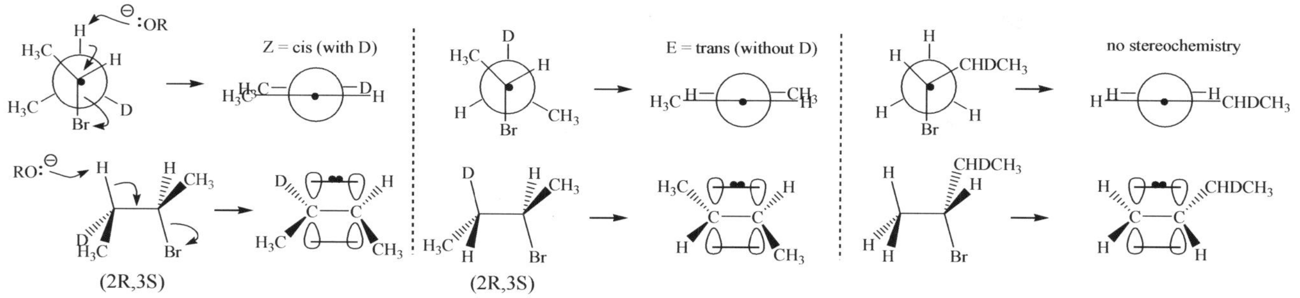

Z = cis (with D)

E = trans (without D)

no stereochemistry

(2R,3S)

(2R,3S)

Problem 16 (p 271) – Sodium dialkylamides are extremely basic and can be sterically bulky (a nitrogen equivalent of t-butoxide). They are often used in reactions with dibromoalkanes, RBr_2, to induce a double E2 reaction (notice there are two leaving groups). We show two different dibromoalkanes below. What reaction is expected and what would be the products after workup? Remember the pK_a of a terminal alkyne is about 25 and the pK_a of a regular amine is about 37 (that's why a final workup step is necessary).

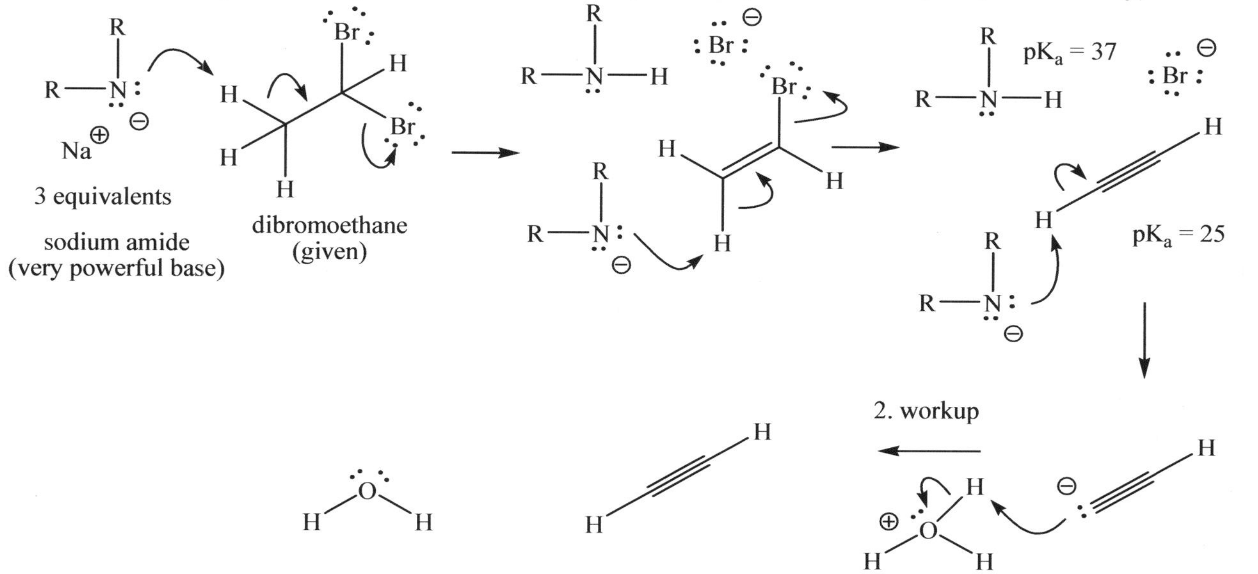

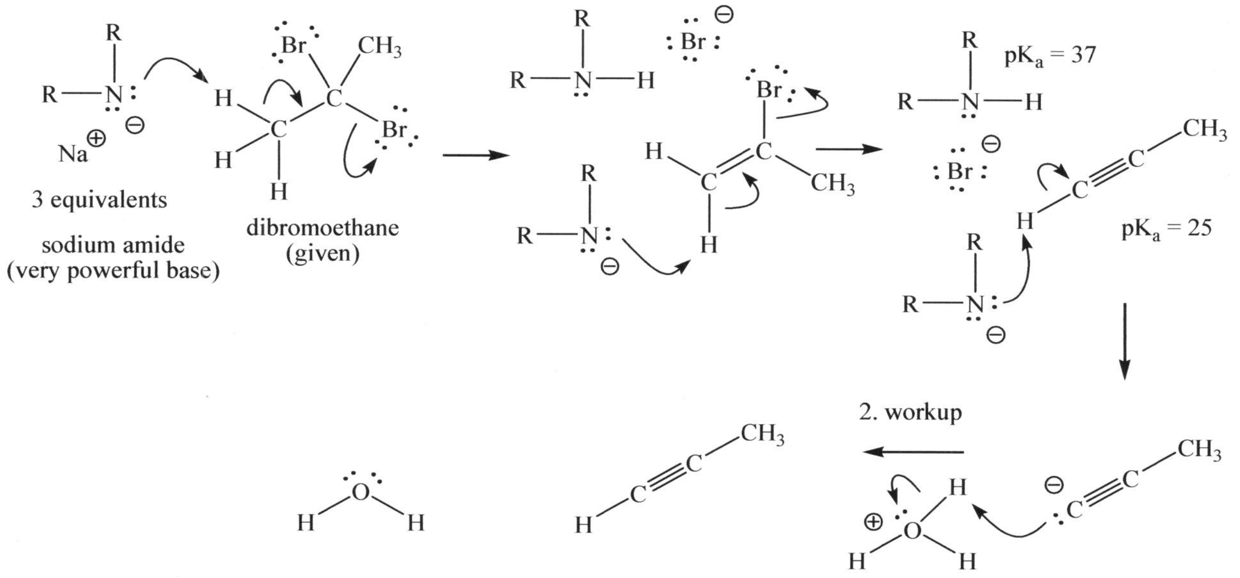

3 equivalents

sodium amide (very powerful base)

dibromoethane (given)

$pK_a = 37$

$pK_a = 25$

2. workup

---

3 equivalents

sodium amide (very powerful base)

dibromoethane (given)

$pK_a = 37$

$pK_a = 25$

2. workup

Problem 17 (p 272) – Propose a reaction using our strong nucleophile/base reagents with one of our C1-C6 R-Br compounds (on pp. 1-2) to make each of the following structures.

a
Br
NaOH
OH

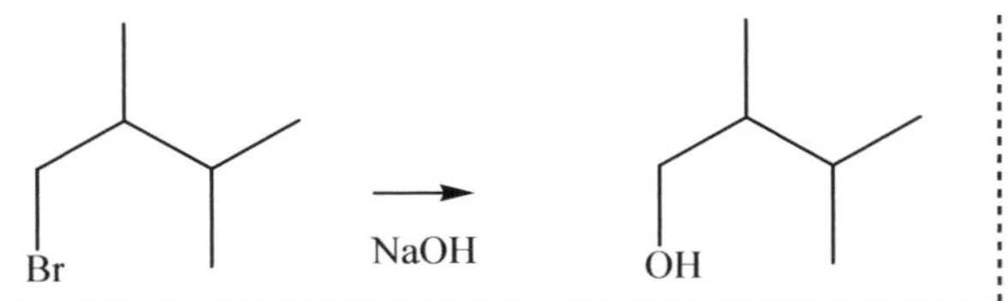

b
Br
O
O
O
O

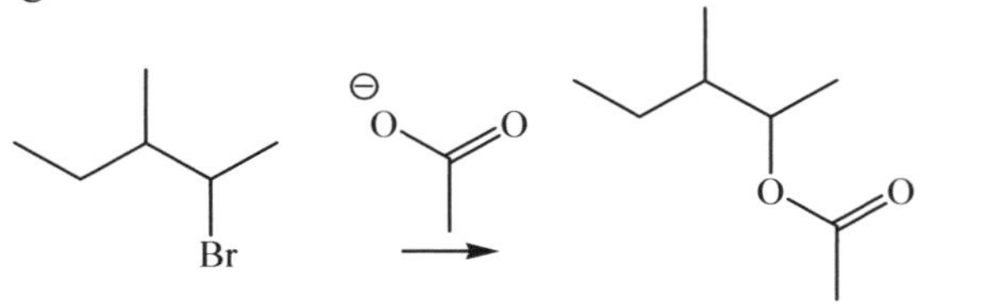

c
Br
1. NaN_3
2. $LiAlH_4$
3. workup
NH_2

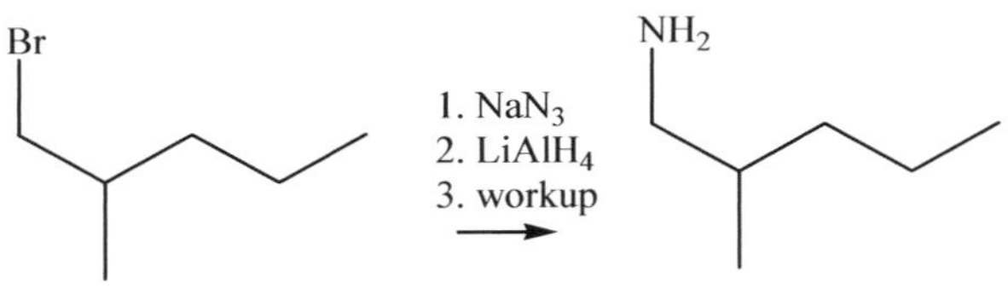

d
Br
NaSH
SH

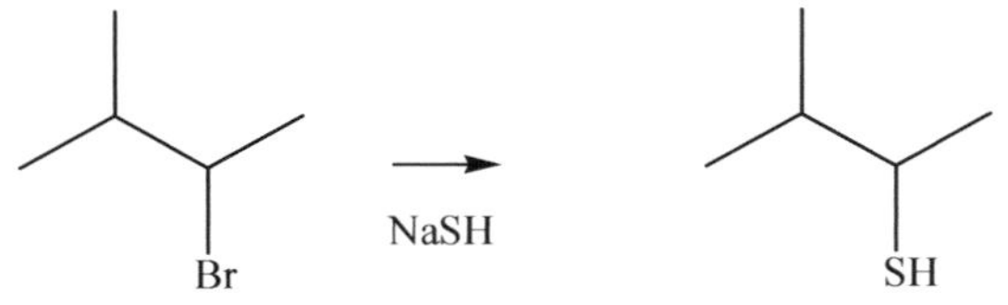

e
Br
O
$Na^{\oplus}$
O

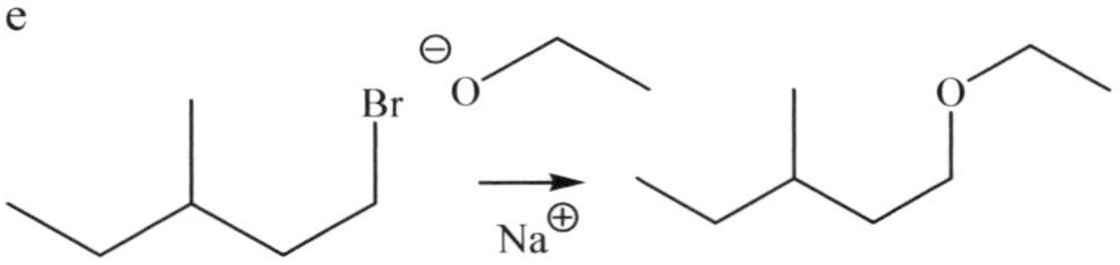

f
Br
$Na^{\oplus}$

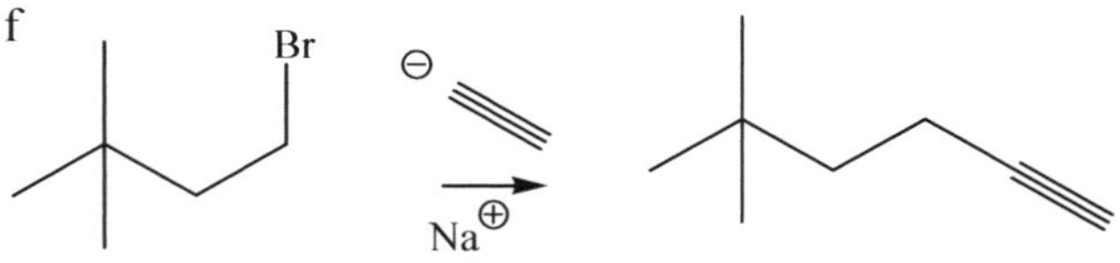

i
Br
C
N
$Na^{\oplus}$
N
C

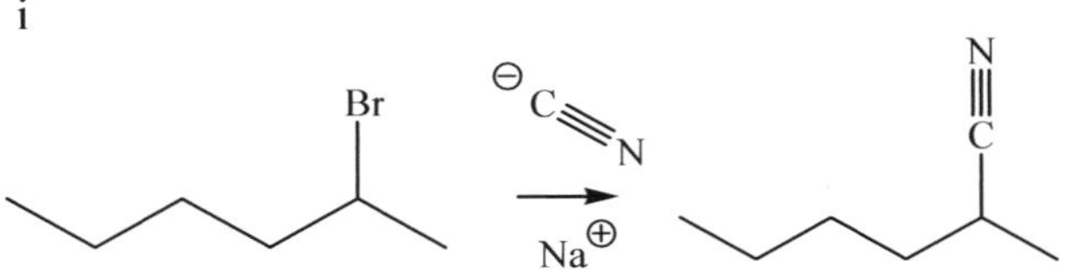

j
Br
S
$Na^{\oplus}$
S

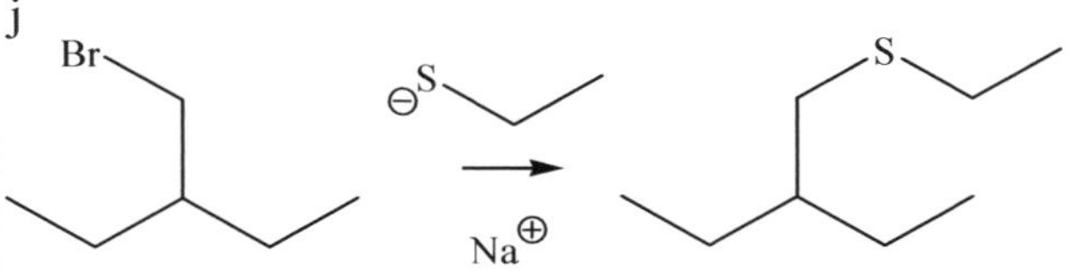

k
Br
1. $LiAlD_4$
2. workup
D

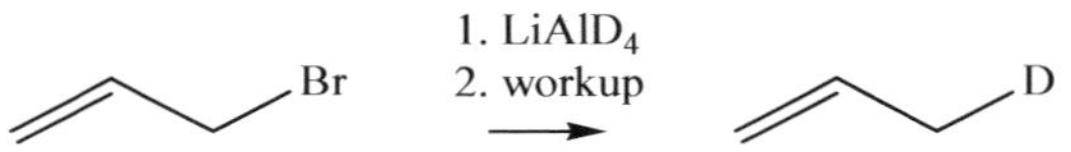

l
O
Br
$K^{\oplus}$

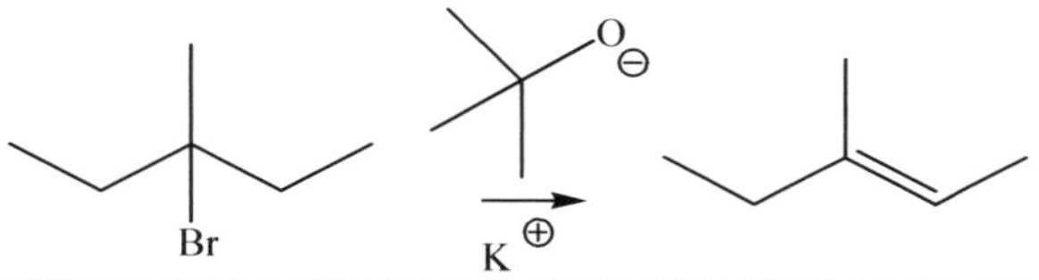

Problem 18 (p 273) – What are the possible products of the following reactions? What is the major product(s) and what is the minor product(s)? There are 40 possible combinations.

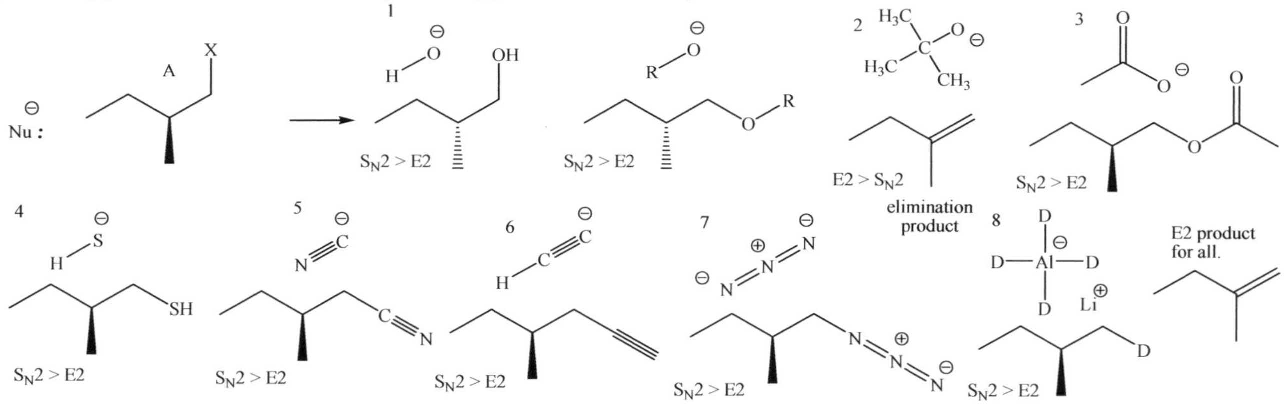

Nu :
A
X

1
H O
OH
$S_N2 > E2$

R O
R
O
$S_N2 > E2$

2
H_3C
C
O
H_3C
CH_3
$E2 > S_N2$
elimination product

3
O
O
O
O
$S_N2 > E2$

4
H S
SH
$S_N2 > E2$

5
N C
C N
$S_N2 > E2$

6
H C C
$S_N2 > E2$

7
N N N
N N N
$S_N2 > E2$

8
D
D Al D
D
$Li^{\oplus}$
D
$S_N2 > E2$

E2 product for all.

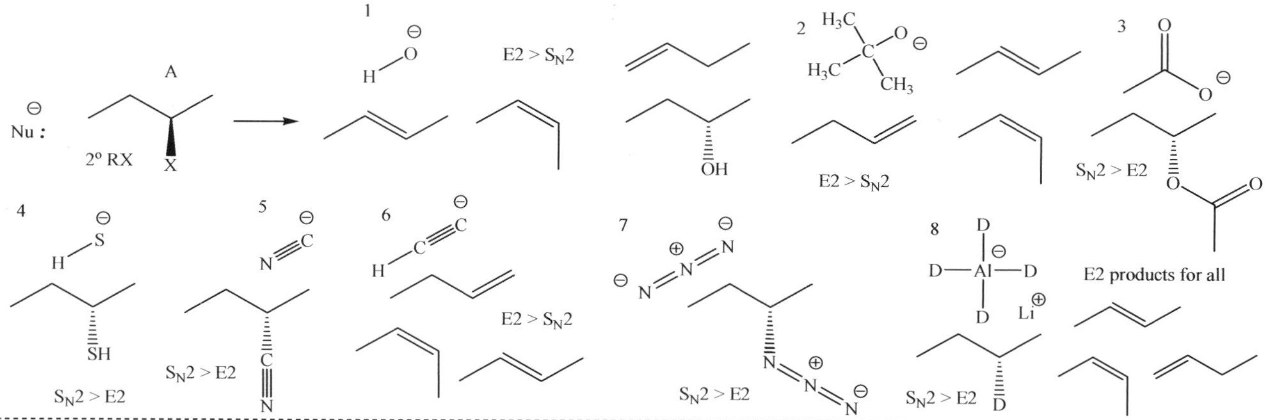

A

$\ominus$ Nu : 2° RX

1 E2 > S_N2

2 E2 > S_N2

3 S_N2 > E2

4 S_N2 > E2

5 S_N2 > E2

6 E2 > S_N2

7 S_N2 > E2

8 S_N2 > E2

E2 products for all

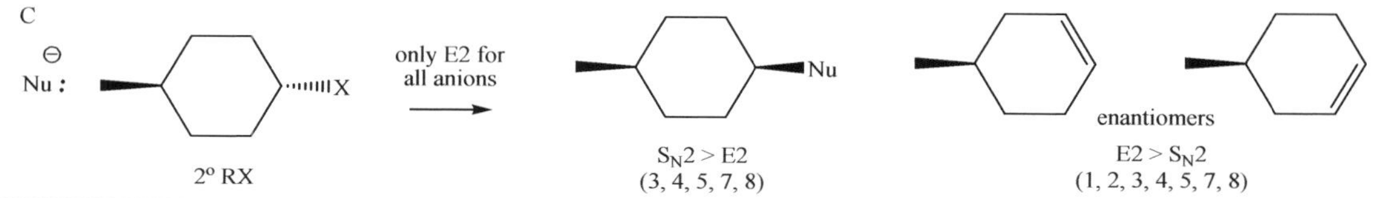

C

$\ominus$ Nu : 2° RX

only E2 for all anions

S_N2 > E2
(3, 4, 5, 7, 8)

enantiomers
E2 > S_N2
(1, 2, 3, 4, 5, 7, 8)

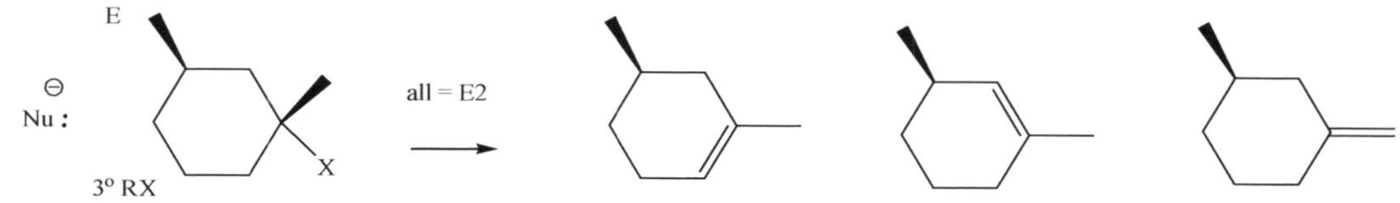

E

$\ominus$ Nu : 3° RX

all = E2

Problem 19 (p 277) – The bond energy depends on charge effects in the anions too. Can you explain the differences in bond energies below? (Hint: Where is the charge more delocalized? Think back to our acid/base topic.)

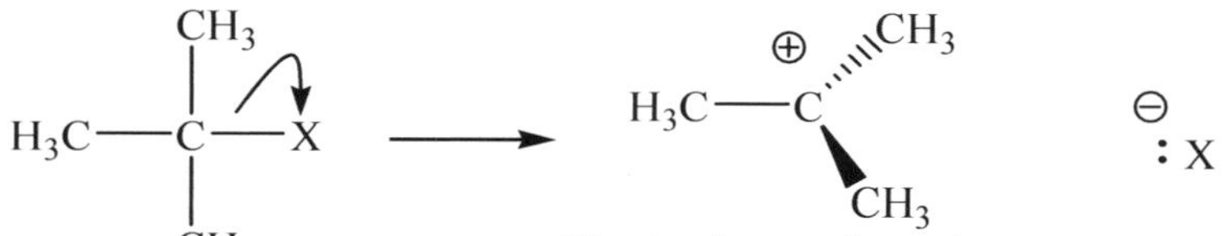

The tertiary carbocation is the same for all examples.

stability of anions = $I^{\ominus} > Br^{\ominus} > Cl^{\ominus}$

greater charge delocalization is more stable, and larger iodide anion is more delocalized than bromide than chloride

Problem 20 (p 278) – Draw in all of the mechanistic steps in an S_N1 reaction of 2R-bromobutane with a. water, b. methanol and c. ethanoic acid. Add in necessary details (3D stereochemistry, curved arrows, lone pairs, formal charge). What are the final products?

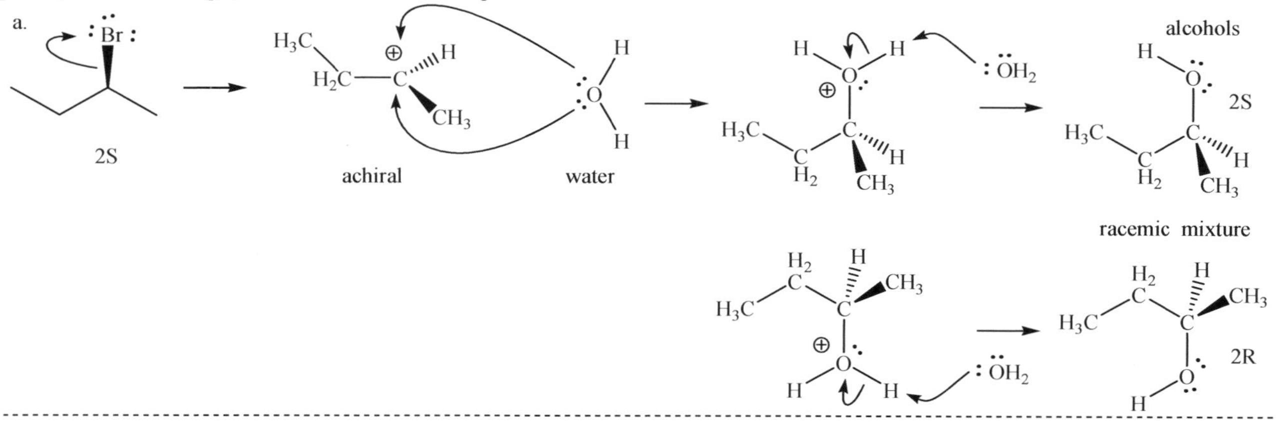

a.

2S

achiral

water

alcohols

2S

racemic mixture

2R

b.

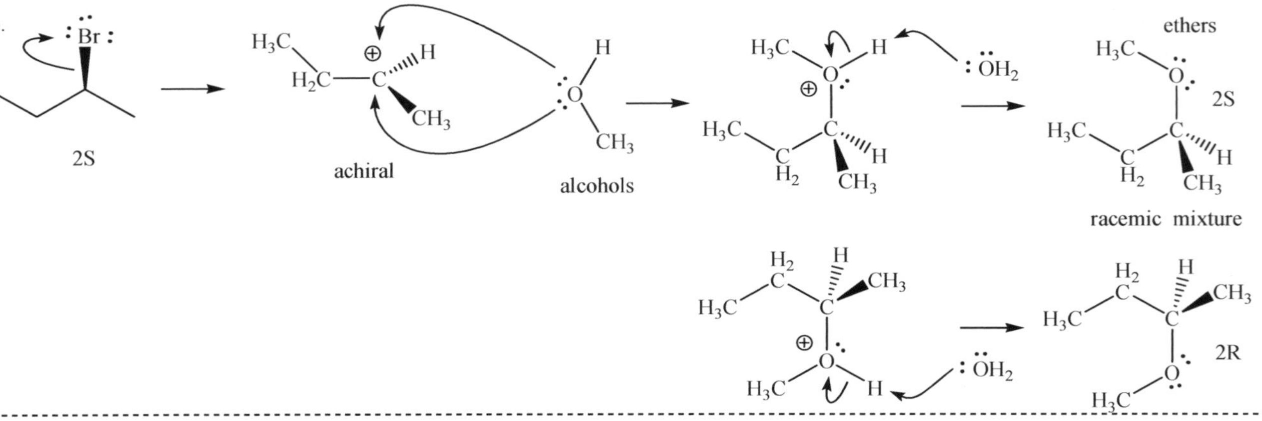

2S

achiral

alcohols

ethers

2S

racemic mixture

2R

c.

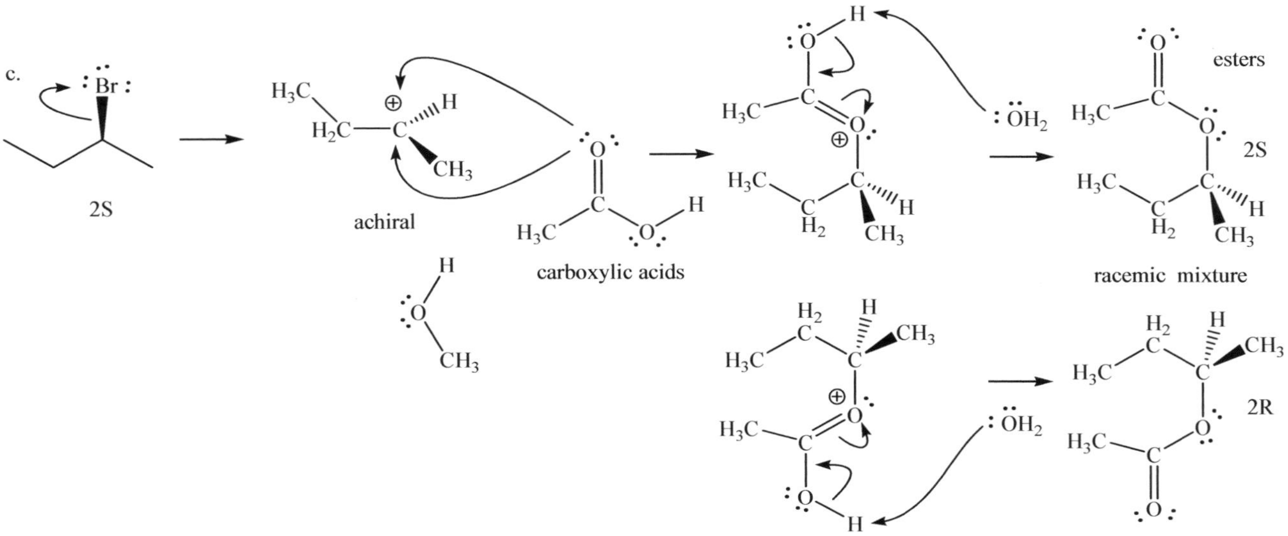

2S

achiral

carboxylic acids

esters

2S

racemic mixture

2R

Problem 21 (p 279) – Reconsider problem 15 and draw in all of the mechanistic steps and show all possible products of the E1 reaction of 2R-bromobutane with any of the weak bases, water, methanol or ethanoic acid (use water in your work). Add in necessary details (3D stereochemistry, curved arrows, lone pairs, formal charge).

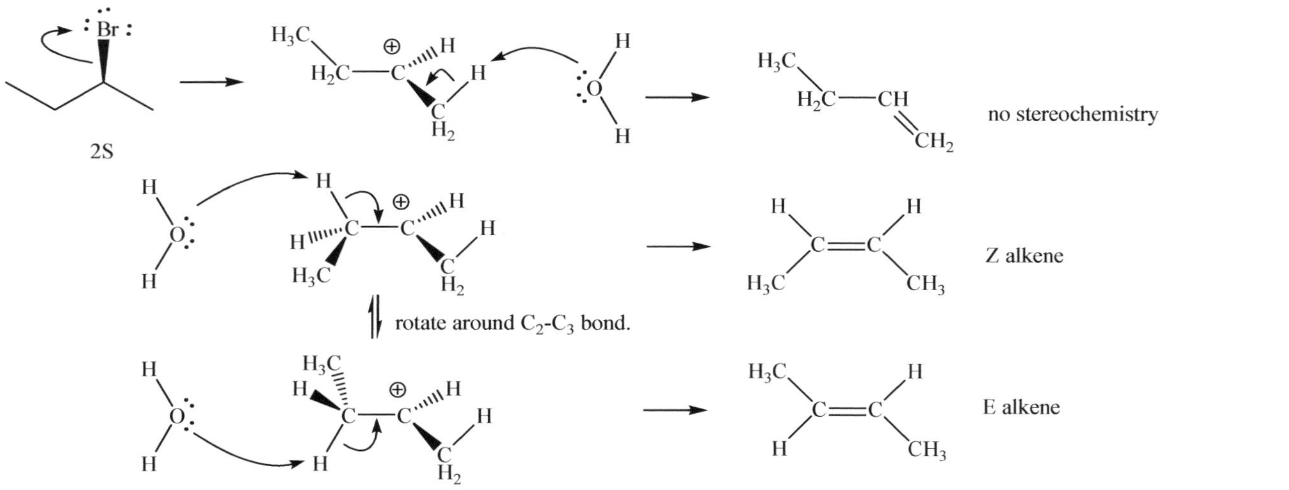

2S

no stereochemistry

Z alkene

rotate around C_2-C_3 bond.

E alkene

Problem 22 (p 280) – What are the likely S_N1 and E1 products of the initial carbocation and the rearrangement carbocations from “a”? Assume water is the nucleophile.

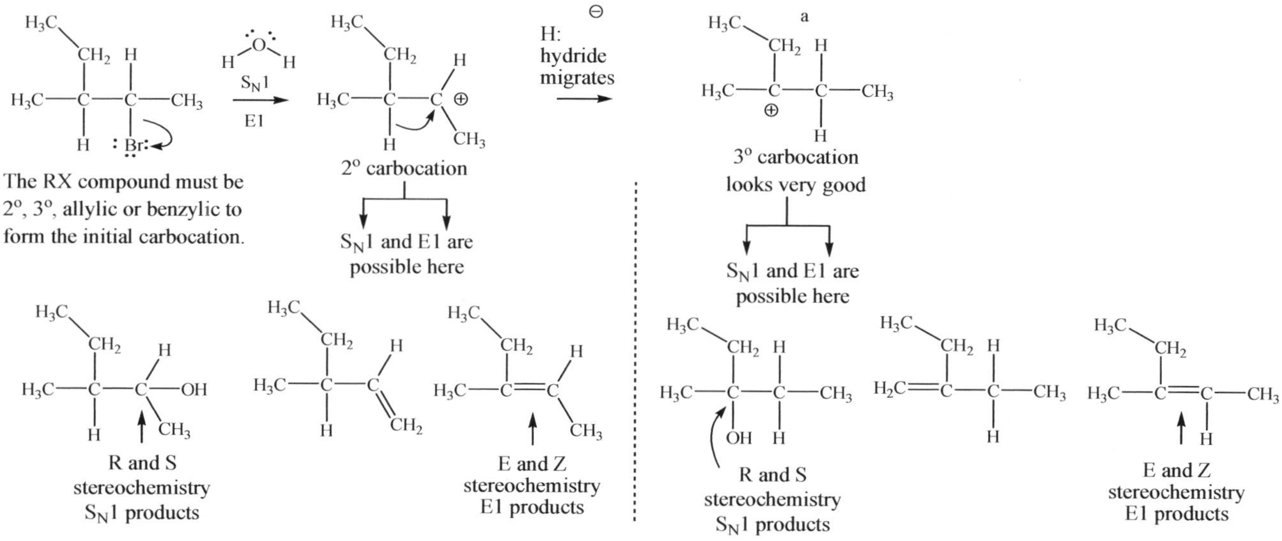

S_N1 / E1

H: hydride migrates

a

2° carbocation

3° carbocation looks very good

The RX compound must be 2°, 3°, allylic or benzylic to form the initial carbocation.

S_N1 and E1 are possible here

S_N1 and E1 are possible here

R and S stereochemistry S_N1 products

E and Z stereochemistry E1 products

R and S stereochemistry S_N1 products

E and Z stereochemistry E1 products

Problem 23 (p 281) – Write out your own mechanism for all reasonable products from the given R-X compound in water (2-halo-3-methylbutane).

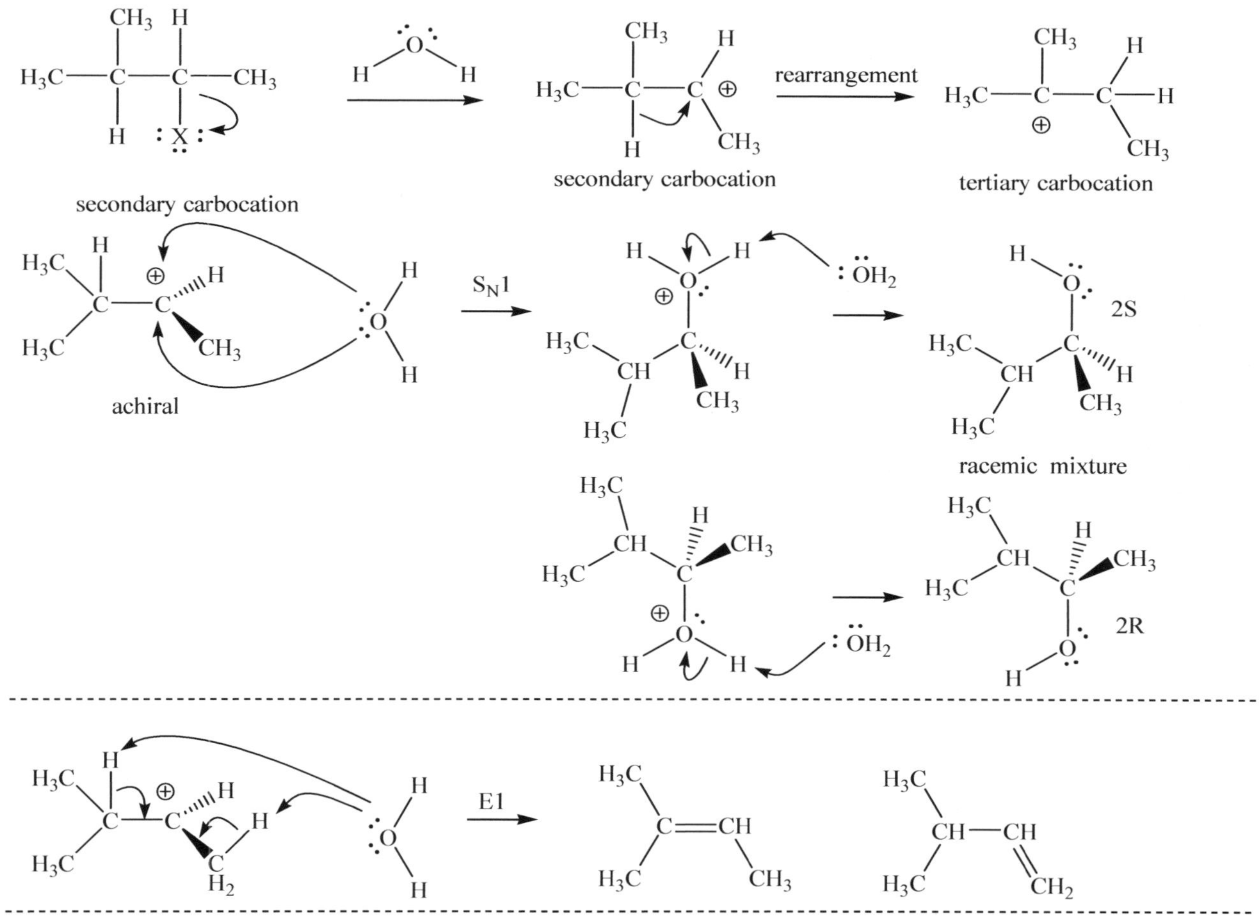

rearrangement

secondary carbocation

secondary carbocation

tertiary carbocation

achiral

S_N1

2S

racemic mixture

2R

E1

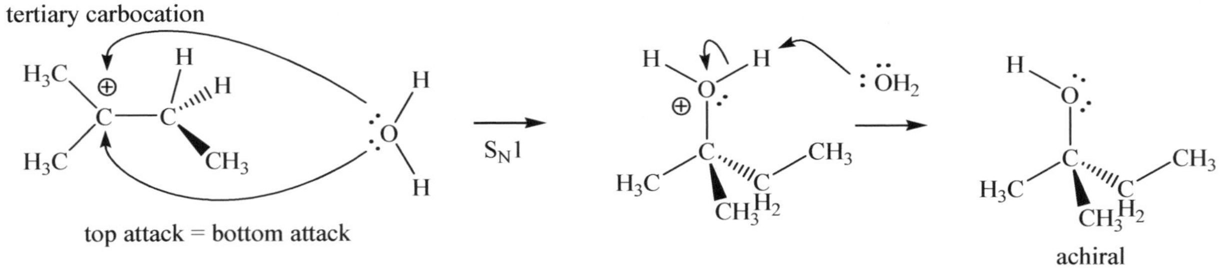

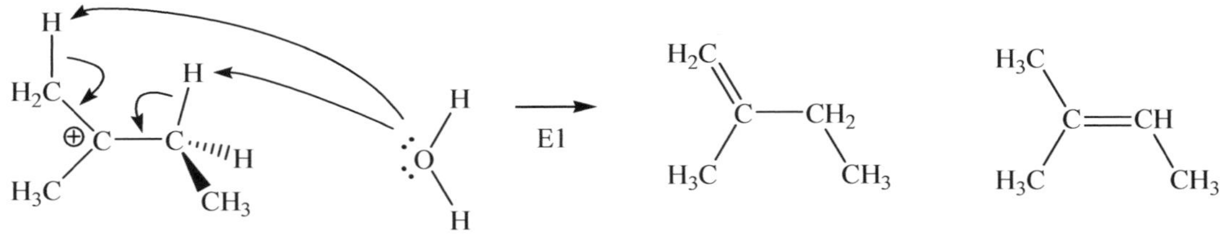

tertiary carbocation

top attack = bottom attack

S_N1

achiral

E1

Problem 24 (p 282) – Lanosterol is the first steroid skeletal structure on the way to cholesterol and other steroids in our bodies. It is formed in a spectacular cyclization of protonated squalene oxide. The initially formed 3° carbocation rearranges 4 times before it undergoes an E1 reaction to form lanosterol. Add in the arrows and formal charge to show the rearrangements and the final E1 reaction.

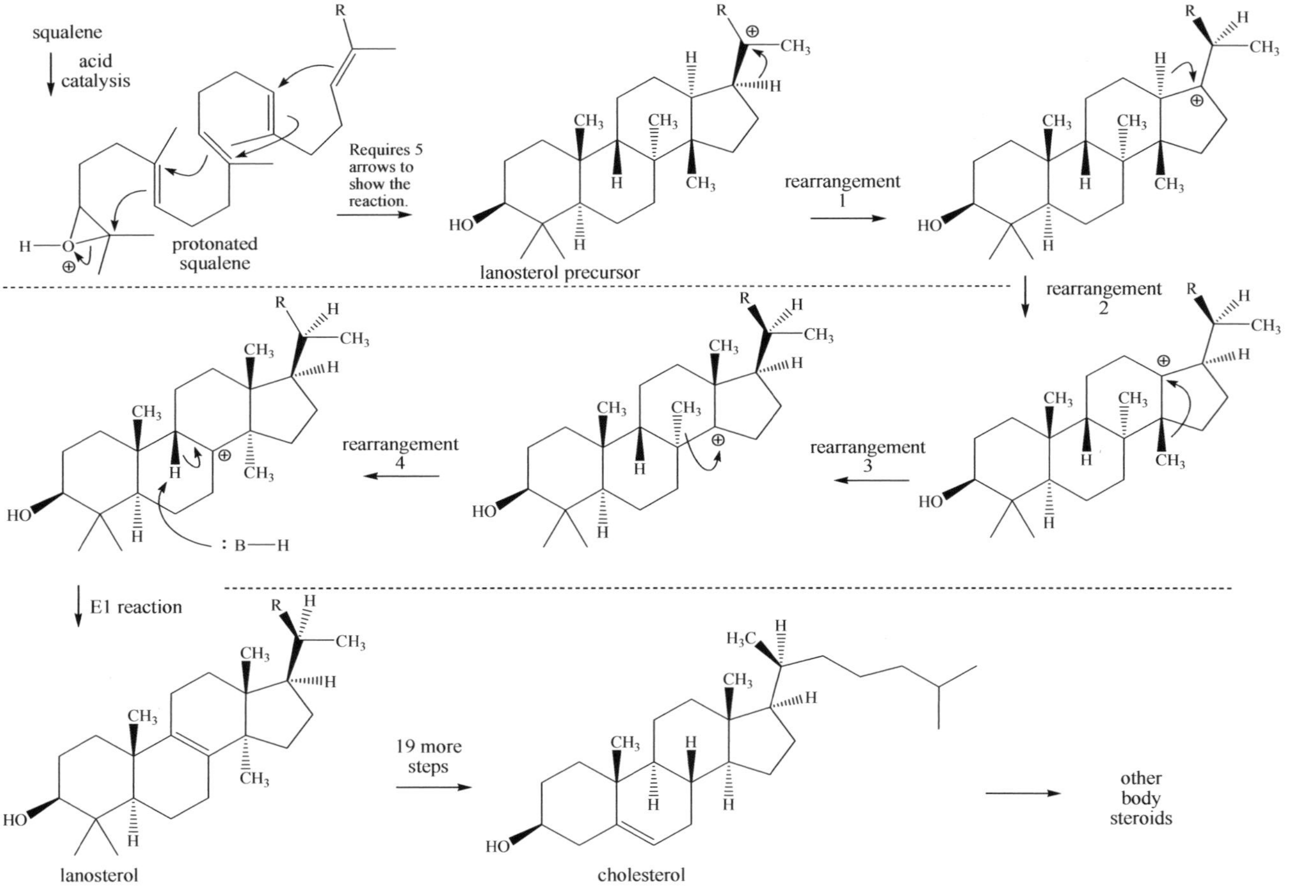

squalene ↓ acid catalysis

protonated squalene

Requires 5 arrows to show the reaction.

lanosterol precursor

rearrangement 1

rearrangement 2

rearrangement 3

rearrangement 4

↓ E1 reaction

lanosterol

19 more steps

cholesterol

other body steroids

Problem 26 (p 288) – Write a detailed arrow-pushing mechanism for each of the following transformations.

a.

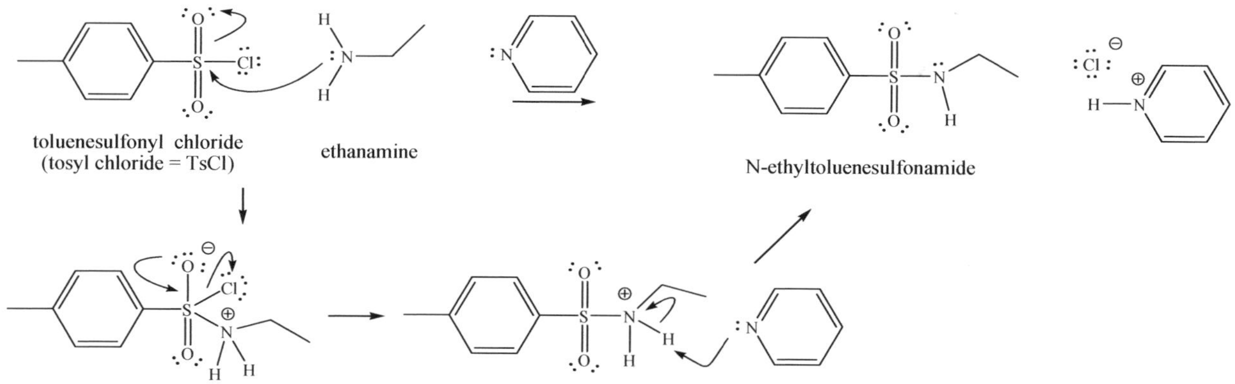

toluenesulfonyl chloride (tosyl chloride = TsCl)

ethanamine

N-ethyltoluenesulfonamide

b.

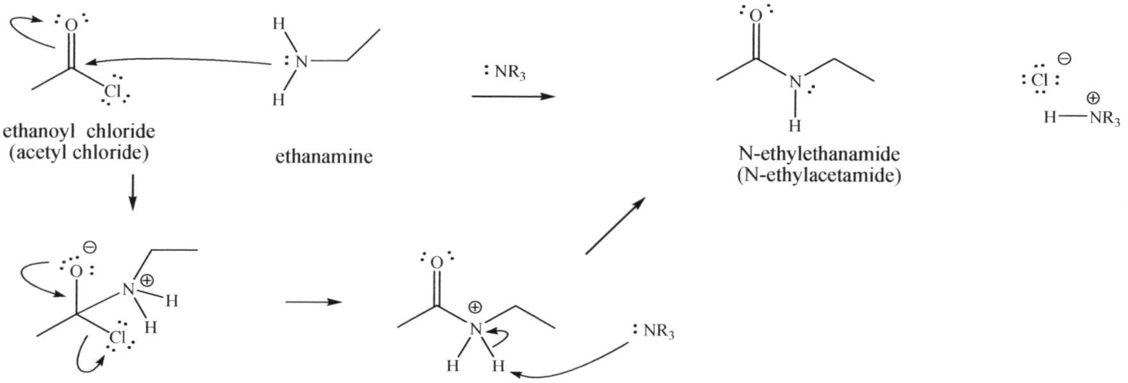

: NR_3

ethanoyl chloride (acetyl chloride)

ethanamine

N-ethylethanamide (N-ethylacetamide)

H—NR_3

: NR_3

Problem 27 (p 289) – We can now make the following molecules. Propose a synthesis for each. (Tosylates formed from alcohols and tosyl chloride/pyridine via acyl substitution reaction, convert "OH" from poor leaving group into a very good leaving group, like bromide). We will only use this for secondary alcohols where rearrangement is a possibility.

All of these syntheses are accomplished from the alcohol + toluenesulfonyl chloride/pyridine.

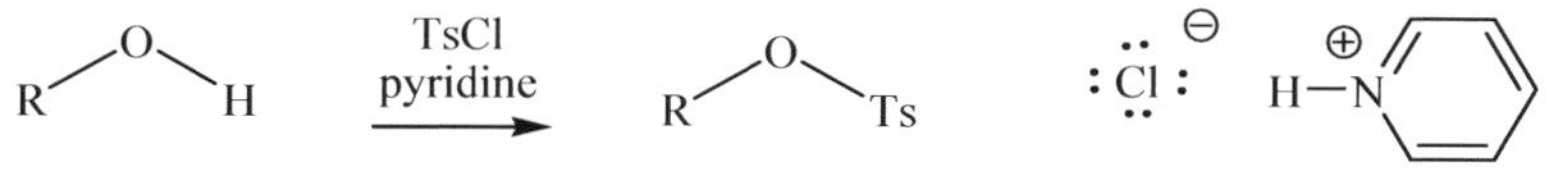

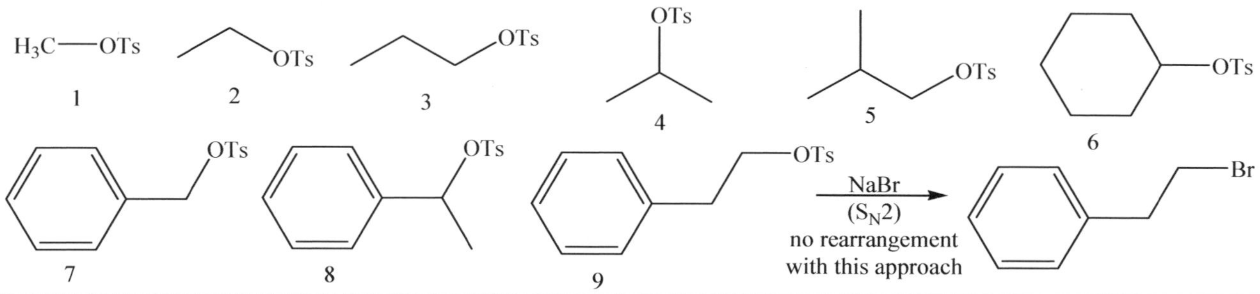

R–O–H →(TsCl, pyridine) R–O–Ts + :Cl:⊖ H–N⊕(pyridine)

H_3C—OTs (1), 2, 3, 4, 5, 6, 7, 8, 9

9 →(NaBr, (S_N2)) Br product

no rearrangement with this approach

Problem 28 (p 290) – Write a mechanism for the following reaction.

H—O—SO_3H

Δ

2° carbocation

rearrangement

tetrasubstituted alkene is best

3° carbocation

Problem 29 (p 291) (R)-2-butanol retains its optical activity indefinitely in aqueous base (^{-}OH), but is rapidly converted to optically inactive 2-butanol (racemic) when in contact with dilute sulfuric acid ($H_2SO_4 + H_2O \rightarrow H_3O^+ + HSO_4^-$). Explain with detailed mechanisms.

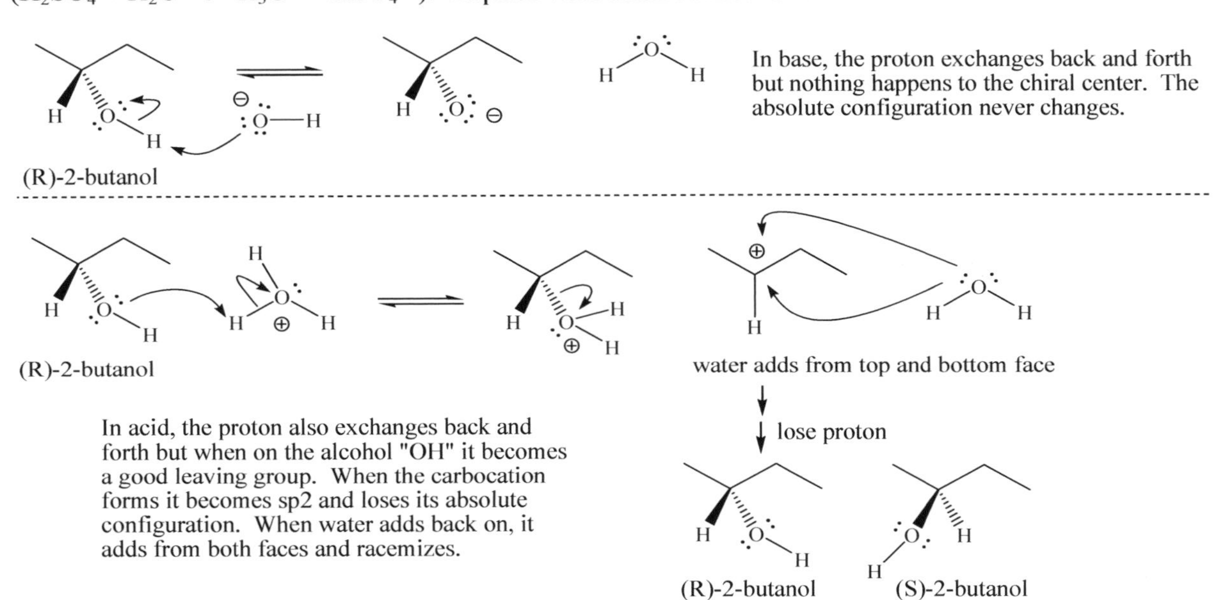

(R)-2-butanol
In base, the proton exchanges back and forth but nothing happens to the chiral center. The absolute configuration never changes.
(R)-2-butanol
water adds from top and bottom face
lose proton
In acid, the proton also exchanges back and forth but when on the alcohol "OH" it becomes a good leaving group. When the carbocation forms it becomes sp2 and loses its absolute configuration. When water adds back on, it adds from both faces and racemizes.
(R)-2-butanol
(S)-2-butanol

Problem 30 (p 294) – Suggest a mechanism for each of the following transformations. Predict major/minor products in each part.

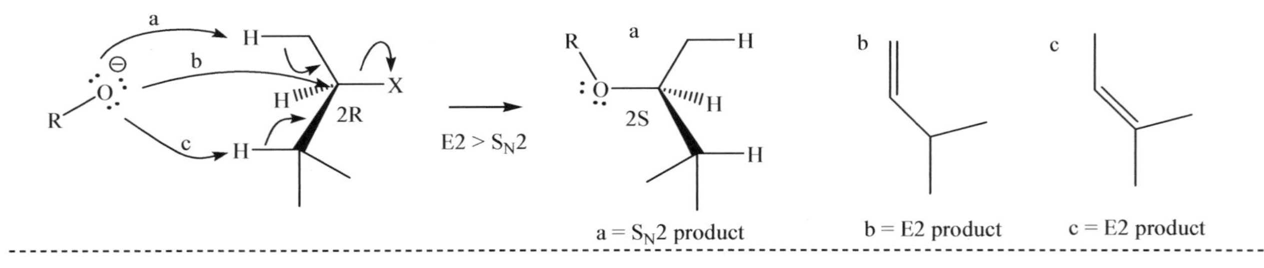

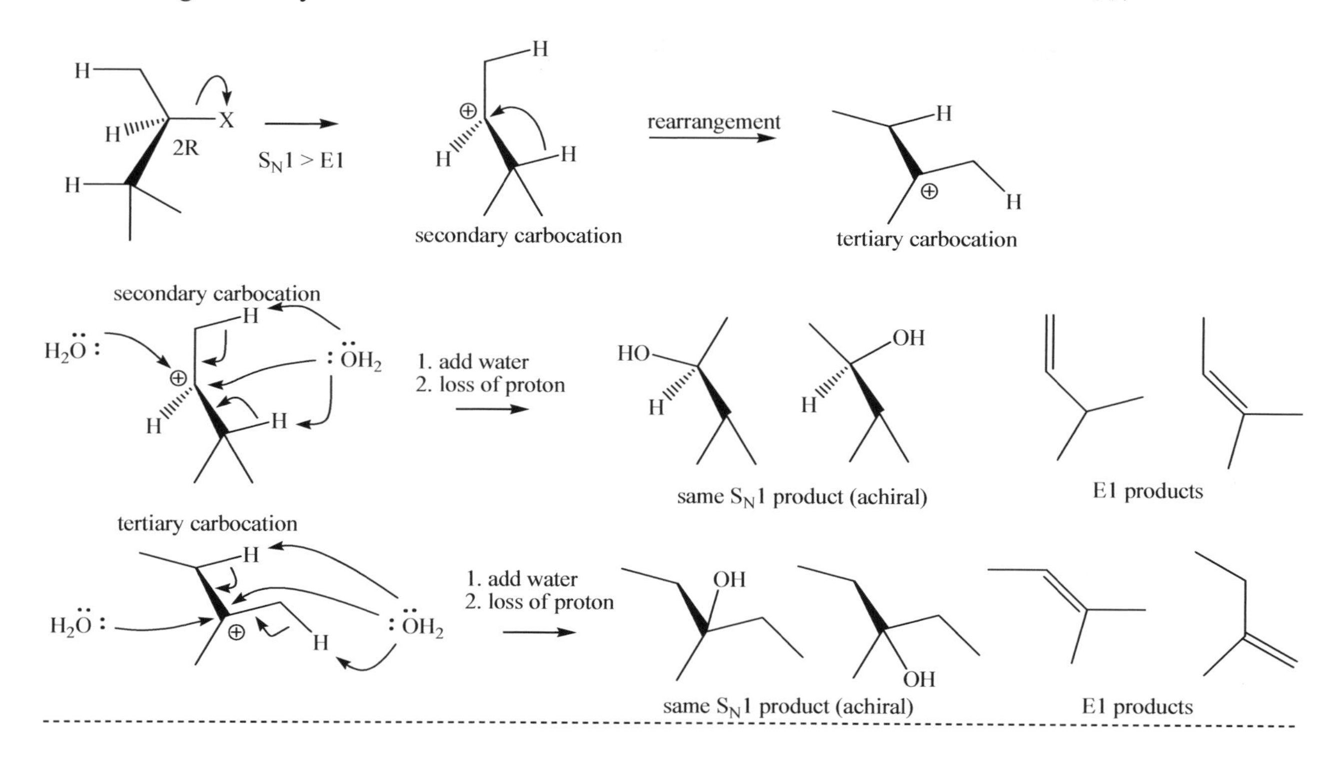

Problem 31 (p 294) - Write out all $C_6H_{13}Br$ isomers. There are 17, not counting stereoisomers, and they are listed on the next page (numbered 1-17).

1. Isomers that can react fastest in S_N2 reactions
   All primary RBr compounds (1,2,3,4,5,7,8), except 1° neopentyl (6).

2. Isomers that give E2 reaction but not S_N2 with sodium methoxide
   All tertiary RBr compounds (15,16,17) and secondary neopentyl (14).

3. Isomers that react fastest in S_N1 reactions
   All tertiary RBr compounds (15,16,17).

4. Isomers that can react by all four mechanisms, S_N2, E2, S_N1 and E1 (What are the necessary conditions?)
   All secondary RBr compounds (9-13), except 2° neopentyl (14) which cannot react by S_N2.

5. Isomers that might rearrange to more stable carbocation in reactions with methanol.
   Secondary RBr with a tertiary or quaternary vicinal neighbor carbon (12,13,14).

6. Isomers that are completely unreactive with methoxide/methanol
   All primary neopentyl RBr compounds (6).

7. Isomers that are completely unreactive with methanol, alone.
   All primary RBr compounds (1,2,3,4,5,6,7,8).

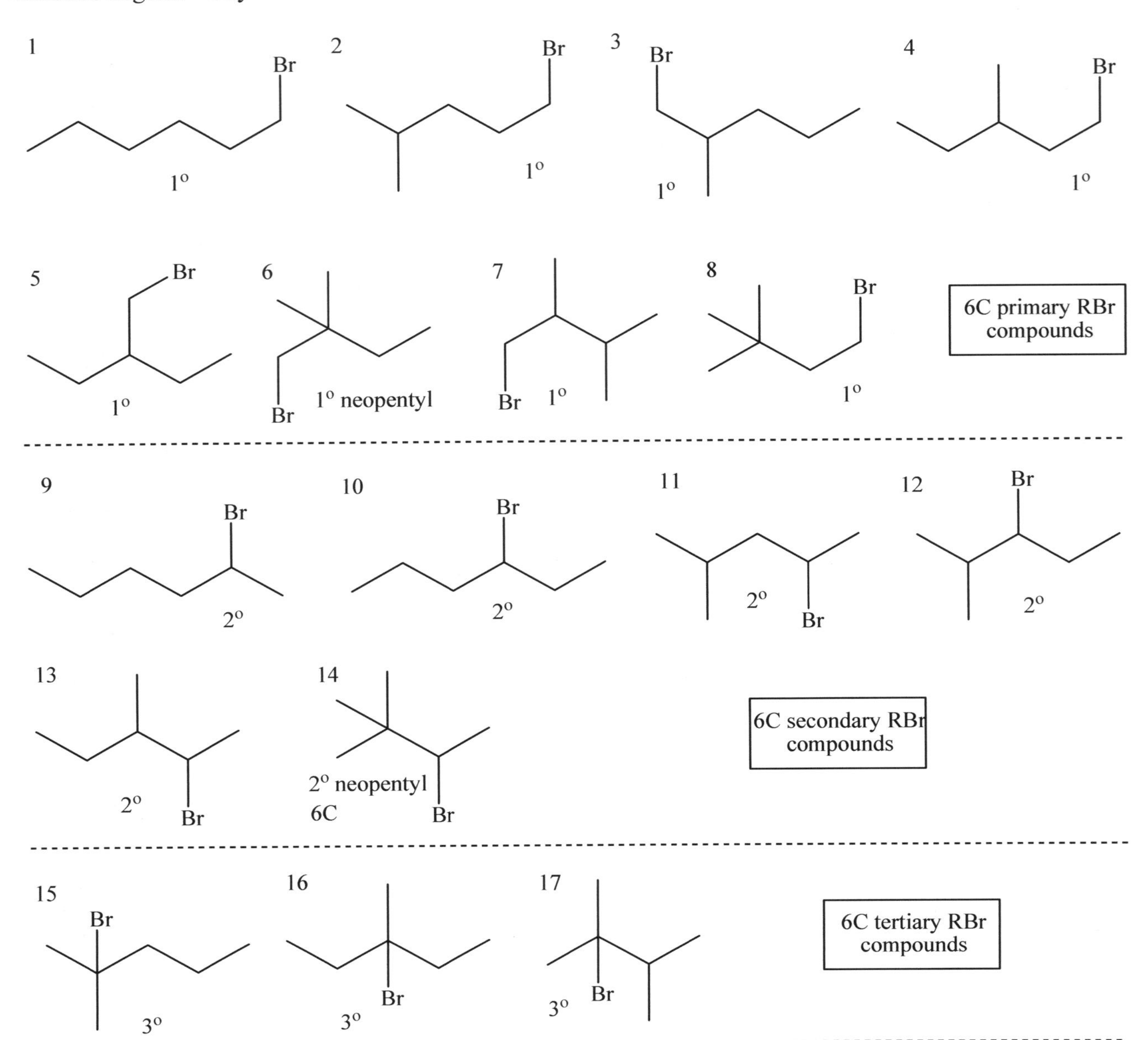

1
Br
1°
2
Br
1°
3
Br
1°
4
Br
1°
5
Br
1°
6
Br
1° neopentyl
7
Br
1°
8
Br
1°
6C primary RBr compounds
9
Br
2°
10
Br
2°
11
2°
Br
12
Br
2°
13
2°
Br
14
2° neopentyl
6C
Br
6C secondary RBr compounds
15
Br
3°
16
Br
3°
17
3°
Br
6C tertiary RBr compounds

Chapter 10

Problem 1 (p 303) - Why don't the reaction conditions (hν and/or Δ) break C-H or C-C bonds? What are the respective bond energies of those bonds? Use the following structures to explain your answer. (Use the bond energy table.)

The weakest bonds break first. If Br_2 is present then this bondswill break first, forming free radicals. C-C and C-H bonds are much stronger and will not break until other free radicals are present. To decide if those bonds will break compare the bond energies that are breaking with the bond energies that are forming. If the reaction is exothermic ($\Delta H < 0$), then that reaction will likely occur. If the reaction is endothermic($\Delta H > 0$) then it is likely that some other reaction step will outcompete that possibility.

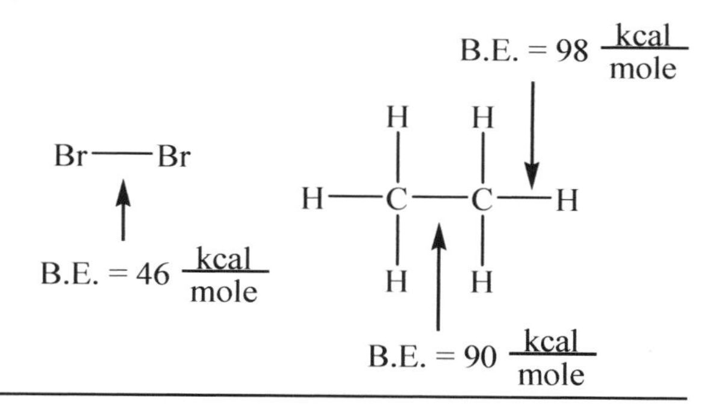

Problem 2 (p 308) - How do the results above explain the observations below? Potential Energy vs. Product of Reaction diagrams are on the next page (PE vs POR diagrams).

a. The energy released in the overall propagation sequence for fluorination is = 108 kcal/mole, an extremelly large value that would likely result in an explosion unless the reactants were diluted with another inert gas that could help dissipate the heat generated. If the reaction was run under controlled conditions there would be essentially no selectivity for which C-H positions reacted because fluorine atoms are highly reactive at any type of C-H bond.

b. The energy released in the overall propagation sequence for chlorination is = -28 kcal/mole. This is a large value that would likely still have to be controlled, but is much less dangerous than fluorination. The high reactivity of chlorine atoms does not lead to very selective reactivity with different types of C-H bond in the alkane reactant.

c. The energy released in the overall propagation sequence for bromination is = -12 kcal/mole. This is exothermic enough to make the reaction work, but the endothermic first step makes the reaction more sluggish and selective in which C-H positions will react. This makes the reaction products more predictable, which is a good thing. The downside is that some positions will not be reactive enough to make significant product using this approach. Other synthetic approaches will be necessary for those products.

d. The energy absorbed in the overall propagation sequence for iodination is = +10 kcal/mole. This will suck energy out of the system and make the reaction fail. Other synthetic approaches will be necessary to make iodoalkanes (like S_N2 reactions with iodide on bromoalkanes in acetone, or HI can be added to alkenes).

Problem 3 (p 310) - Set up a table using methane, similar to the above calculations for ethane. Show each step of the free radical substitution mechanism using X_2 = F_2, Cl_2, Br_2, I_2. Analyze the energetics for each step. Use X_2 for your general examples to provide an arrow pushing mechanism for each step of free radical halogenation of methane. Calculate a ΔH_{rxn} for each step of your mechanisms, using the actual bond energies.

Step 1 - Initiation

$$X—X \xrightarrow[\Delta \text{ (heat)}]{h\nu \text{ (light) or}} X\cdot \quad \cdot X$$

	Calculated ΔH_{rxns}			
	F_2	Cl_2	Br_2	I_2
ΔH_{rxn} (bond energy) =	+38	+58	+46	+36

All of these bonds are easy to break and the energy to do so is supplied by light or heat.

Step 2a - Propagation

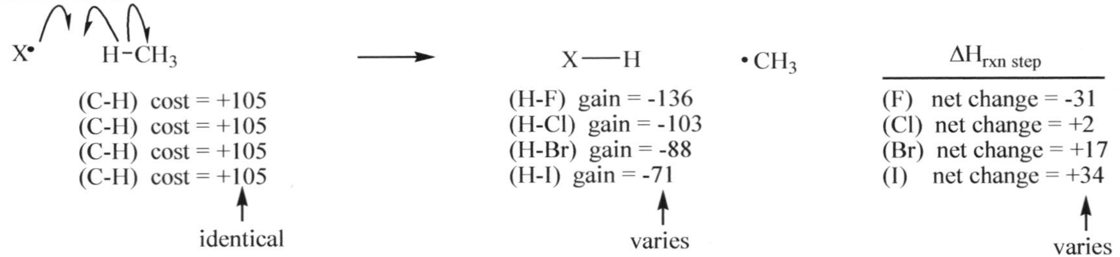

		$\Delta H_{rxn\ step}$
(C-H) cost = +105	(H-F) gain = -136	(F) net change = -31
(C-H) cost = +105	(H-Cl) gain = -103	(Cl) net change = +2
(C-H) cost = +105	(H-Br) gain = -88	(Br) net change = +17
(C-H) cost = +105	(H-I) gain = -71	(I) net change = +34
identical	varies	varies

Step 2b – Propagation

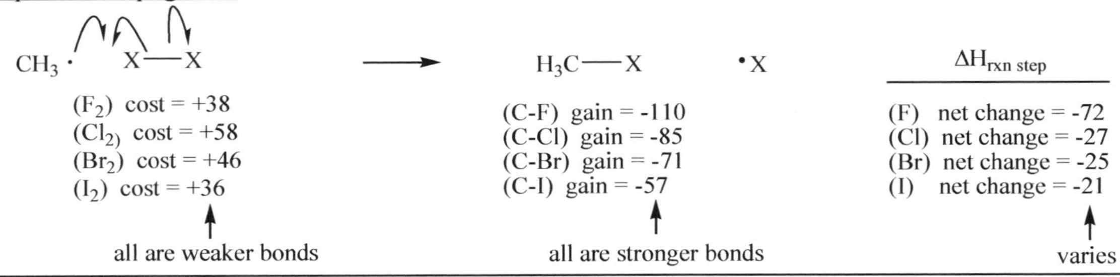

		$\Delta H_{rxn\ step}$
(F_2) cost = +38	(C-F) gain = -110	(F) net change = -72
(Cl_2) cost = +58	(C-Cl) gain = -85	(Cl) net change = -27
(Br_2) cost = +46	(C-Br) gain = -71	(Br) net change = -25
(I_2) cost = +36	(C-I) gain = -57	(I) net change = -21
all are weaker bonds	all are stronger bonds	varies

Step 2a + 2b

$\Delta H_{rxn\ steps\ 2a\ +\ 2b}$

(F) both steps = -31 - 72 = -103 (exothermic)
(Cl) both steps = +2 - 27 = -25 (exothermic)
(Br) both steps = +17 - 25 = -8 (exothermic)
(I) both steps = +34 - 21 = +13 (endothermic)

Step 3 – Termination

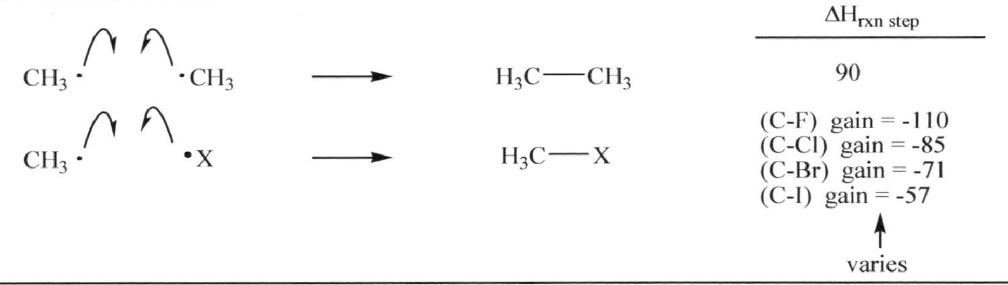

$\Delta H_{rxn\ step}$

90

(C-F) gain = -110
(C-Cl) gain = -85
(C-Br) gain = -71
(C-I) gain = -57

varies

Problem 4 (p 3112)- Use this data to predict the relative amounts of chlorinated product and brominated product for propane and 2-methylpropane. (Amount = [#H]x[relative rate]). Which halogen is more selective?

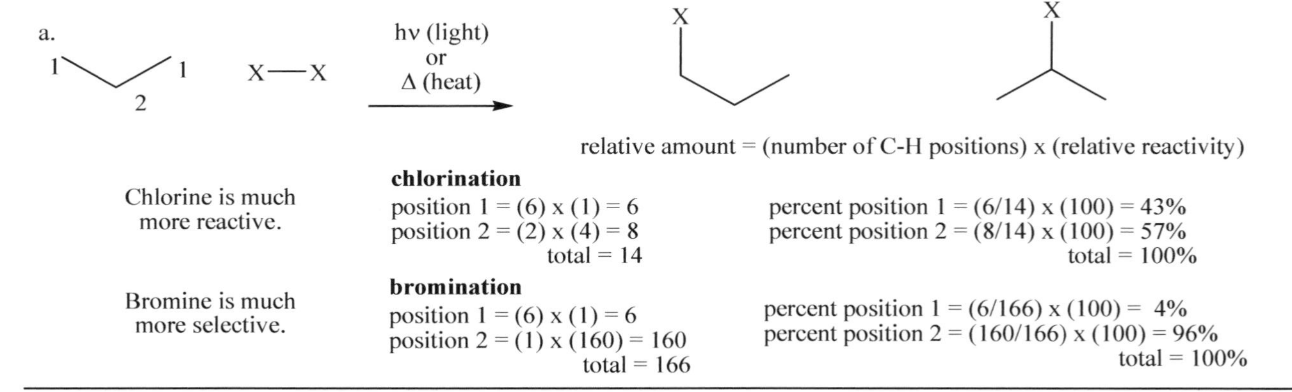

relative amount = (number of C-H positions) x (relative reactivity)

Chlorine is much more reactive.

chlorination

position 1 = (6) x (1) = 6
position 2 = (2) x (4) = 8
total = 14

percent position 1 = (6/14) x (100) = 43%
percent position 2 = (8/14) x (100) = 57%
total = 100%

Bromine is much more selective.

bromination

position 1 = (6) x (1) = 6
position 2 = (1) x (160) = 160
total = 166

percent position 1 = (6/166) x (100) = 4%
percent position 2 = (160/166) x (100) = 96%
total = 100%

b.

1 1 2 1 X—X hν (light) or Δ (heat) X X

relative amount = (number of C-H positions) x (relative reactivity)

Chlorine is much more reactive.

chlorination
position 1 = (9) x (1) = 9
position 2 = (1) x (5) = 5
total = 14

percent position 1 = (9/14) x (100) = 64%
percent position 2 = (5/14) x (100) = 36%
total = 100%

Bromine is much more selective.

bromination
position 1 = (9) x (1) = 9
position 2 = (1) x (1600) = 1600
total = 1609

percent position 1 = (9/1609) x (100) = 0.5%
percent position 2 = (1600/1609) x (100) = 99.5%
total = 100%

Problem 5 (p 312) - How many different types of hydrogen atoms are in each of the following molecules? Which type of hydrogen atom in each molecule is most reactive? How did you decide this? What is the maximum number of monosubstituted products that could form if each type of hydrogen atom was substituted in each molecule? Use the given table of relative rates (partly real, partly made up data) to show the product distribution for chlorine and bromine. We won't consider stereoisomers here. Using the most reactive hydrogen atom in the first molecule, write out a mechanism for formation of the major reaction product, using bromine as the halogen reactant. (Amount = [#H]x[relative rate]).

relative rates for chlorination		
tertiary C-H	=	5.1
secondary C-H	=	4.0
primary C-H	=	1.0
tertiary allylic/benzylic C-H	=	100
secondary allylic/benzylic C-H	=	70
primary allylic/benzylic C-H	=	20

Consider vinyl and phenyl C-H to have a relative rate of 0.

relative rates for bromination		
tertiary C-H	=	1600
secondary C-H	=	80
primary C-H	=	1.0
tertiary allylic/benzylic C-H	=	1,000,000
secondary allylic/benzylic C-H	=	40,000
primary allylic/benzylic C-H	=	16,000

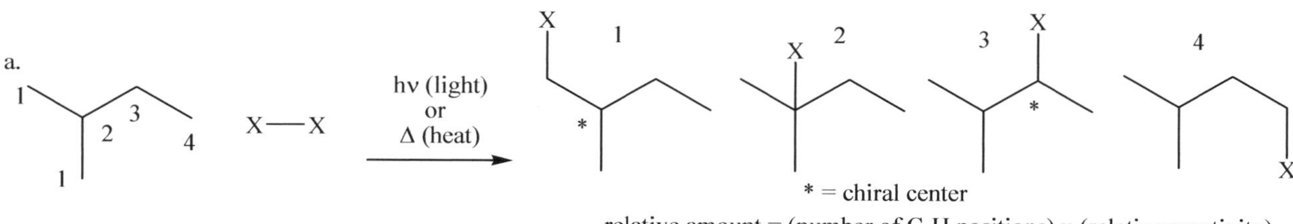

relative amount = (number of C-H positions) x (relative reactivity)

The tertiary C-H position is the most reactive position.

chlorination
position 1 = (6) x (1) = 6
position 2 = (1) x (5) = 5
position 3 = (2) x (4) = 8
position 4 = (3) x (1) = 3
total = 22

percent position 1 = (6/22) x (100) = 27%
percent position 2 = (5/22) x (100) = 23%
percent position 3 = (8/22) x (100) = 36%
percent position 4 = (3/22) x (100) = 14%
total = 100%

bromination
position 1 = (6) x (1) = 6
position 2 = (1) x (1600) = 1600
position 3 = (2) x (80) = 160
position 4 = (3) x (1) = 3
total = 1769

percent position 1 = (6/1769) x (100) = 0%
percent position 2 = (1600/1769) x (100) = 90%
percent position 3 = (160/1769) x (100) = 9%
percent position 4 = (3/1769) x (100) = 0%
total = 99%

---

b.

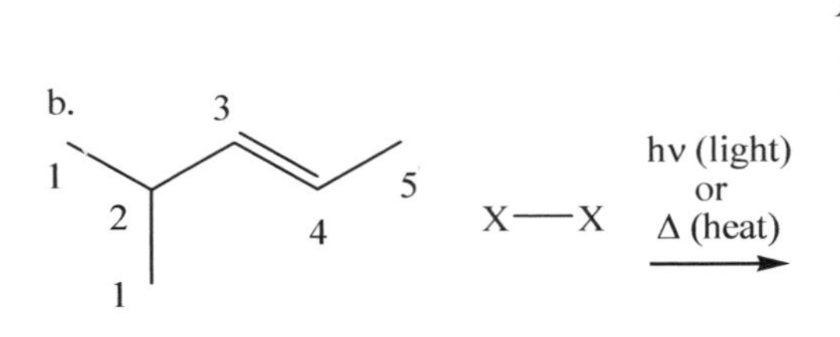

X 1 * | X 2 | X 3

4 X | * = chiral center | 5 X

The tertiary allylic C-H position is the most reactive position. The double bond can move as a result of resonance in some intermediates (not shown here).

relative amount = (number of C-H positions) x (relative reactivity)

chlorination

position 1 = (6) x (1) = 6
position 2 = (1) x (100) = 100
position 3 = (1) x (0) = 0
position 4 = (1) x (0) = 0
position 5 = (3) x (70) = 70
total = 176

percent position 1 = (6/176) x (100) = 3%
percent position 2 = (100/176) x (100) = 57%
percent position 3 = (0/176) x (100) = 0%
percent position 4 = (0/176) x (100) = 0%
percent position 5 = (70/176) x (100) = 40%
total = 100%

bromination

position 1 = (6) x (1) = 6
position 2 = (1) x (1,000,000) = 1,000,000
position 3 = (1) x (0) = 0
position 4 = (1) x (0) = 0
position 5 = (3) x (40,000) = 120,000
total = 1,120,006

percent position 1 = (6/1,120,006) x (100) = 0%
percent position 2 = (1,000,000/1,120,006) x (100) = 89%
percent position 3 = (0/1,120,006) x (100) = 0%
percent position 4 = (0/1,120,006) x (100) = 0%
percent position 2 = (120,000/1,120,006) x (100) = 11%
total = 100%

---

c.

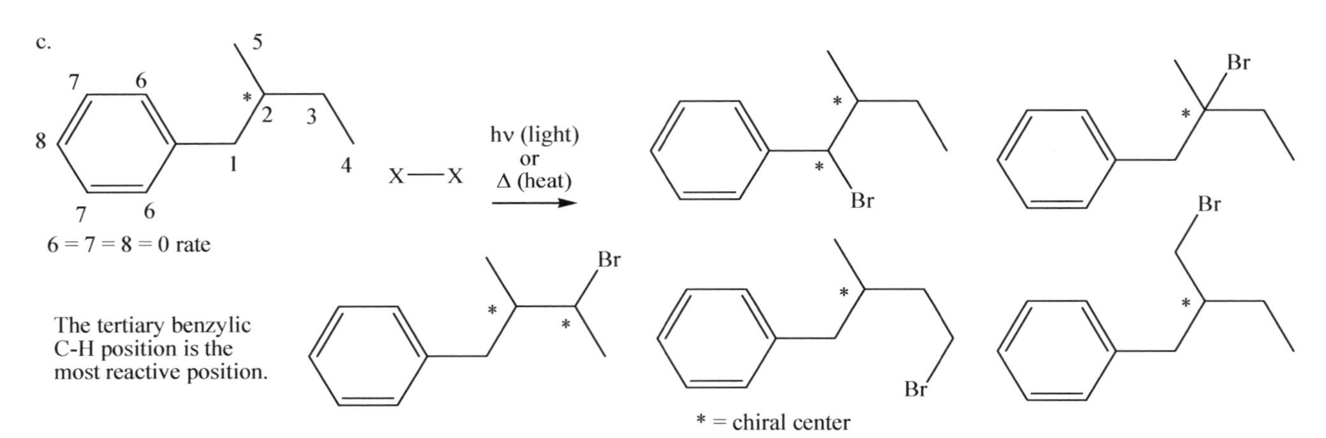

6 = 7 = 8 = 0 rate

The tertiary benzylic C-H position is the most reactive position.

* = chiral center

relative amount = (number of C-H positions) x (relative reactivity)

chlorination

position 1 = (2) x (70) = 140
position 2 = (1) x (5) = 5
position 3 = (2) x (4) = 8
position 4 = (3) x (1) = 3
position 5 = (3) x (1) = 3
total = 159

percent position 1 = (140/159) x (100) = 88%
percent position 2 = (5/159) x (100) = 3%
percent position 3 = (8/159) x (100) = 5%
percent position 4 = (3/159) x (100) = 2%
percent position 5 = (3/159) x (100) = 2%
total = 100%

6 = 7 = 8 = 0 rate

bromination

position 1 = (2) x (40,000) = 80,000
position 2 = (1) x (1,600) = 1,600
position 3 = (2) x (80) = 160
position 4 = (3) x (1) = 3
position 5 = (3) x (1) = 3
total = 81,766

percent position 1 = (80,000/81,766) x (100) = 98%
percent position 2 = (1,600/81,766) x (100) = 2%
percent position 3 = (160/81,766) x (100) = 0%
percent position 4 = (3/81,766) x (100) = 0%
percent position 2 = (3/81,766) x (100) = 0%
total = 100%

6 = 7 = 8 = 0 rate

---

1. **Initiation**: cleavage of weakest bond using light or heat (the halogen bond)

:Br—Br: → hν (light) or Δ (heat) → :Br• •Br:

BE (cost) = +46

$\Delta H_{rxn} = +46 \frac{kcal}{mole}$

2a. **Propagation**: halogen atom abstracts a hydrogen atom from a C-H bond, forming a carbon free radical

attack at the weakest C-H bond

H—Br:

BE (cost) = +92

BE (gain) = -88

$\Delta H_{rxn} = (-88) - (-92)$

$\Delta H_{rxn} = +4 \frac{kcal}{mole}$

2b. **Propagation**: carbon free radical abstracts a halogen atom from a halogen molecule, forming a halogen atom

:Br—Br: → R—C(H)(H)—Br •Br:

BE (cost) = +46

BE (gain) = -67

$\Delta H_{rxn} = (-67) - (-46)$

$\Delta H_{rxn} = -21 \frac{kcal}{mole}$

$\Delta H_{2a + 2b} = -17 \frac{kcal}{mole}$

3. **termination**: combination of two reactive free radicals into one stable bond, ending two chain reaction sequences

R• •Br: → R—Br

BE (gain) = -67

$\Delta H_{rxn} = -67$

R• •R → R—R

BE (gain) = -71

$\Delta H_{rxn} = -71$

Problem 6 (p 316) Predict the products and the approximate relative amounts of each (as a percent). Include stereoisomers, if present (enantiomers, diastereomers, meso, cis/trans). Assume the relative reactivities are the same as listed at the end of the problem. Write an arrow-pushing mechanism for the major product formed in part f. Calculate ΔH_{step} for each step of your mechanism.

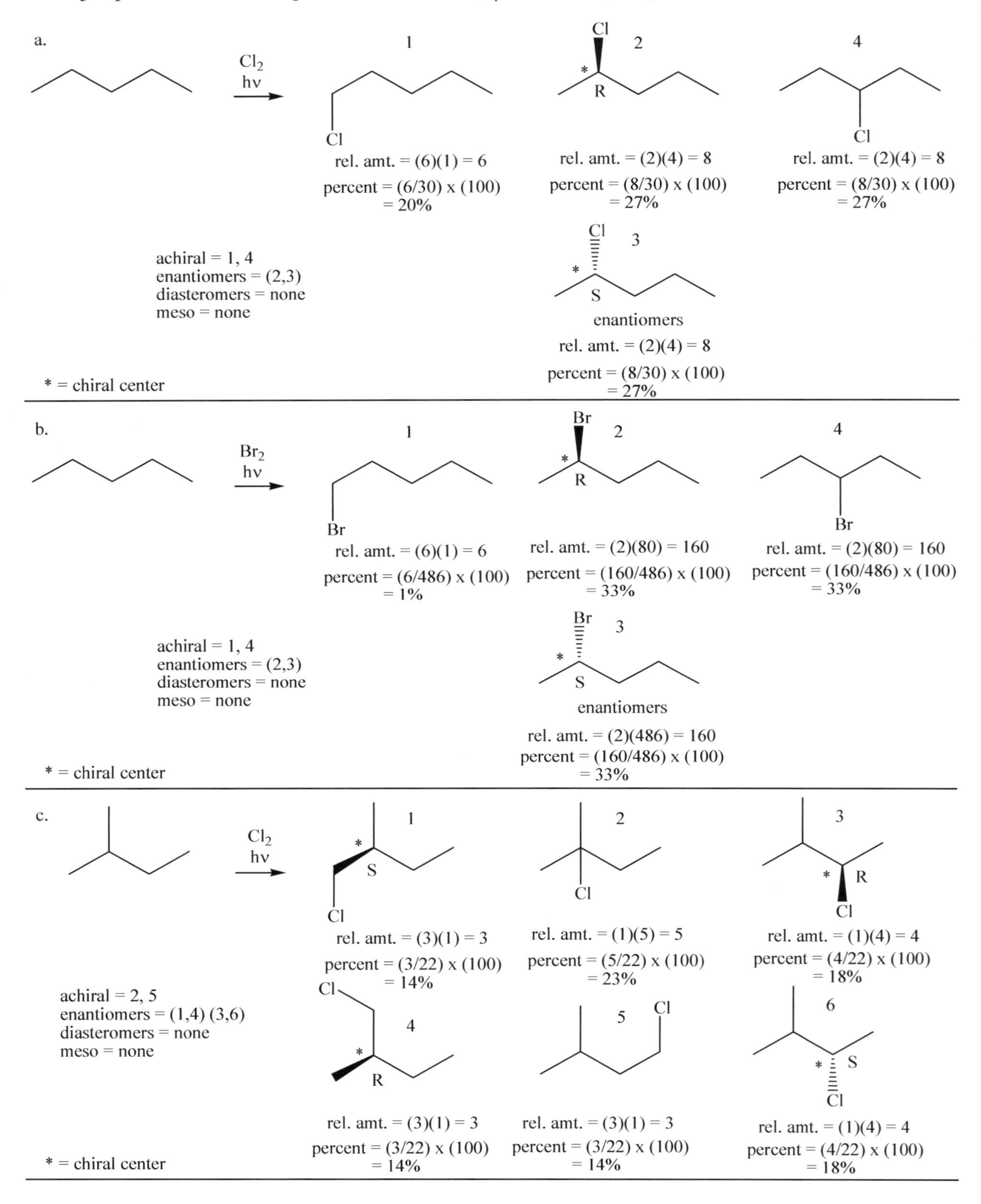

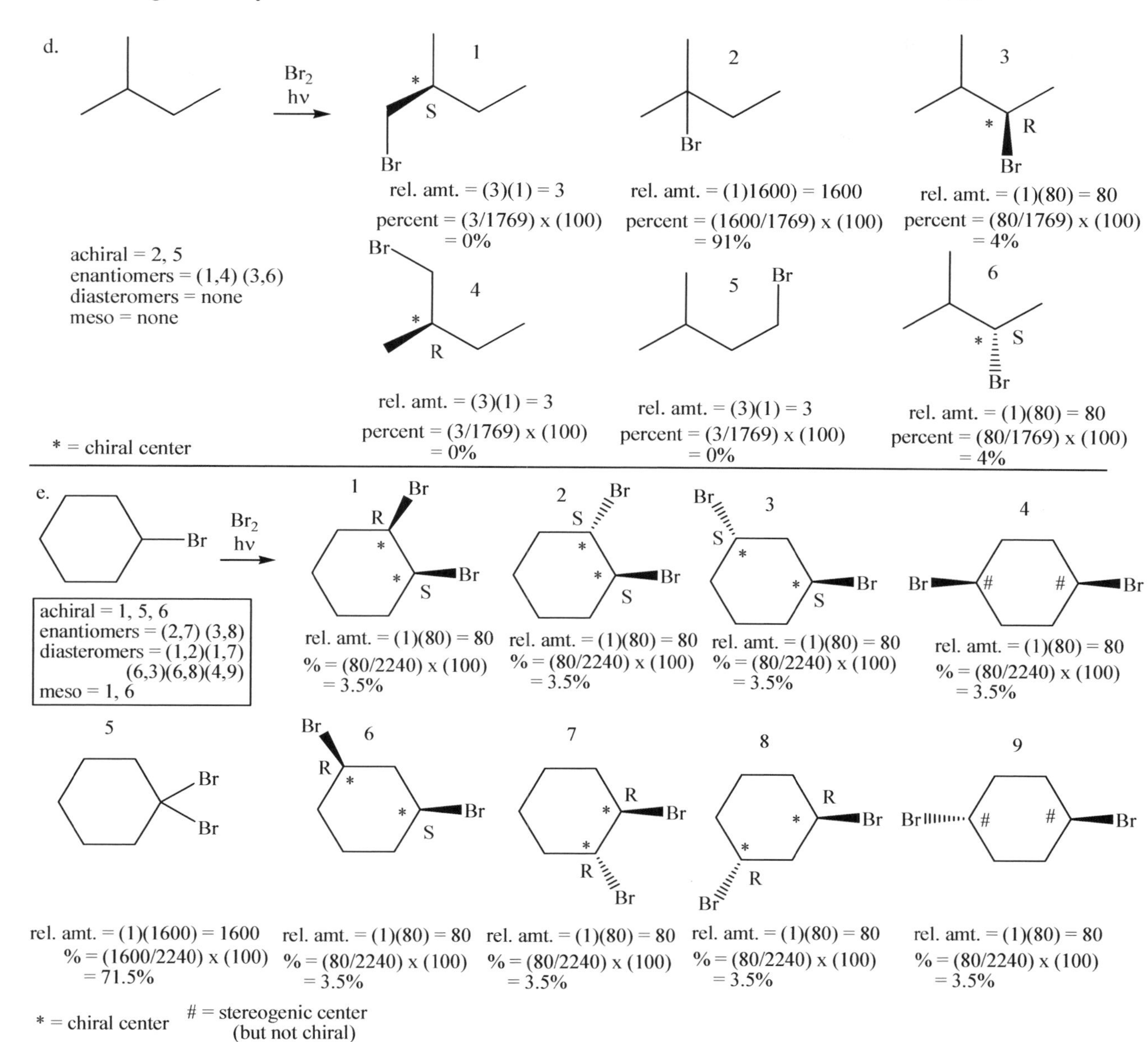
d.
Br2
hν
1
2
3
Br
Br
Br
S
R
rel. amt. = (3)(1) = 3
percent = (3/1769) x (100)
= 0%
rel. amt. = (1)1600) = 1600
percent = (1600/1769) x (100)
= 91%
rel. amt. = (1)(80) = 80
percent = (80/1769) x (100)
= 4%
achiral = 2, 5
enantiomers = (1,4) (3,6)
diasteromers = none
meso = none
Br
4
R
5
Br
6
S
Br
rel. amt. = (3)(1) = 3
percent = (3/1769) x (100)
= 0%
rel. amt. = (3)(1) = 3
percent = (3/1769) x (100)
= 0%
rel. amt. = (1)(80) = 80
percent = (80/1769) x (100)
= 4%
* = chiral center
e.
Br
Br2
hν
achiral = 1, 5, 6
enantiomers = (2,7) (3,8)
diasteromers = (1,2)(1,7)
(6,3)(6,8)(4,9)
meso = 1, 6
rel. amt. = (1)(80) = 80
% = (80/2240) x (100)
= 3.5%
rel. amt. = (1)(1600) = 1600
% = (1600/2240) x (100)
= 71.5%
* = chiral center
# = stereogenic center
(but not chiral)

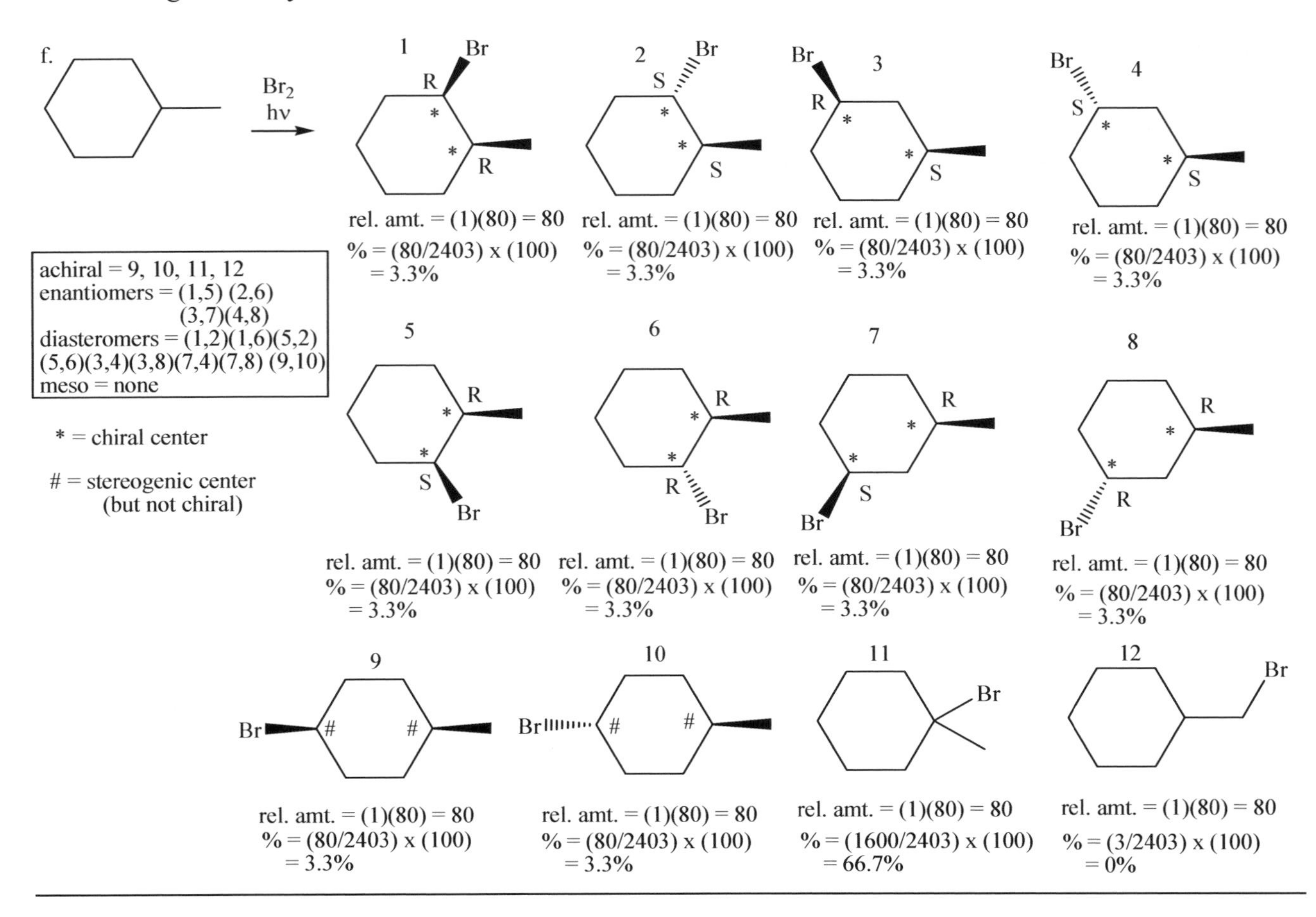

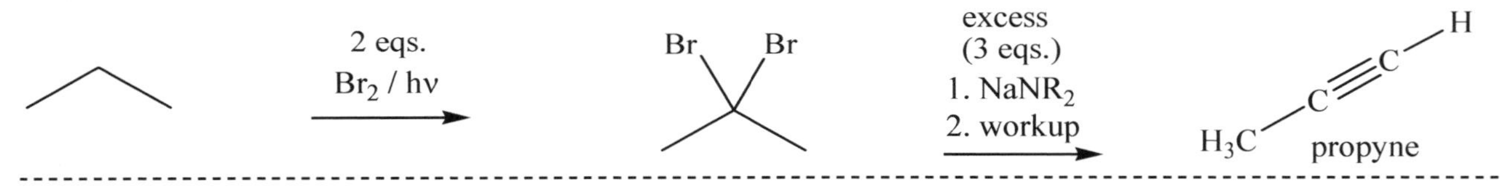

Problem 7 (p 320)– Supply all missing mechanism arrows and any formal charge to complete the following free radical addition reaction.

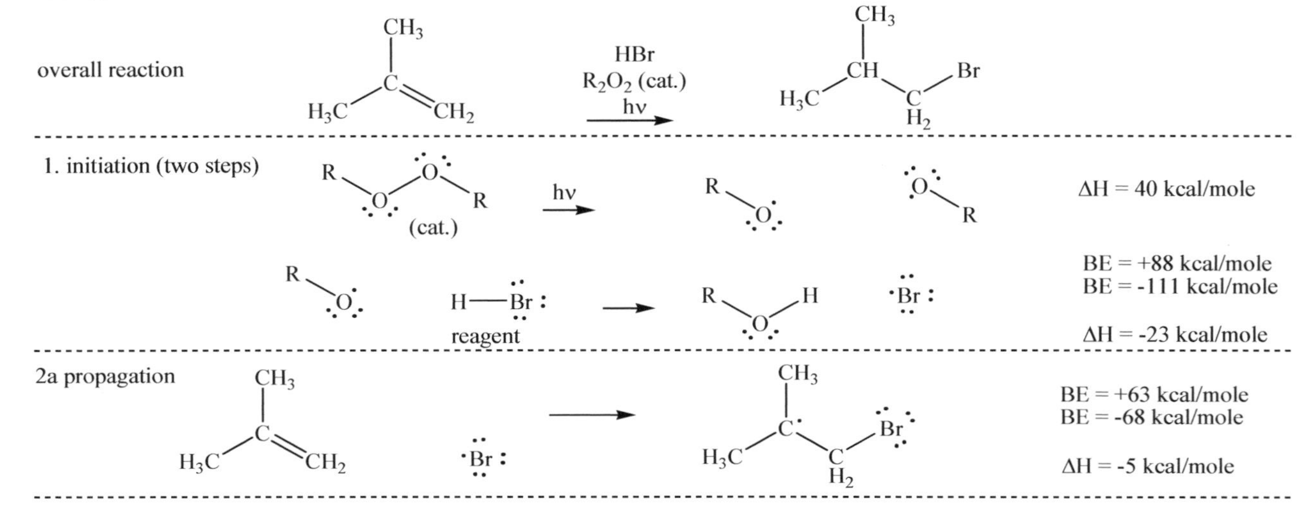

2b propagation

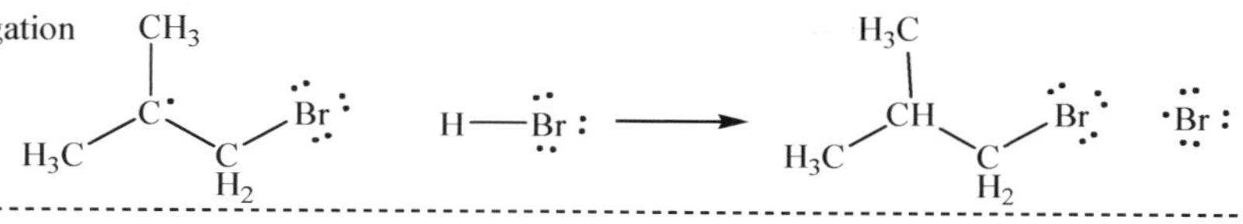

$(CH_3)_2\dot{C}{-}CH_2{-}Br + H{-}Br \rightarrow (CH_3)_2CH{-}CH_2{-}Br + \cdot Br$

BE = +88 kcal/mole
BE = -98 kcal/mole
ΔH = -10 kcal/mole

ΔH = -15 both steps (2a + 2b)

Problem 8 (p 320)– Write a complete mechanism for the following free radical addition reaction.

overall reaction

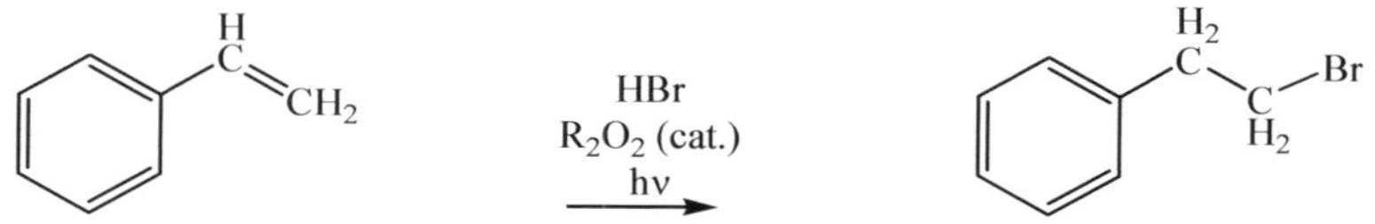

$C_6H_5{-}CH{=}CH_2 \xrightarrow[h\nu]{HBr,\ R_2O_2\ (cat.)} C_6H_5{-}CH_2{-}CH_2{-}Br$

1. initiation (two steps)

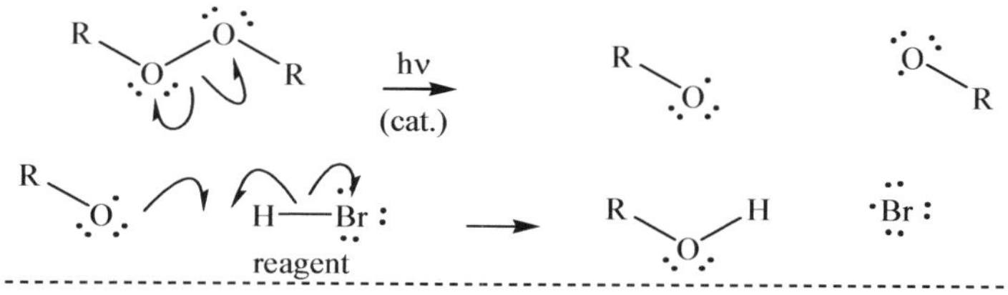

$R{-}O{-}O{-}R \xrightarrow[(cat.)]{h\nu} R{-}O\cdot + \cdot O{-}R$ ΔH = 40 kcal/mole

$R{-}O\cdot + H{-}Br$ (reagent) $\rightarrow R{-}O{-}H + \cdot Br$

BE = +88 kcal/mole
BE = -111 kcal/mole
ΔH = -23 kcal/mole

2a propagation

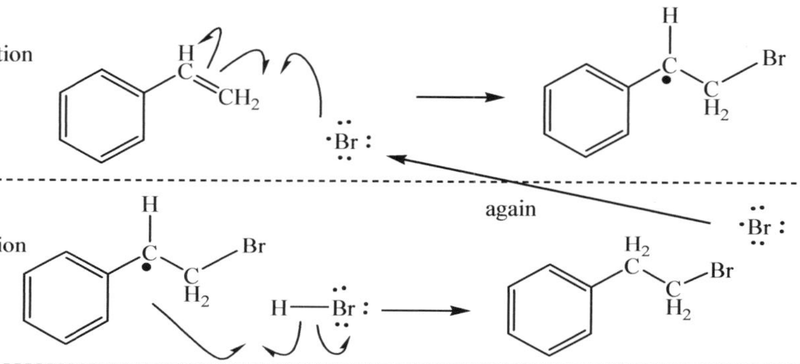

$C_6H_5{-}CH{=}CH_2 + \cdot Br \rightarrow C_6H_5{-}\dot{C}H{-}CH_2{-}Br$

BE = +63 kcal/mole
BE = -68 kcal/mole
ΔH = -5 kcal/mole

again

2b propagation

$C_6H_5{-}\dot{C}H{-}CH_2{-}Br + H{-}Br \rightarrow C_6H_5{-}CH_2{-}CH_2{-}Br + \cdot Br$

BE = +88 kcal/mole
BE = -98 kcal/mole
ΔH = -10 kcal/mole

ΔH = -15 both steps (2a + 2b)

Problem 9 (p 322)– Propose a reasonable synthesis for the following molecules from the given starting materials.

propane used as an example compound

Br_2, hν

Br

K⊕ ⊖O

Br_2, hν, ROOR

Br

NaOH

OH

2 Br_2, hν

Br_2, hν

Br

NaOCH₃

O

Br Br

1. 3 eqs. Na^+ R_2N^-

2. workup

Na^+ R_2N^-

Na⊕

O ⊖

NaSH

SH

R–O–H NaH → R–O⊖ Na⊕

NaSCH₃

S

1. NaOH → Na⊕ O⊖

OH

NaCN

C≡N

R–S–H 1. NaOH → R–S⊖ Na⊕

NaN₃

N_3

H

R–N–R

1. n-BuLi

2. O

Li⊕ ⊖

1. $LiAlH_4$

2. workup

NH_2

NaCCR

R

Many other variations are also possible.

Br

O Li⊕ ⊖

O

Chapter 11

Problem 1 (p 327)– What are the oxidation states of each carbon atom below? What is the formal charge of every carbon atom below?

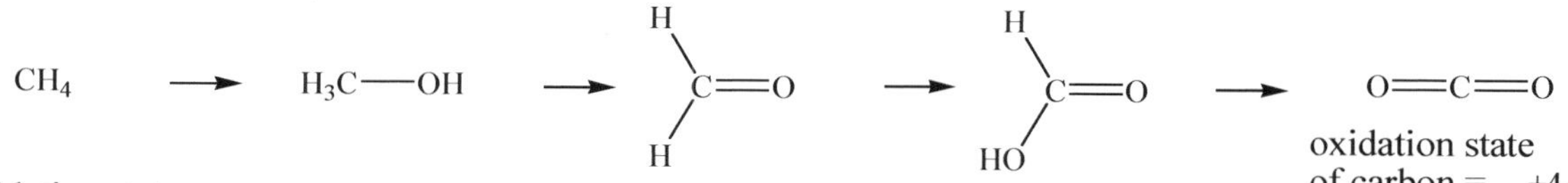

CH_4 → $H_3C—OH$ → $H_2C=O$ → $HC(=O)OH$ → $O=C=O$

oxidation state of carbon = -4 ? oxidation state of carbon = -2 ? oxidation state of carbon = 0 ? oxidation state of carbon = +2 ? oxidation state of carbon = +4 ?

⇅ ± H_2O

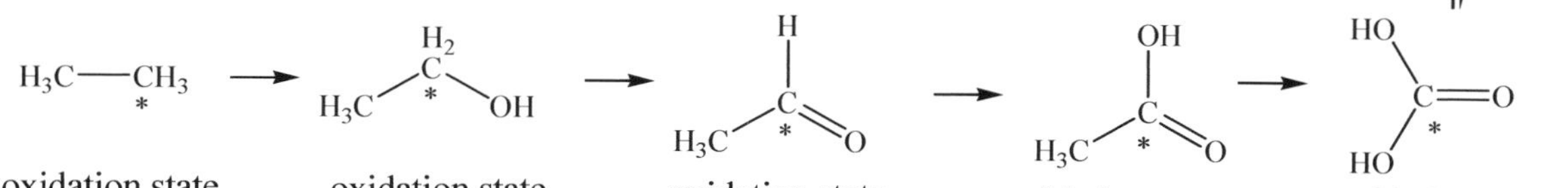

$H_3C—CH_3$ (*) → $H_3C–CH_2(*)–OH$ → $H_3C–C(*)H=O$ → $H_3C–C(*)(OH)=O$ → $(HO)_2C(*)=O$

oxidation state of * carbon = -3 ? oxidation state of * carbon = -1 ? oxidation state of * carbon = +1 ? oxidation state of * carbon = +3 ? oxidation state of * carbon = +4 ?

Problem 2 (p 327) – Which has a higher energy content per "*gram*", glucose or hexan-1-ol? You might also need the molecular weights to solve this problem (given). Speculate why this is the case. The necessary data is in the table above. Use the given reactions and necessary heats of formations.

Given information

$\Delta H_f (CO_2)$ = 94.0 kcal/mole (carbon dioxide)
$\Delta H_f (H_2O)$ = -57.0 kcal/mole (water)
$\Delta H_f (C_6H_{12}O_6)$ = -304.3 kcal/mole (glucose, MW = 180)
$\Delta H_f (C_6H_{12}O)$ = -90.3 kcal/mole (hexanol, MW = 100)

Heats of combustion.

$$C_6H_{12}O_6 + 6\ O_2 \xrightarrow{\Delta H_{rxn} = ?} 6\ CO_2 + 6\ H_2O$$
(glucose)

$$C_6H_{14}O + 8.5\ O_2 \xrightarrow{\Delta H_{rxn} = ?} 6\ CO_2 + 7\ H_2O$$
(hexanol)

$$\Delta H_{rxn} = \Delta H_f(\text{products}) - \Delta H_f(\text{reactants})$$

glucose (carbohydrate)

$$C_6H_{12}O_6 + 6\ O_2 \longrightarrow 6\ CO_2 + 6\ H_2O + \Delta$$

$$\Delta H_{rxn} = \Sigma\ \Delta H_{products} - \Sigma\ \Delta H_{reactants}$$

$$\Delta H_{rxn} = 6(-94.0) + 6(-57.0) - (-304.3) - 6(0)$$
$$= (-564) + (-342) + 304.3 = -601.7$$

$$\frac{\text{kcal}}{\text{gram}} = \frac{-601.7\ \text{kcal}}{180\ \text{g}} = 3.34\ \text{kcal/g}$$

hexanol (approximates a fat)

$$C_6H_{14}O + 9\ O_2 \longrightarrow 6\ CO_2 + 7\ H_2O + \Delta$$

$$\Delta H_{rxn} = \Sigma\ \Delta H_{products} - \Sigma\ \Delta H_{reactants}$$

$$\Delta H_{rxn} = 6(-94.0) + 7(-57.0) - (-90.3) - 6(0)$$
$$= (-564) + (-399) + 90.3 = -872.7$$

$$\frac{\text{kcal}}{\text{gram}} = \frac{-872.7\ \text{kcal}}{100\ \text{g}} = 8.73\ \text{kcal/g}$$

Much more energy is available in hexanol because it is almost completely carbon and hydrogen, both of which make very stable carbon dioxide and water. Glucose (carbohydrates) are already partially burned (an oxygen for every carbon atom) so that energy is not available for combustion. Fatty acid chains are more like hexanol, long strings of (CH_2)s with a methyl at the end.

Problem 3 (p 328) – What are the oxidation states below on the carbon atom and the chromium atom as the reaction proceeds? Which step does the oxidation/reduction occur? (PCC, B: = pyridine and Jones, B: = water)

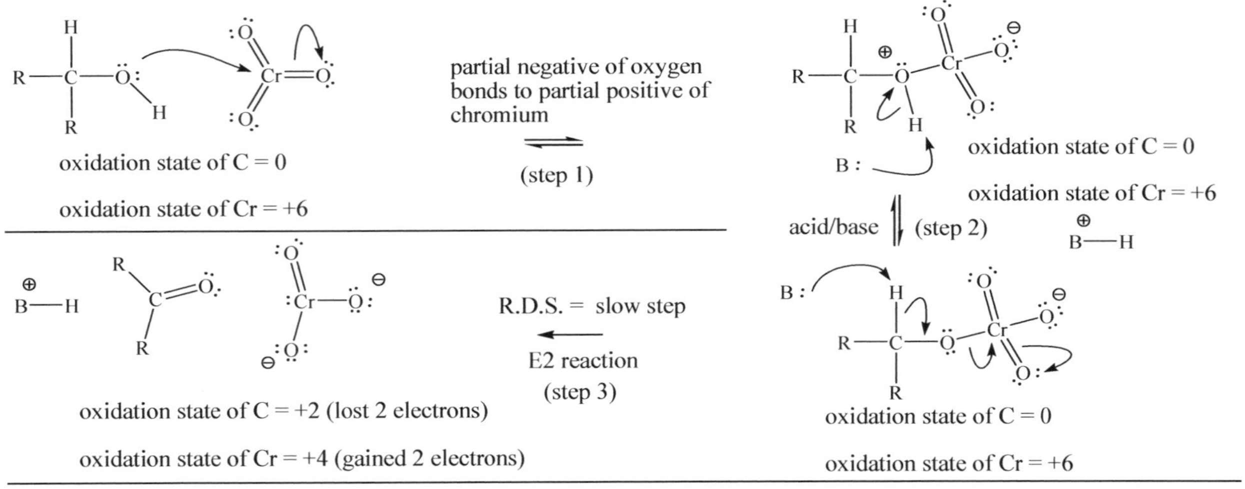

partial negative of oxygen bonds to partial positive of chromium

(step 1)

oxidation state of C = 0

oxidation state of Cr = +6

oxidation state of C = 0

oxidation state of Cr = +6

acid/base (step 2)

oxidation state of C = 0

oxidation state of Cr = +6

R.D.S. = slow step

E2 reaction

(step 3)

oxidation state of C = +2 (lost 2 electrons)

oxidation state of Cr = +4 (gained 2 electrons)

Problem 4 (p 331) – Supply all of the mechanistic details in the sequences below showing 1. the oxidation of a primary alcohol, 2. hydration of the carbonyl group and 3. oxidation of the carbonyl hydrate (Jones conditions). This answer is written in the book (p 313), but recopied here. Can you write it without any hints?

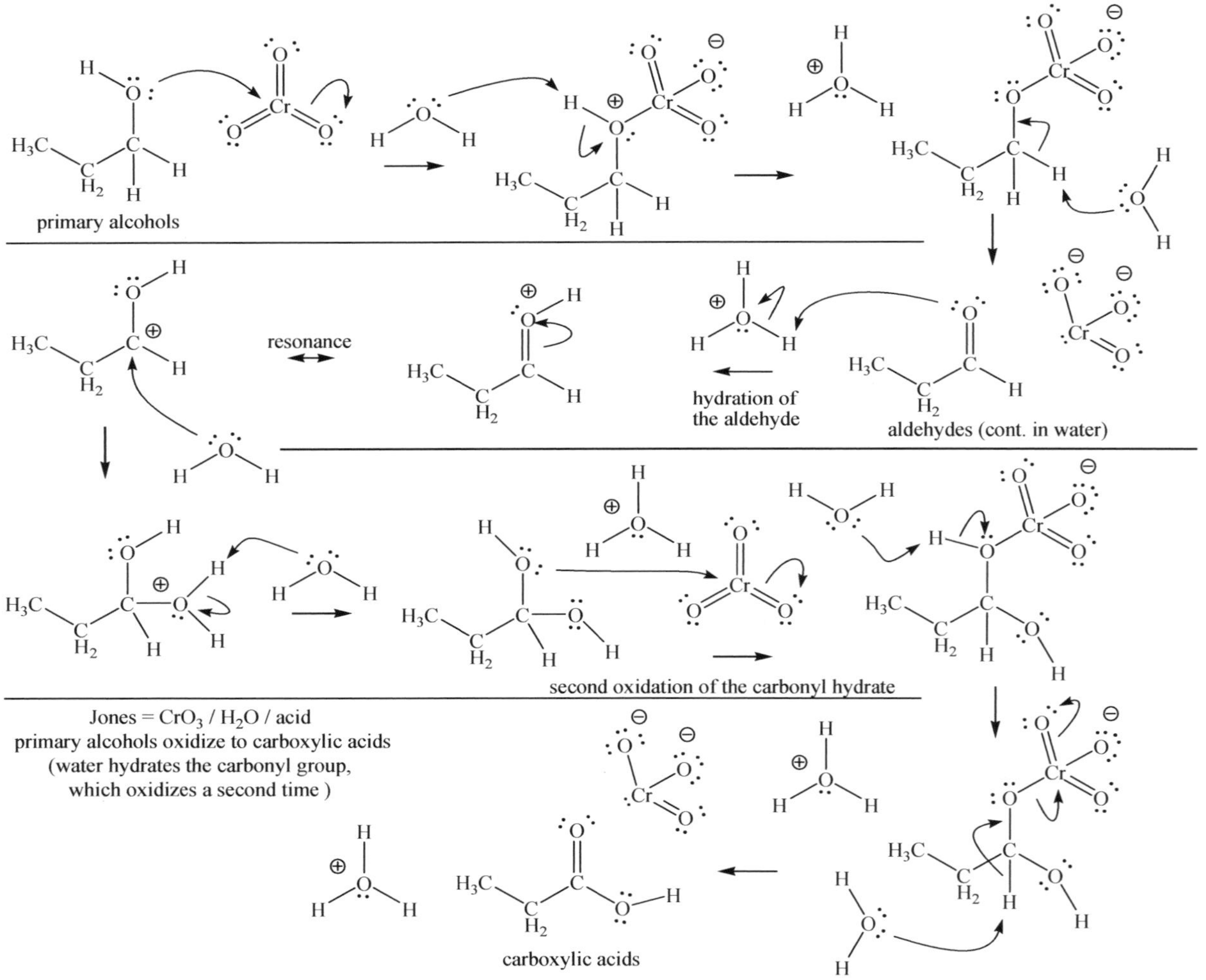

primary alcohols

resonance

hydration of the aldehyde

aldehydes (cont. in water)

second oxidation of the carbonyl hydrate

Jones = CrO_3 / H_2O / acid
primary alcohols oxidize to carboxylic acids
(water hydrates the carbonyl group, which oxidizes a second time)

carboxylic acids

Problem 5 (p 337) – Propose a mechanism for the reverse reaction, hydrolysis of an imine with water to form a carbonyl compound (ketone here) and a primary amine.

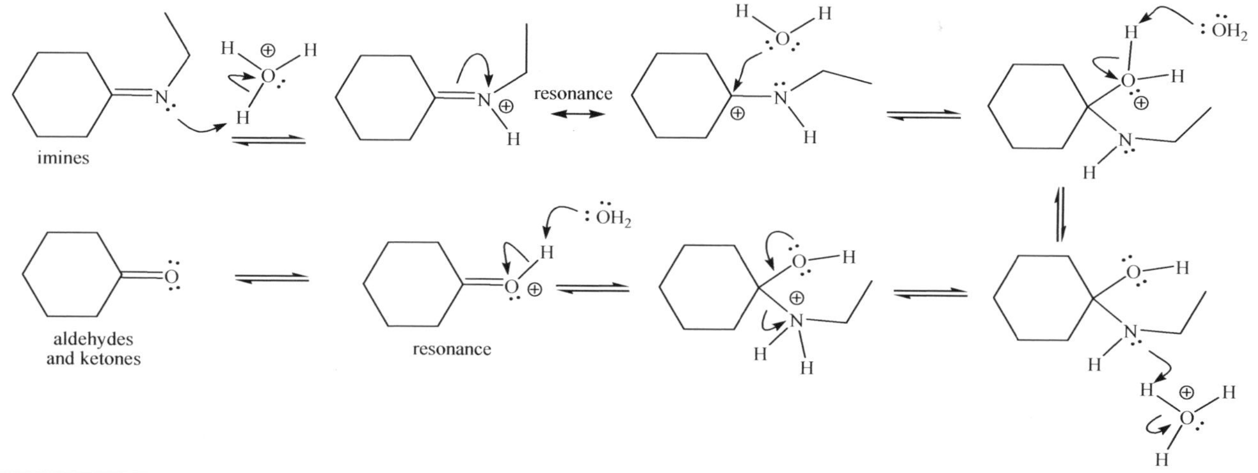

imines

resonance

$:\ddot{O}H_2$

aldehydes and ketones

resonance

Problem 6 (p 337) – Fill in the necessary reagents to accomplish the following transformations.

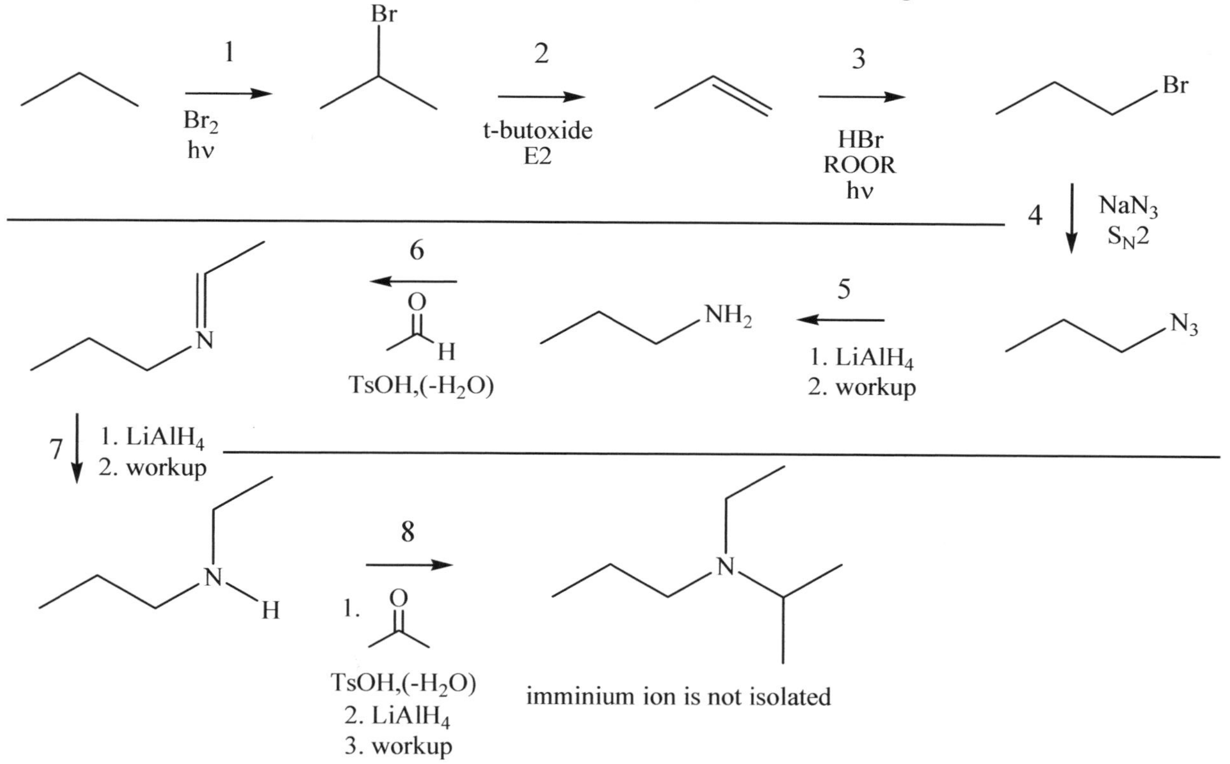

1: Br_2, hν

2: t-butoxide, E2

3: HBr, ROOR, hν

4: NaN_3, S_N2

5: 1. $LiAlH_4$ 2. workup

6: acetaldehyde (CH_3CHO), TsOH,(-H_2O)

7: 1. $LiAlH_4$ 2. workup

8: 1. acetone, TsOH,(-H_2O) 2. $LiAlH_4$ 3. workup

imminium ion is not isolated

Problem 7 (p 337) – Propose a synthetic path to make the triethylamine.

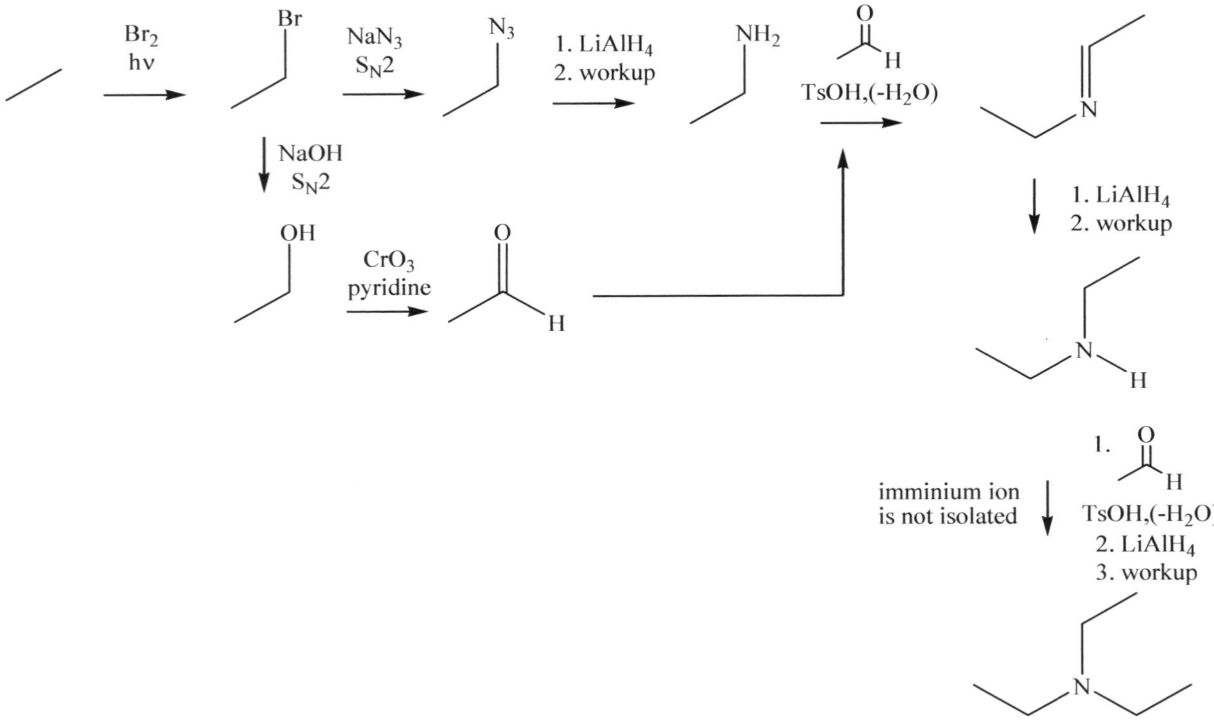

Br_2, hν; NaN_3, S_N2; 1. $LiAlH_4$ 2. workup; NH_2; TsOH,(-H_2O); NaOH, S_N2; OH; CrO_3 pyridine; H; 1. $LiAlH_4$ 2. workup; N; H

imminium ion is not isolated

1. (acetaldehyde) TsOH,(-H_2O)
2. $LiAlH_4$
3. workup

Problem 8 (p 339) - All the functional groups we make are formed by "acyl substitution" reactions. Supply the necessary mechanistic details to complete the mechanisms.

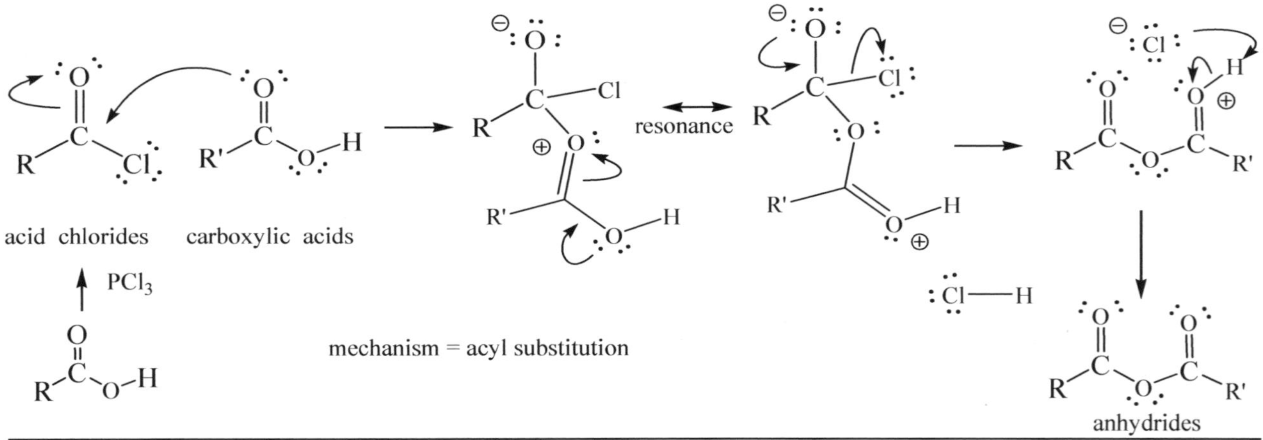

acid chlorides carboxylic acids

PCl_3

resonance

mechanism = acyl substitution

anhydrides

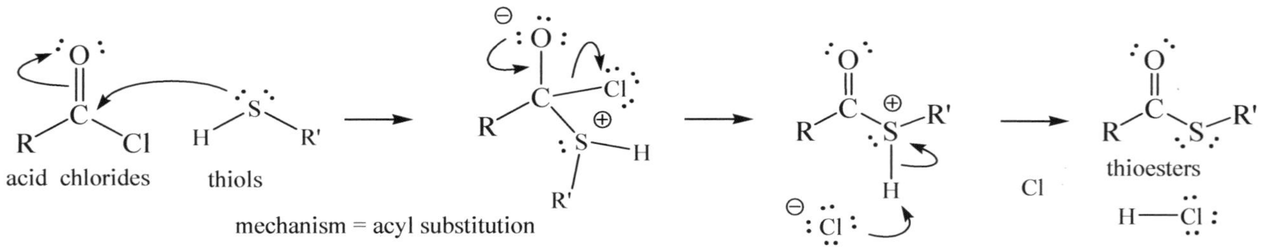

acid chlorides thiols

mechanism = acyl substitution

thioesters

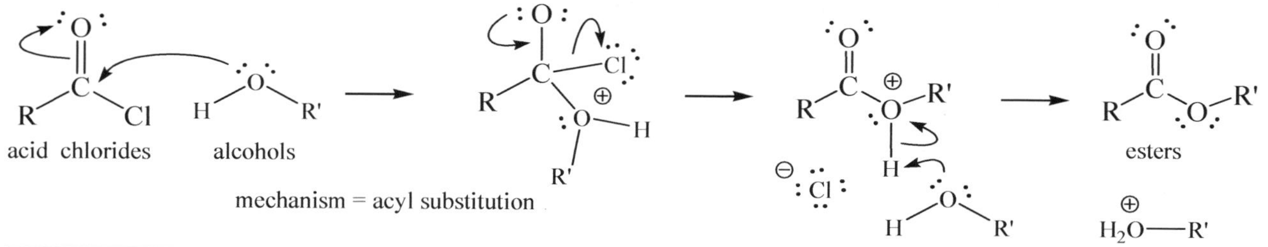

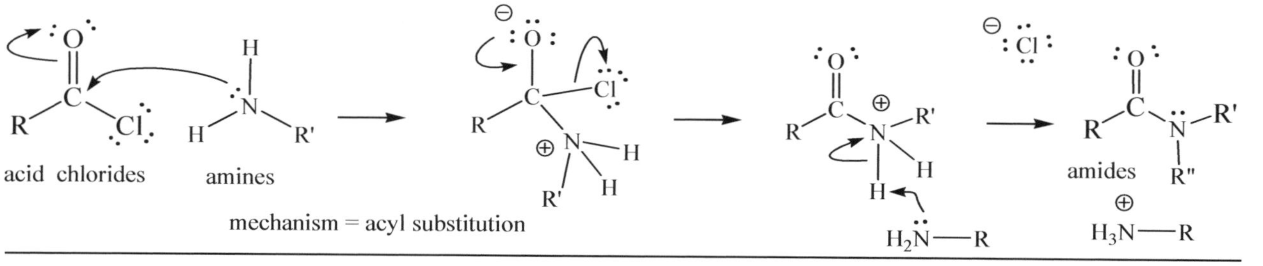

acid chlorides + alcohols → esters + $H_2O—R'$ (with Cl^-)

mechanism = acyl substitution

acid chlorides + amines → amides + $H_3N—R$ (with Cl^-)

mechanism = acyl substitution

Problem 9 (p 340)– Write a pK_a equation for each conjugate acid group referred to above and write the ΔG of the reaction.

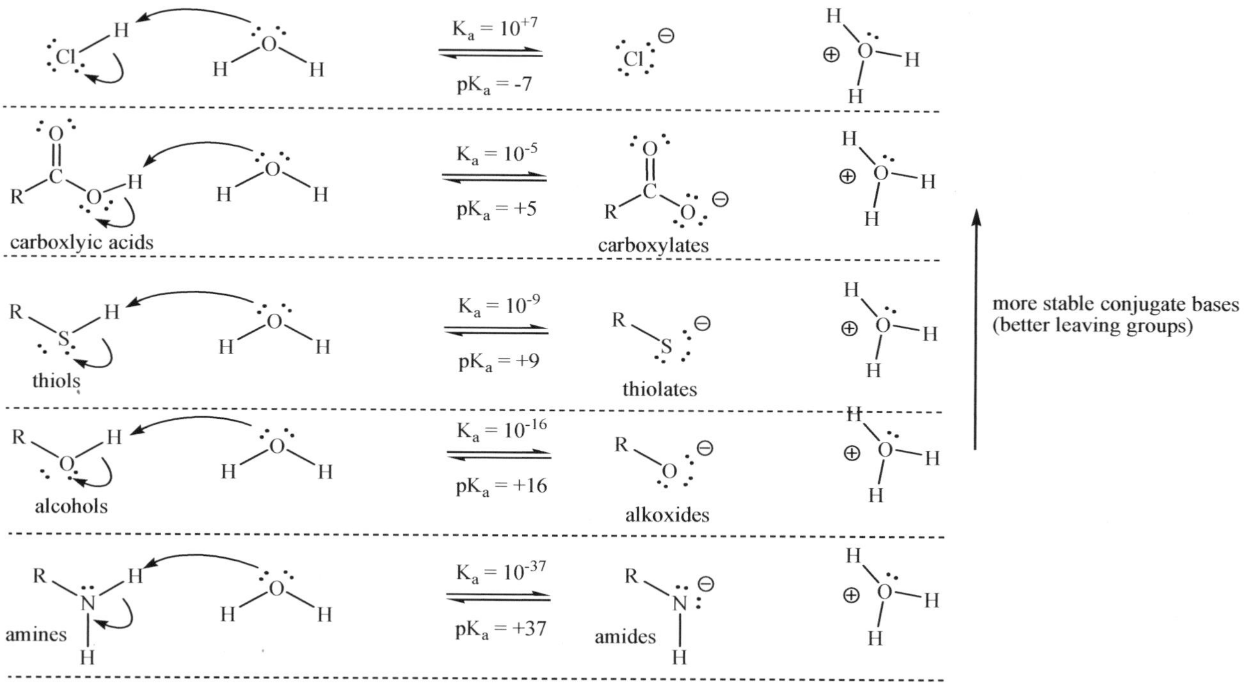

Acid		Equilibrium	Conjugate base	
HCl	H_2O	$K_a = 10^{+7}$, $pK_a = -7$	Cl^-	H_3O^+
carboxlyic acids	H_2O	$K_a = 10^{-5}$, $pK_a = +5$	carboxylates	H_3O^+
thiols	H_2O	$K_a = 10^{-9}$, $pK_a = +9$	thiolates	H_3O^+
alcohols	H_2O	$K_a = 10^{-16}$, $pK_a = +16$	alkoxides	H_3O^+
amines	H_2O	$K_a = 10^{-37}$, $pK_a = +37$	amides	H_3O^+

more stable conjugate bases (better leaving groups)

Problem 10 (p 341)– Which ketone reacts faster with strong nucleophiles (A or B)? Explain your reasoning.

Reaction A is faster because the tetrahedral intermediate is much less sterically crowded. Tetrahedral intermediate B would be much more difficult to form.

Problem 11 (p 343)–
a. How does the reactivity of methanal compare with the reactivity of a simple aldehyde? Explain your reasoning using structures. (Hint: write out the resonance structures and evaluate the partial positive on the carbonyl carbon.)

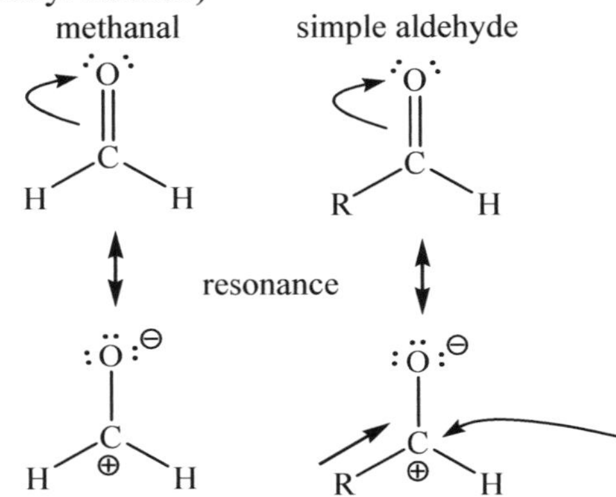

This is really the same comparison made, just above, between aldehydes and ketones. We are really comparing a hydrogen at the carbonyl with an R group. The R group is inductively electron donating and that will reduce the partial positive of the carbonyl. This will make nucleophiles slower in their attack, so a simple aldehyde will be less reactive than methanal. Methanal is a unique aldehyde in that it is the only aldehyde with two hydrogens attached to the carbonyl carbon.

R group provides an extra inductive effect that reduces the partial postive charge of this carbon.

b. How does the reactivity of a methanoate ester compare with the reactivity of a regular simple ester? Explain your reasoning using structures. (Hint: write out the resonance structures and evaluate the partial positive on the carbonyl carbon.)

esters

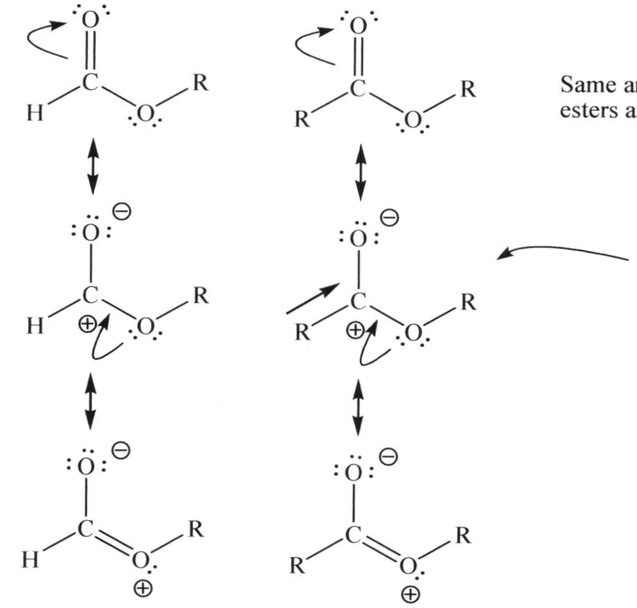

Same argument as above. Methanoate esters are more reactive than simple esters.

R group provides an extra inductive effect that reduces the partial postive charge of this carbon.

c. How does the reactivity of methanamide compare with the reactivity of a regular simple amide? Explain your reasoning using structures. (Hint: write out the resonance structures and evaluate the partial positive on the carbonyl carbon.)

methanamides

amides

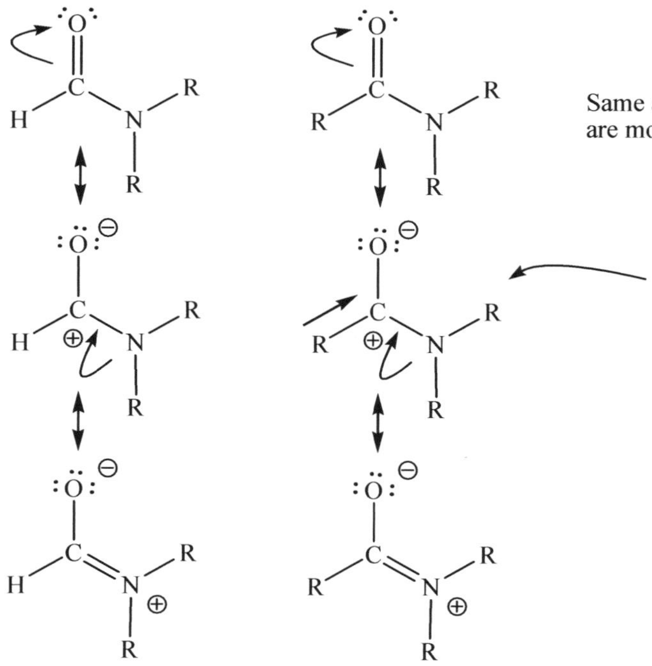

Same argument as above. Methanamides are more reactive than simple amides.

R group provides an extra inductive effect that reduces the partial postive charge of this carbon.

d. Is a thioester more like an oxygen ester or an acid chloride? Explain your answer. What order of reactivity would you predict for these 3 carboxyl functional groups?

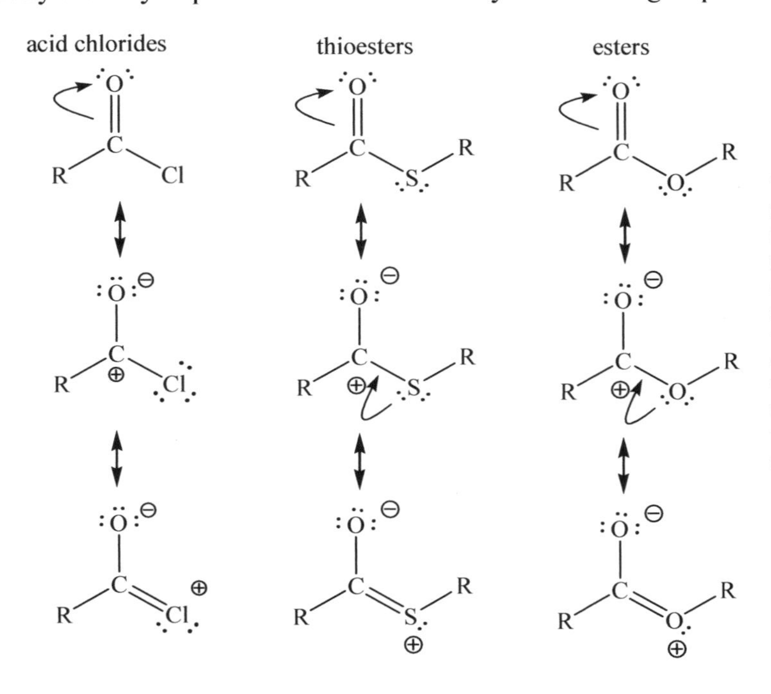

The third resonance structure for the acid chloride and the thioester is not a strong contributor because the 2p (carbon) and 3p (chlorine, sulfur) are different sizes and the electron density is more dispersed with chlorine and sulfur. Because chlorine is a better leaving group, we would expect the acid chloride to be most reactive, followed by the thioester (common in biochemistry) and then the normal ester.

21 functional group summaries (p 344-356)

Chapter 12

Problem 1 (p 358) – THF boils at 66°C and its dipole moment is 1.63 D. The boiling point of diethyl ether is 35°C and its dipole moment is 1.15 D. The boiling point of ethylene oxide is 11°C and its dipole moment is 1.8 D. The boiling point of dimethyl ether is -24°C and its dipole moment is 1.30 D. Propose a possible explanation for the different physical properties of these similar looking ethers.

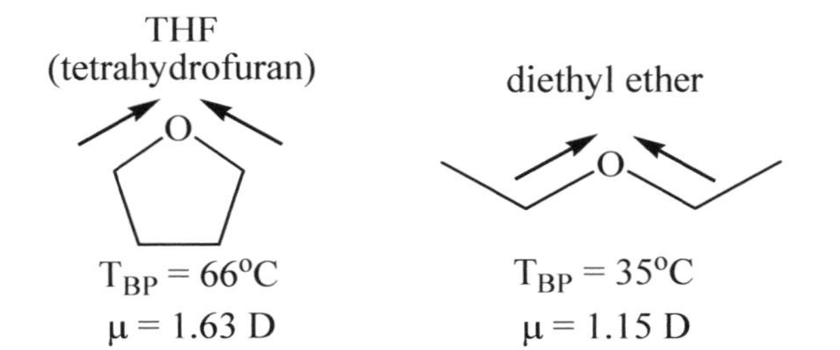

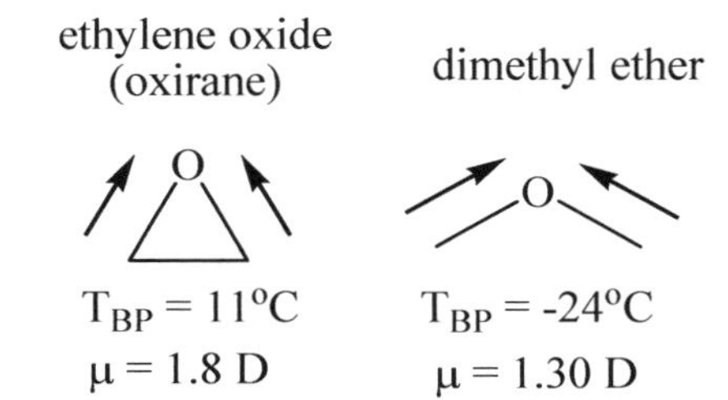

THF has the alkyl branches tied back in a fixed shape. The C-O dipoles are rigidly fixed and the nonpolar branches do not ever get in the way of the front of the oxygen. This makes the molecule more polar, which also gives it a higher boiling point. The ethyl branches of ether are free to rotate through 360° and reduce its ability to interact with other ether molecules or electrophiles.

Ethylene oxide also has the alkyl branches tied back in a fixed shape but the angles are even closer together, leading to a larger molecule diople and boliing point. This makes the molecule more polar, which also gives it a higher boiling point than dimethyl ether.

Problem 2 (p 361) – Nearly all of the epoxides in this problem can be made two ways using the sulfur ylid/carbonyl strategy. Propose a reasonable synthesis for each of the following epoxides using the sulfur ylid approach.

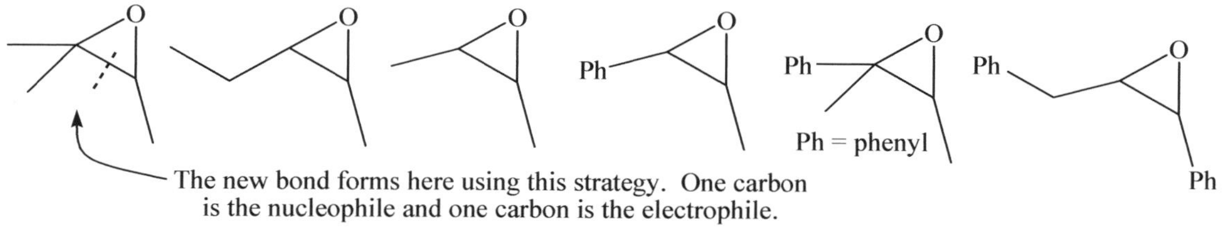

Ph
Ph
Ph
Ph = phenyl
Ph

The new bond forms here using this strategy. One carbon is the nucleophile and one carbon is the electrophile.

Each carbon of the epoxide originally comes from an RBr compound. There will be 2 approaches for all the epoxides shown. Only the synthesis of the first epoxide is provided.

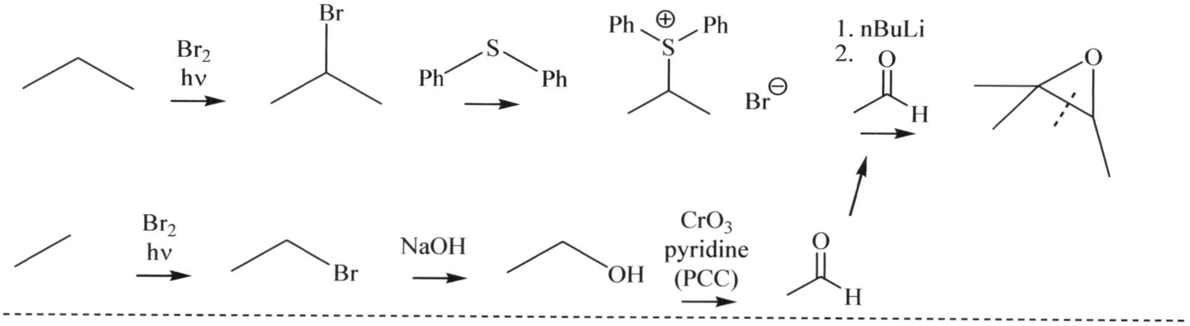

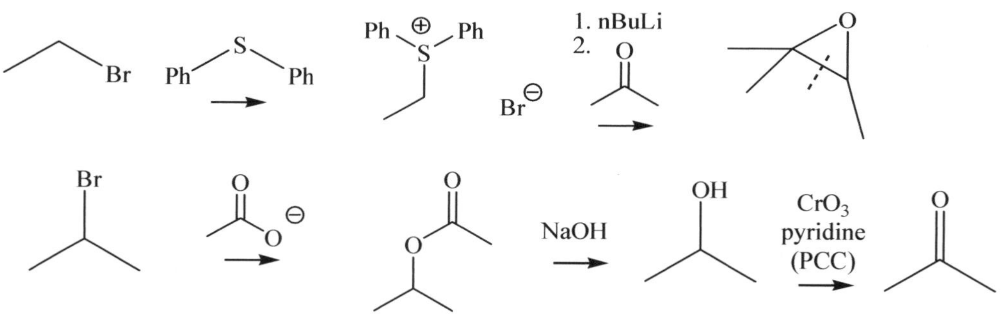

Br_2 hν; Br; Ph–S–Ph; Ph ⊕ Ph S; Br⊖; 1. nBuLi 2. O H

Br_2 hν; Br; NaOH; OH; CrO_3 pyridine (PCC); O H

Br; Ph–S–Ph; Ph ⊕ Ph S; Br⊖; 1. nBuLi 2. O

Br; O O⊖; O O; NaOH; OH; CrO_3 pyridine (PCC); O

We still cannot make cycloalkene epoxide (like cyclohexene oxide), but once we study alkenes we will be able to make almost any needed epoxide.

Problem 3 (361) – Propose a mechanism for how the bromohydrin of cyclohexene (given for now, but made from cyclohexene and Br_2/H_2O in our next topic) could make cyclohexene oxide. Consider what is missing in the product epoxide from the starting bromohydrin, and the reaction conditions that make it happen.

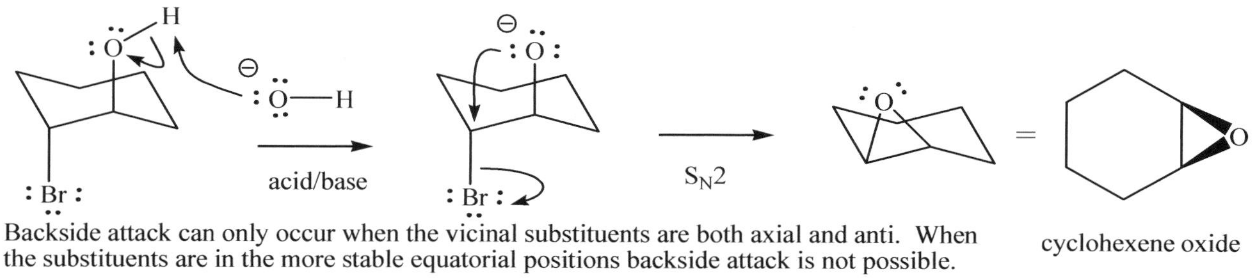

H; :O:; ⊖ :O–H; acid/base; :Br:; ⊖ :O:; :Br:; S_N2; O; =; O

Backside attack can only occur when the vicinal substituents are both axial and anti. When the substituents are in the more stable equatorial positions backside attack is not possible.

cyclohexene oxide

Problem 4 (p 362) – A side product of magnesium organometallic reactions is coupling of two R groups (R-R). Propose a possible mechanism for how this could happen.

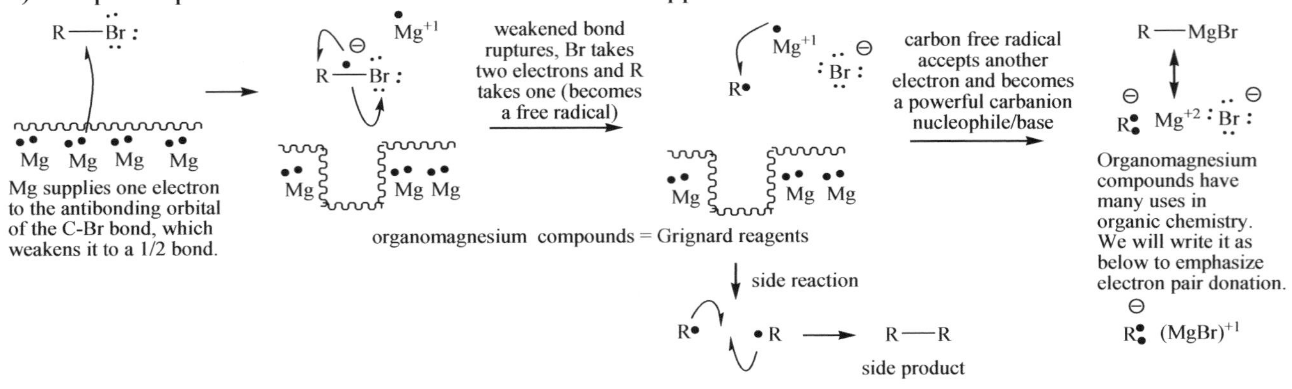

Problem 5 (363) – Propose a reasonable synthesis to make the following organomagnesium compounds (use a single reaction arrow). Pick one of your reactions and write a simplistic mechanism for the reaction (2 electron transfer).

Some examples of simple organometallic reagents 'makeable' from our R-Br list (lots will be possible)

Problem 6 (p 366) – Complete the mechanistic details in each of the following reactions and write the expected product. Workup means to neutralize with acid (H_3O^+).

1 ... 2. workup

2 ... 2. workup (protonate)

3 ... 2. workup (protonate)

4 ... 2. workup

5 ... 2. workup (protonate)

6 ... 2. workup (protonate)

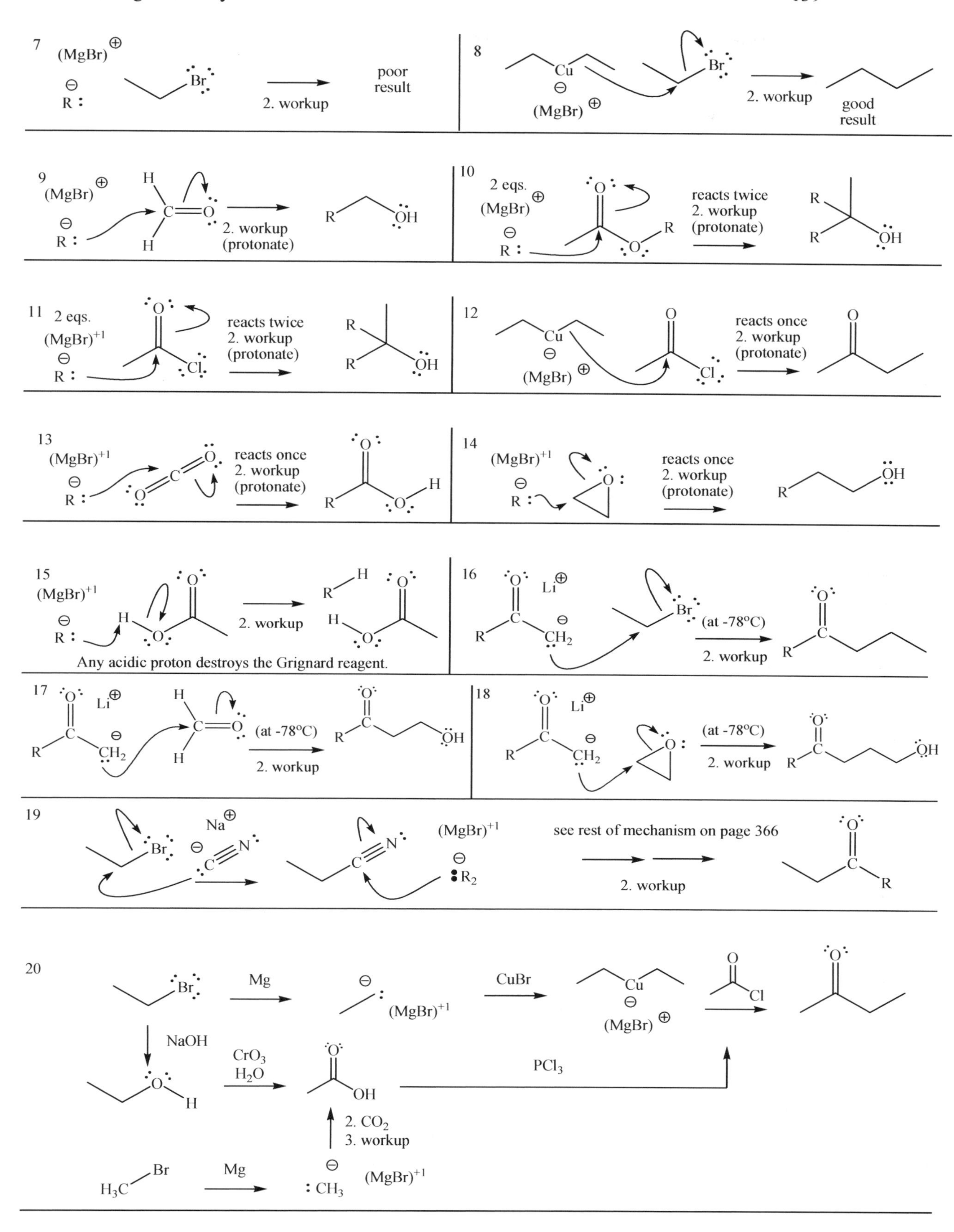
7
(MgBr)
R :
Br
2. workup
poor result
8
Cu
(MgBr)
Br
2. workup
good result
9
(MgBr)
R :
H
C=O
H
2. workup (protonate)
R
OH
10
2 eqs.
(MgBr)
R :
O
O
R
reacts twice
2. workup
(protonate)
R
R
OH
11
2 eqs.
(MgBr)+1
R :
O
Cl
reacts twice
2. workup
(protonate)
R
R
OH
12
Cu
(MgBr)
O
Cl
reacts once
2. workup
(protonate)
O
13
(MgBr)+1
R :
O=C=O
reacts once
2. workup
(protonate)
O
R
O
H
14
(MgBr)+1
R :
O
reacts once
2. workup
(protonate)
R
OH
15
(MgBr)+1
R :
H
O
O
2. workup
R
H
H
O
O
Any acidic proton destroys the Grignard reagent.
16
O
Li
R
C
CH2
Br
(at -78°C)
2. workup
O
C
R
17
O
Li
R
C
CH2
H
C=O
H
(at -78°C)
2. workup
O
R
C
OH
18
O
Li
R
C
CH2
O
(at -78°C)
2. workup
O
C
R
OH
19
Br
Na
C≡N
C≡N
(MgBr)+1
R2
see rest of mechanism on page 366
2. workup
O
C
R
20
Br
Mg
(MgBr)+1
CuBr
Cu
(MgBr)
O
Cl
O
NaOH
O
H
CrO3
H2O
O
OH
PCl3
2. CO2
3. workup
H3C
Br
Mg
CH3
(MgBr)+1

Problem 7 (p 367) – Use $LiAlD_4$ with the given molecules and show expected products when followed with a workup step. Write arrow-pushing mechanisms.

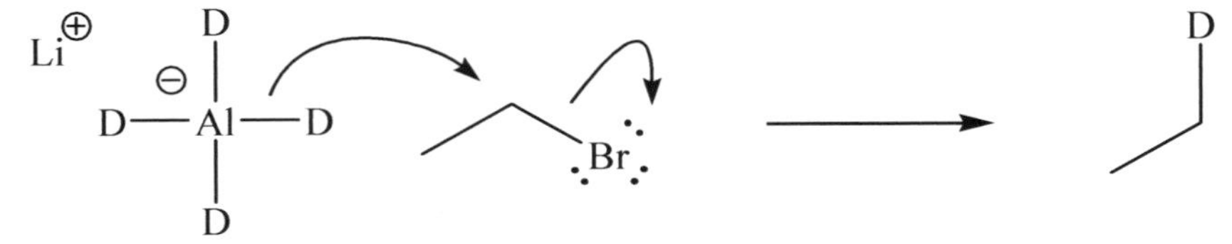

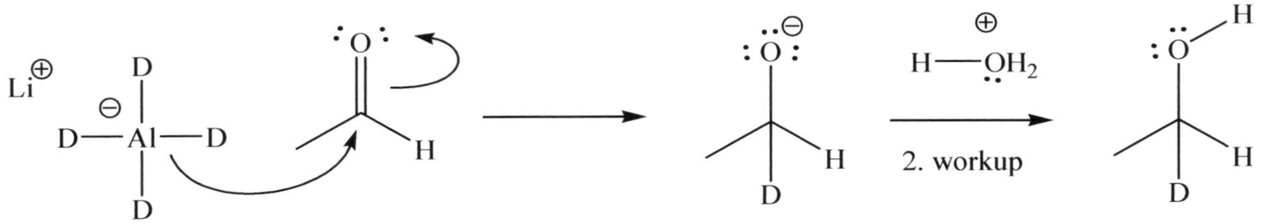

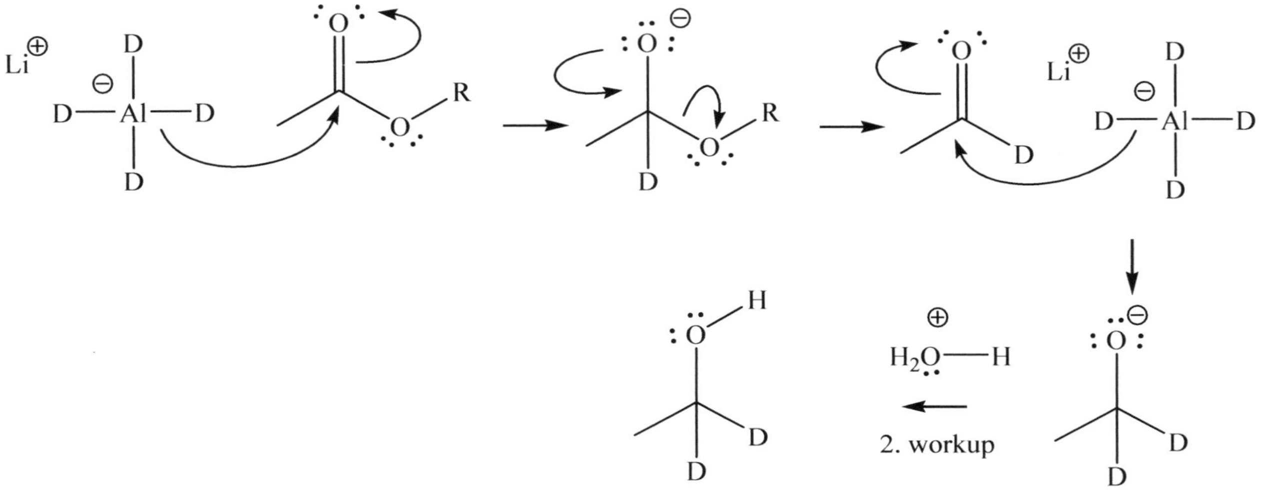

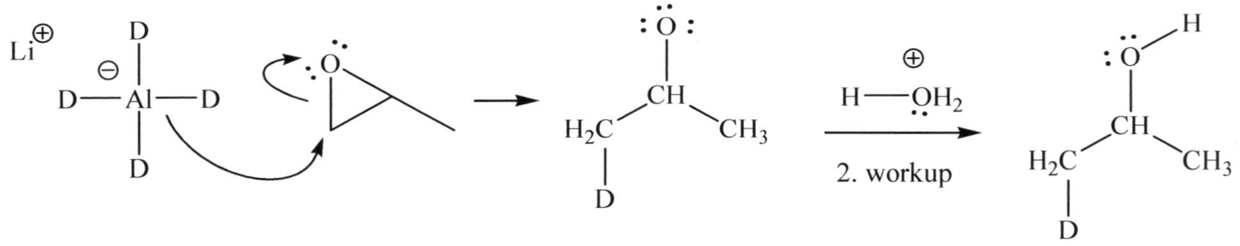

Li⊕ ⊖ D—Al—D (D, D) + Br → D

H—OH₂ ⊕, 2. workup

H₂O—H ⊕, 2. workup

H₂C CH CH₃ D, H—OH₂ ⊕, 2. workup, H₂C CH CH₃ D

Problem 8 (p 368) – A few examples of different strategies to the same target molecules are shown. Show short reaction sequences to make the following combinations work. No mechanisms are necessary. Show each step with the necessary reagents over each arrow.

Reaction 1 - few possible approaches (one part nucleophile and one part electrophile)

1. Mg 2. O (H, H) 3. workup — Br

1. Mg 2. O (epoxide) 3. workup — Br

NaOH — Br

1. $LiAlH_4$ or $NaBH_4$ 2. workup — O, H

1. only $LiAlH_4$ (2 eqs) 2. workup — O, O, R

Reaction 2 – a few possible approaches (one part nucleophile and one part electrophile)

1. Mg 2. O (H) 3. workup — Br

1. $LiAlH_4$ or $NaBH_4$ 2. workup — O

1. $LiAlH_4$ or $NaBH_4$ 2. workup — O

1. Mg 2. O (H) 3. workup — Br, CH_3

1. Mg 2. O (epoxide) 3. workup — Br

1. $CH_3CO_2^{\ominus}$ 2. NaOH — Br

Reaction 3 – a few possible approaches (one part nucleophile and one part electrophile)

1. Mg 2. O 3. workup — Br

1. Mg 2. O, O, R 3. workup — 2 eqs. Br, CH_3

1. $LiAlH_4$ or $NaBH_4$ 2. workup — O

1. Mg 2. O 3. workup — Br, CH_3

1. Mg 2. O (epoxide) 3. workup — Br

Reaction 4 – a few possible approaches (one part nucleophile and one part electrophile)

1. Mg 2. CO_2 3. workup — Ph, Br

CrO_3 / H_2O (Jones) — Ph, O, H

H_2SO_4 / H_2O — Ph, O, O, R

CrO_3 / H_2O (Jones) — Ph, OH

H_2SO_4 / H_2O — Ph, C≡N

1. NaOH / H_2O 2. workup — Ph, O, O, R

Reaction 5 – a few possible approaches (one part nucleophile and one part electrophile)

1. Mg 2. O 3. workup — Br, CH_3

1. Mg 2. O, O, R 3. workup — 2 eqs. Br

H_2O — Br

1. Mg 2. O 3. workup — Br

1. Mg 2. O (epoxide) 3. workup — Br, CH_3

Problem 9 (p 372) – Provide your own cuprate approaches for coupling at bonds 4, 5 and 6. Use the key if you have to. We will use C4 and C5 target molecules in just a bit to develop explicit synthetic strategies, using our available reactions. Possible syntheses via 1, 2 and 3 in book.

? Possibilities? (via cuprates) → TM

1 2 3 4 5 6

Sources of carbon for this example

CH_4 Br

Bond 4

4

Br

$H_3C-Cu^{\ominus}-CH_3$ $(MgBr)^{\oplus}$ ← 0.5 eqs CuBr ← $H_3C-MgBr$ ← Mg ← H_3C-Br

Br ← ? PBr_3 ← HO ← Grignard 1. Mg 2. $H_2C{=}O$ 3. workup ← Br ← 1. TsCl pyridine 2. NaBr ← OH

Br 1. Mg 2. (epoxide) 3. workup → OH → CrO_3 pyridine (PCC) → H, O ↑ 1. Mg 2. 3. workup

CH_4 → Br_2 hv → Br CH_3 → $H_2P-S-PH_2$ → CH_3 $H_2P-S^{\oplus}-PH_2$ → 1. nBuLi 2. $H_2C{=}O$ → O (epoxide)

Br_2 hv → Br

CH_4 → Br_2 hv → Br CH_3 → NaOH → OH CH_3 → CrO_3 pyridine (PCC) → O CH_2

Bond 5

5

Br

$Cu^{\ominus}$ $(MgBr)^{\oplus}$ ← 0.5 eqs CuBr ← MgBr ← Mg ← Br

Br ← ? 1. TsCl pyridine 2. NaBr ← OH ← ? Grignard 1. Mg 2. H O 3. workup ← Br ← Br_2 hv ←

made above

Bond 6

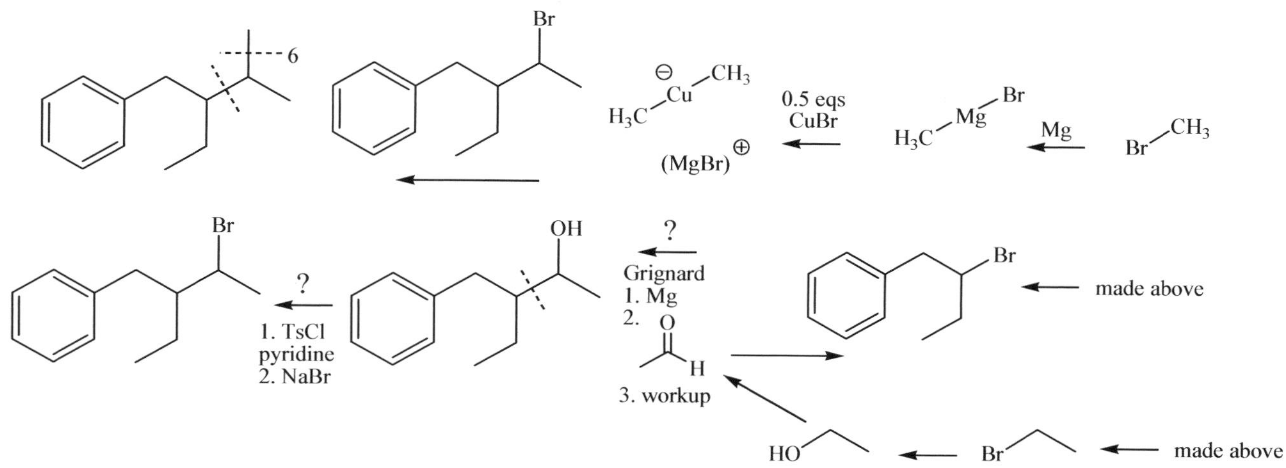

Br
6
H_3C Cu CH_3 (MgBr)⊕ ← 0.5 eqs CuBr ← H_3C Mg Br ← Mg ← Br CH_3

Br ← ? 1. TsCl pyridine 2. NaBr ← OH ← ? Grignard 1. Mg 2. O H 3. workup ← Br ← made above

HO ← Br ← made above

Problem 10 (p 376) – Show the expected product and provide a mechanism for each of the following reactions.

a

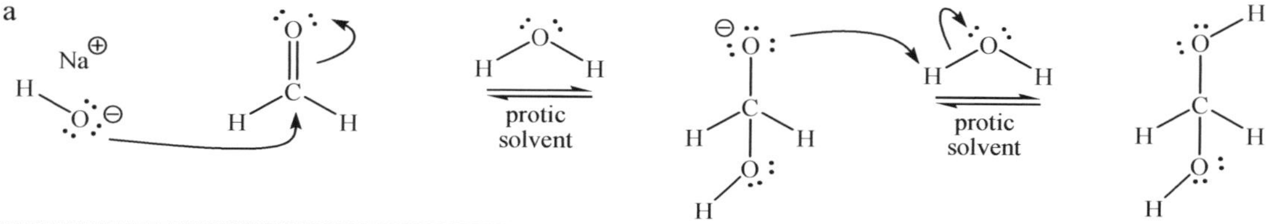

Na⊕ H O⊖ + H C(=O) H ⇌ (H–O–H, protic solvent) ⇌ (H–O–H, protic solvent)

b

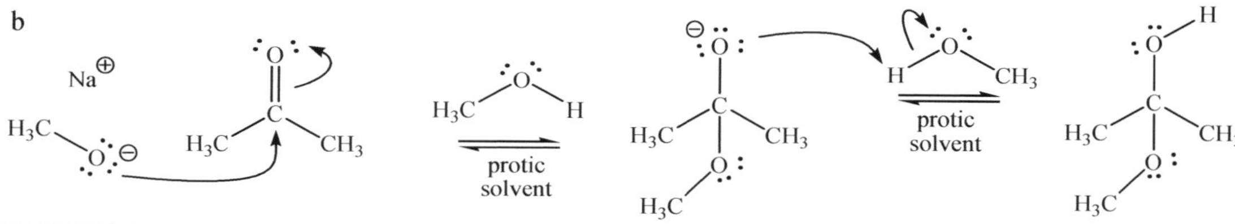

Na⊕ H_3C O⊖ + H_3C C(=O) CH_3 ⇌ (H_3C–O–H, protic solvent) ⇌ (H–O–CH_3, protic solvent)

c

Na⊕ N≡C⊖ + H_3C C(=O) H → aprotic solvent → 2. workup (add a proton)

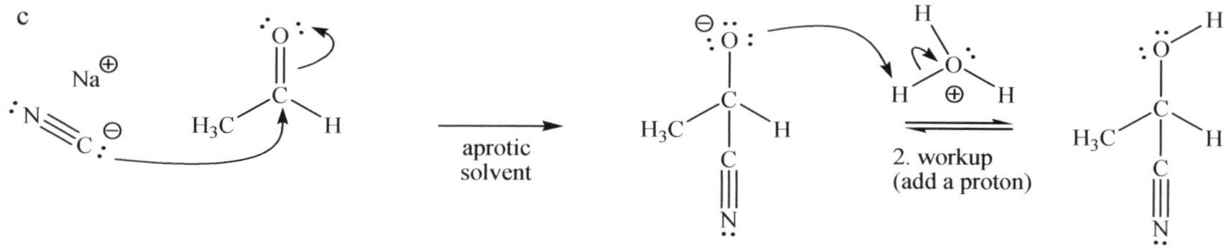

d

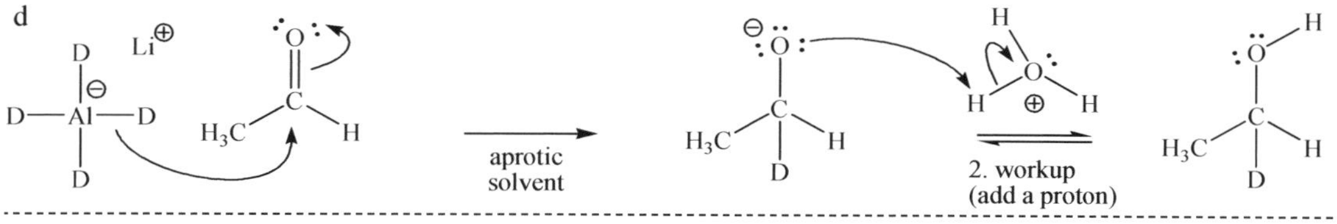

Li⊕ D–Al⊖(D)(D)–D + H_3C C(=O) H → aprotic solvent → 2. workup (add a proton)

e

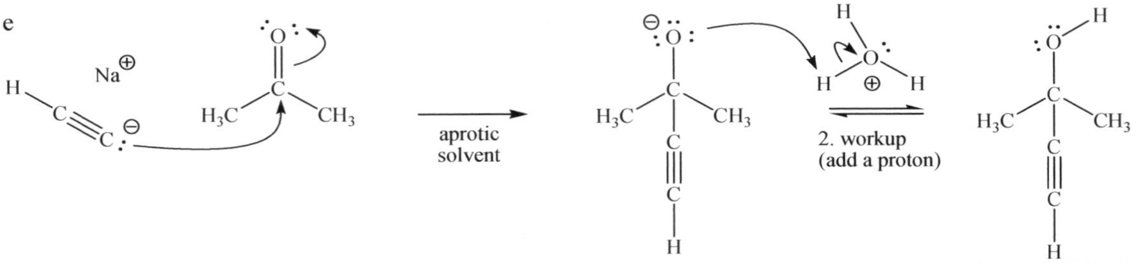

Na⊕ H–C≡C:⊖ + H_3C–C(=O)–CH_3 → (aprotic solvent) → alkoxide ⇌ (2. workup (add a proton), H_3O^+) → H_3C–C(OH)(C≡C–H)–CH_3

f

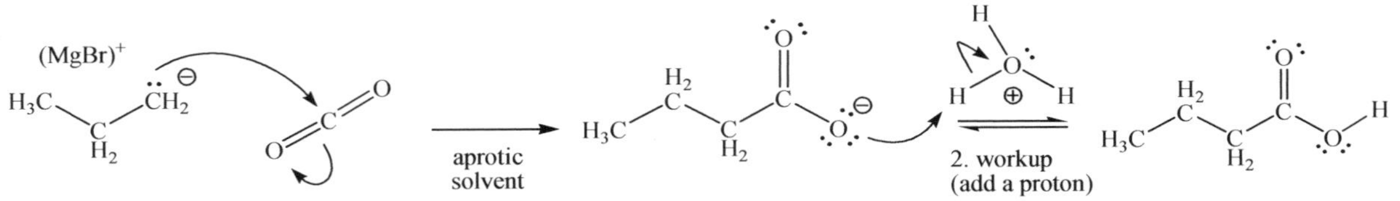

$(MgBr)^+$ H_3C–CH_2–CH_2^- + O=C=O → (aprotic solvent) → carboxylate ⇌ (2. workup (add a proton)) → H_3C–CH_2–CH_2–C(=O)–OH

g

$(MgBr)^+$ H_3C–CH_2–CH_2^- + H_3C–C≡N: → (aprotic solvent) → imine anion ⇌ (2. workup (add a proton)) → imine → protonated imine → water addition → proton transfer → loss of NH_3 → resonance → ketone + NH_4^+

resonance

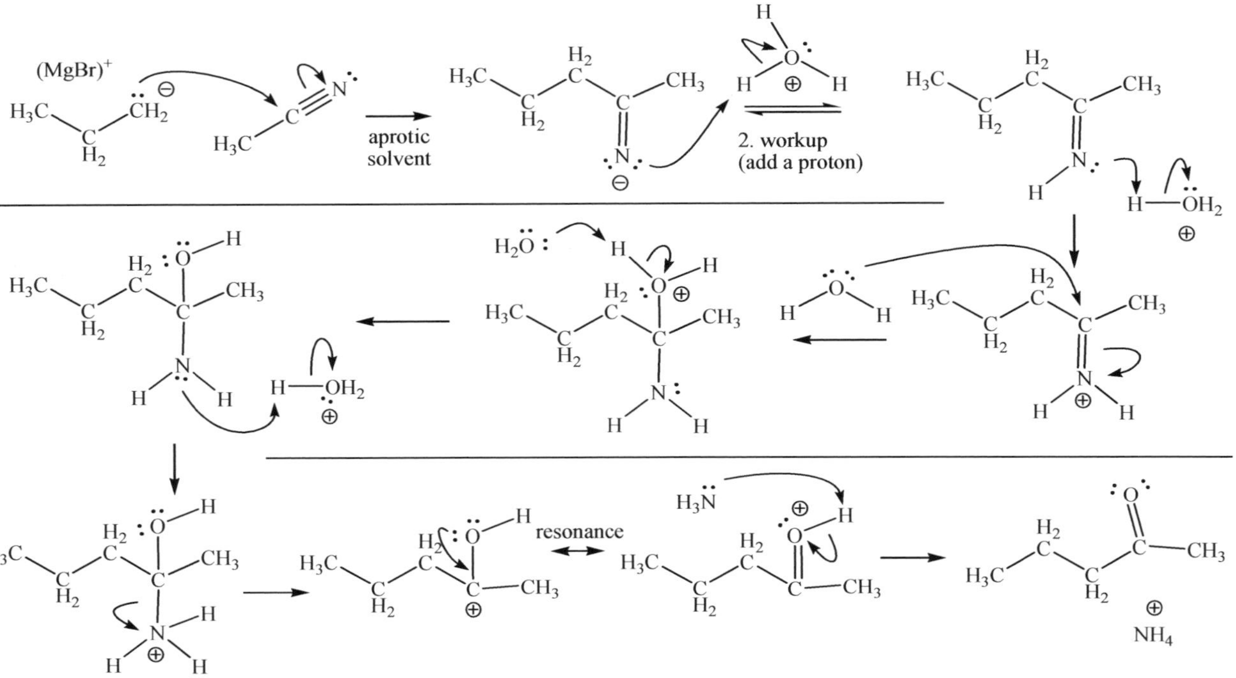

h. Base hydrolysis of a nitrile to form a primary amide.

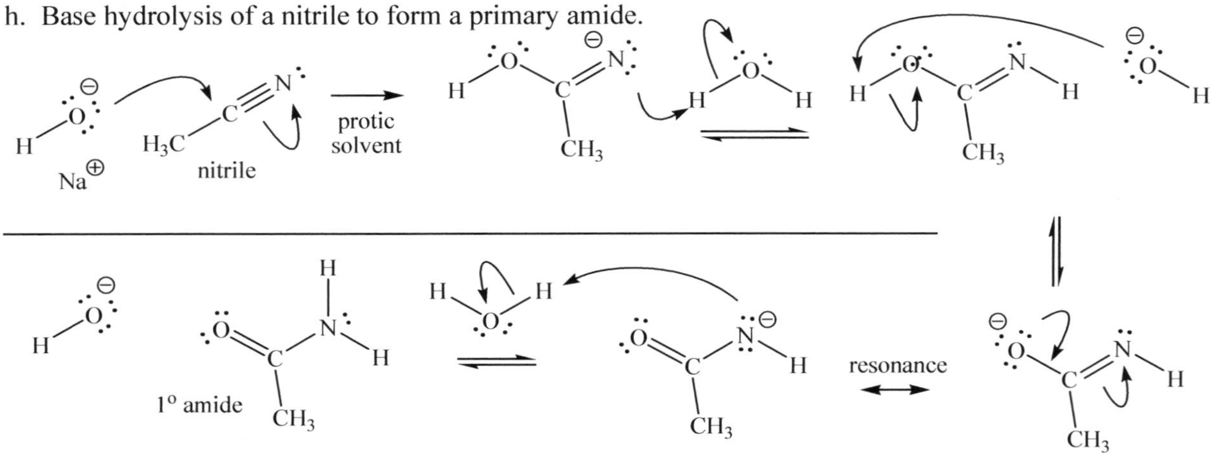

H–O:⊖ Na⊕ + H_3C–C≡N: (nitrile) → (protic solvent) → ⇌ → resonance → ⇌ 1° amide

nitrile

1° amide

Problem 11 (p 378) – Predict the expected product and provide a mechanism for each of the following reactions.

a 2 equivalents

$(MgBr)^+$

ethyl methanoate
(ethyl formate)

aprotic
solvent

ketone
(more reactive)

2. workup
(add a proton)

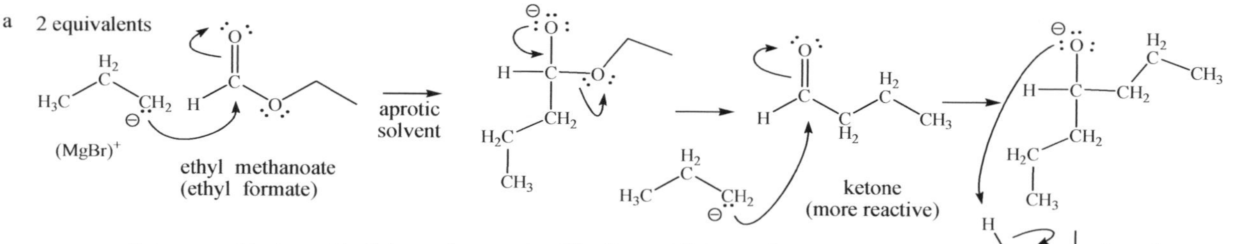

Esters react twice with Grignard reagents. The first addition makes a tetrahedral intermediate that collapses, forming a ketone. As we just saw, ketones are more reactive than esters so the ketone reacts before any other ester can, leading to a non-reactive alkoxide. The alkoxide will get protonated in the workup, forming an alcohol with two "R" groups the same.

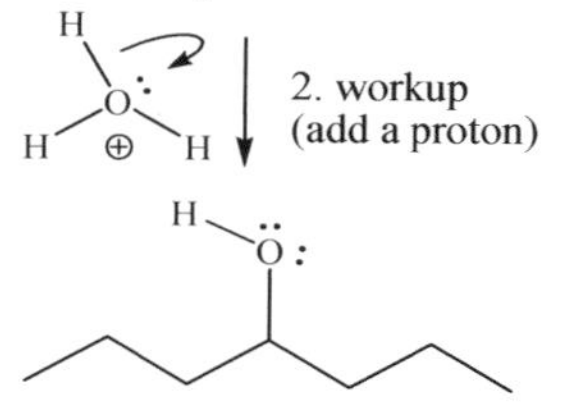

b
2 equivalents

ethyl ethanoate
(ethyl acetate)

protic
solvent

2. workup
(add a proton)

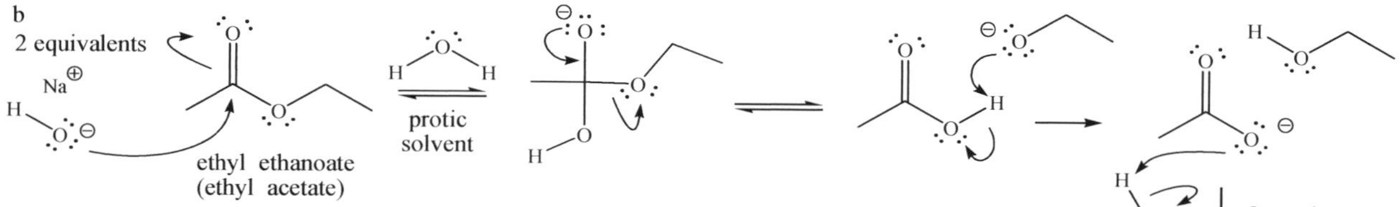

This is base hydrolysis of an ester. It also goes by "saponification" (soap making) because they used to make soap from animal fat mixed with ashes containing NaOH and KOH that would convert the triglycerides into the conjugate bases of fatty acid chains, which would make soap cakes to wash off grease and grime. Of course, it was pretty harsh on the skin, so taking baths wasn't all that popular.

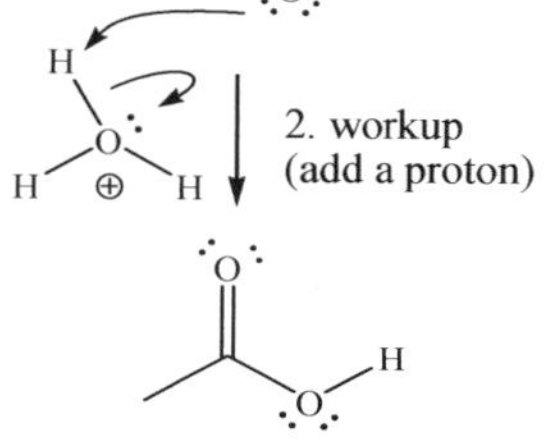

c 2 equivalents of hydride

ethyl propanoate

aprotic
solvent

2. workup
(add a proton)

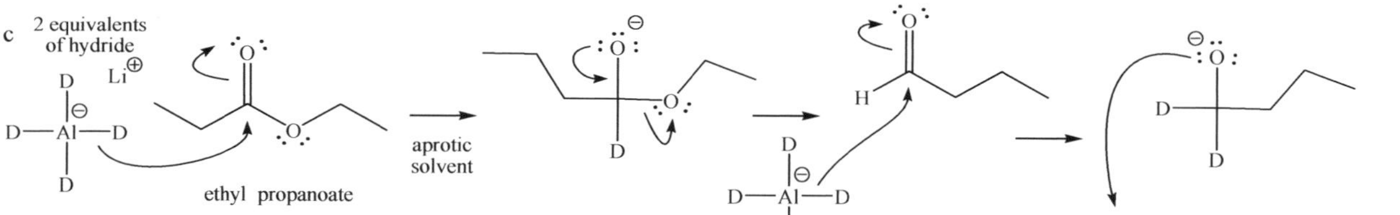

lithium aluminium hydride,
very powerful nucleophilic
hydrogen

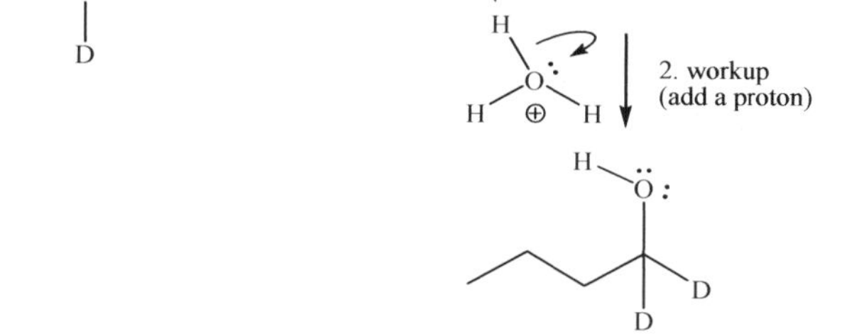

Problem 12 (p 379) – Show the expected product and provide a mechanisms for each of the following reactions.

a

protic
solvent

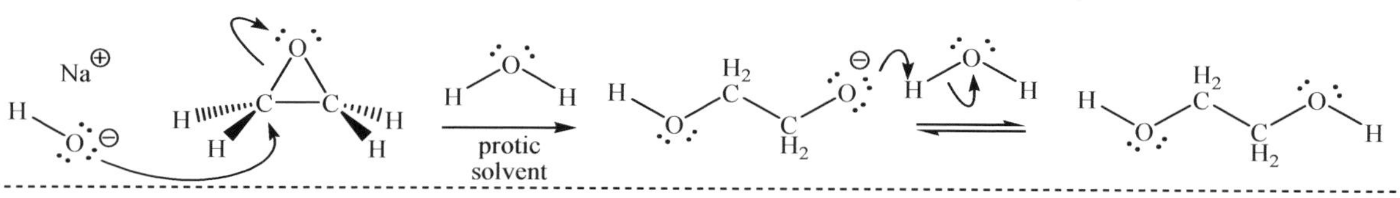

b

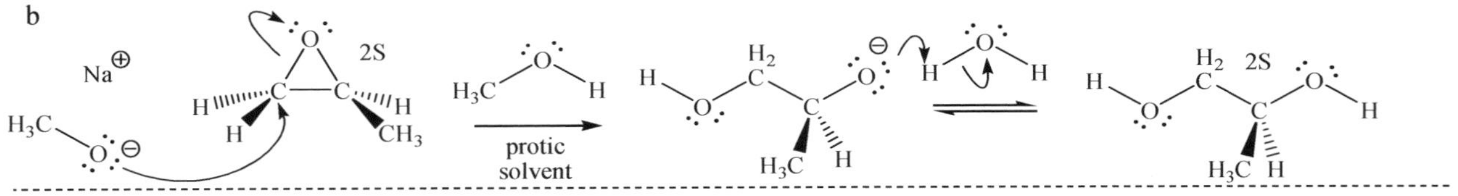

protic solvent

c

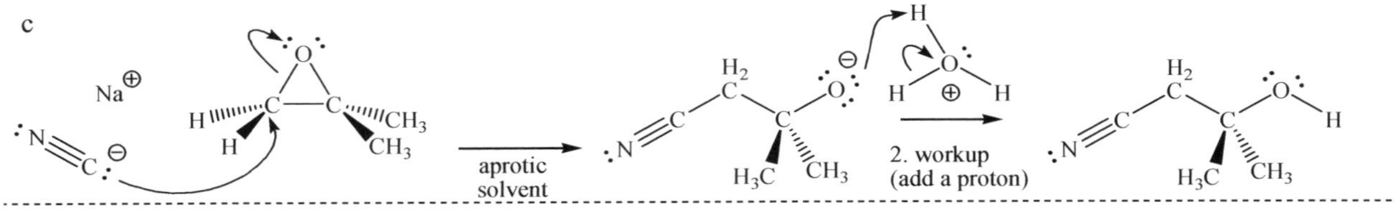

aprotic solvent

2. workup (add a proton)

d

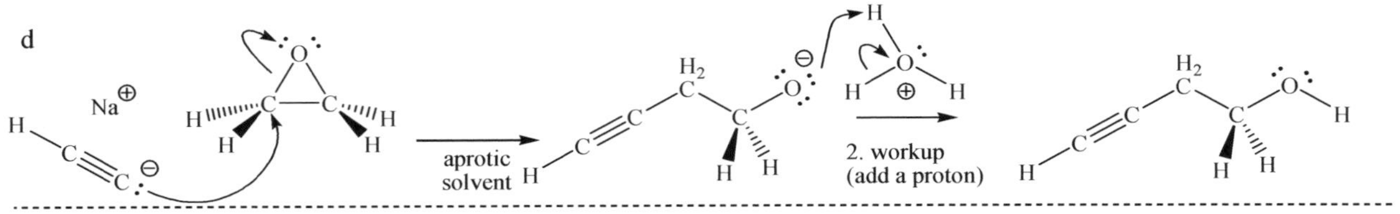

aprotic solvent

2. workup (add a proton)

e

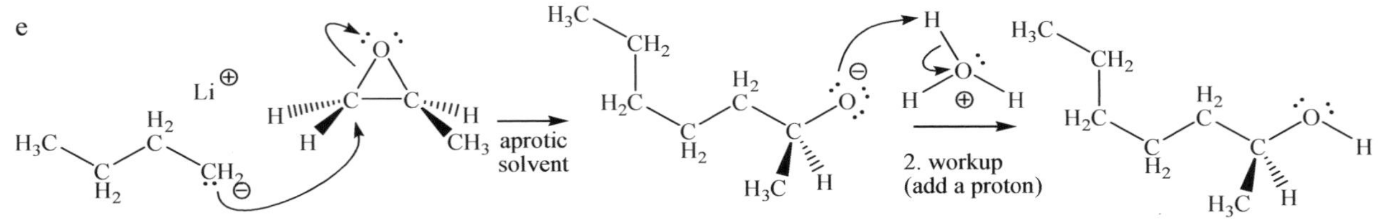

aprotic solvent

2. workup (add a proton)

f

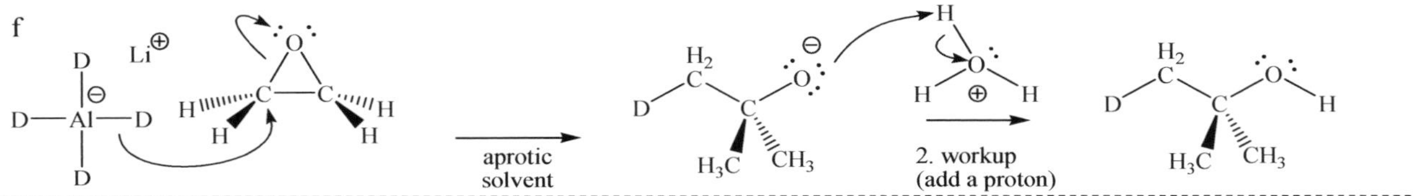

aprotic solvent

2. workup (add a proton)

Problem 13 (p 382) – Show the enolate of propanone reactions with the following electrophiles: a. bromoethane, b. methanal and c. ethylene oxide. Assume a final workup step where necessary.

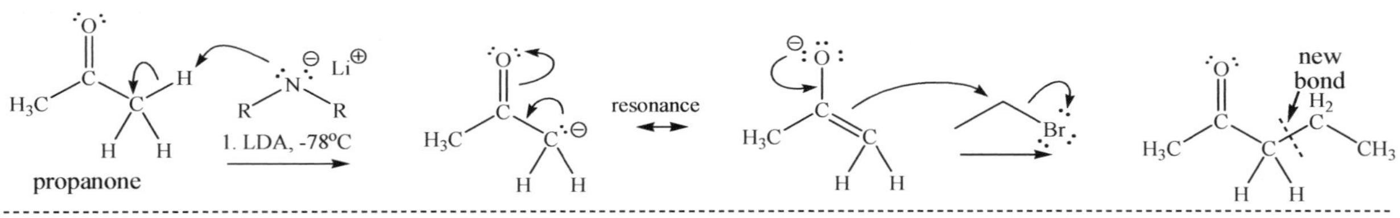

propanone

1. LDA, -78°C

resonance

new bond

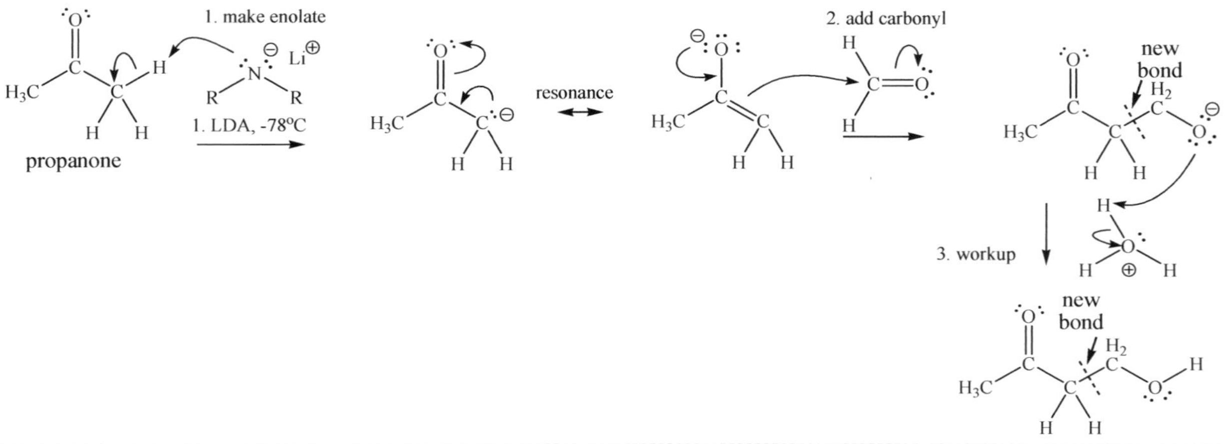

1. make enolate

propanone

1. LDA, -78°C

resonance

2. add carbonyl

new bond

3. workup

new bond

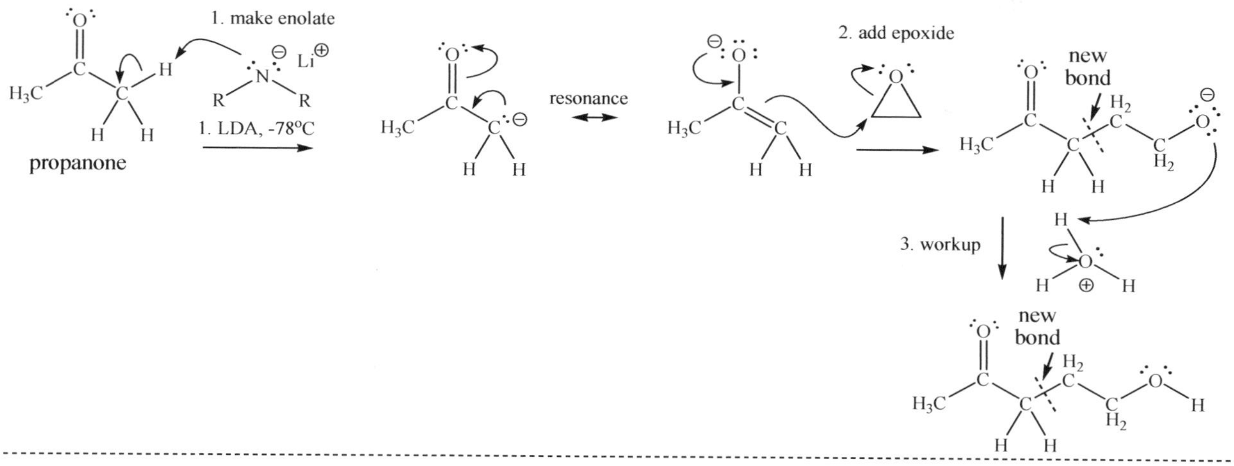

O
C
H_3C C H
H H
propanone
1. make enolate
⊖ Li⊕
N
R R
1. LDA, -78°C
O
H_3C C C ⊖
H H
resonance
⊖ O
H_3C C C
H H
2. add epoxide
O
O new bond
H_3C C C H_2 C C H_2 O ⊖
H H
3. workup
H
O
H ⊕ H
O new bond
H_3C C C H_2 C C H_2 O H
H H

Problem 14 (p 383) – Predict the product and show a mechanism for the following reaction of LDA with cyclohexeneoxide, followed by neutralization (workup).

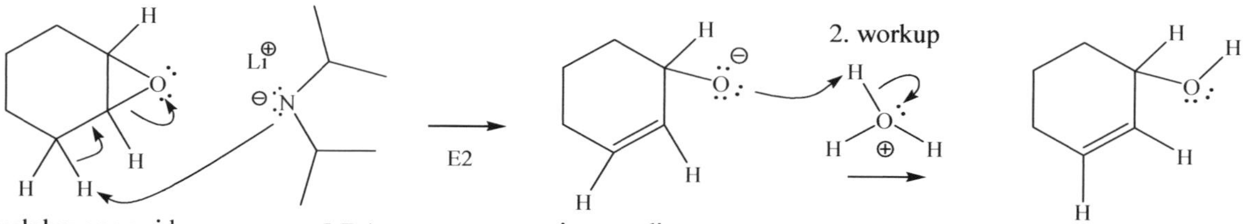

H
O
H
H H
cyclohexene oxide
Li⊕
⊖ N
LDA
E2
H
O ⊖
H
H
intermediate
2. workup
H
O
H ⊕ H
H H
O
H
H
product = allylic alcohol

Very strong acids used in our course.

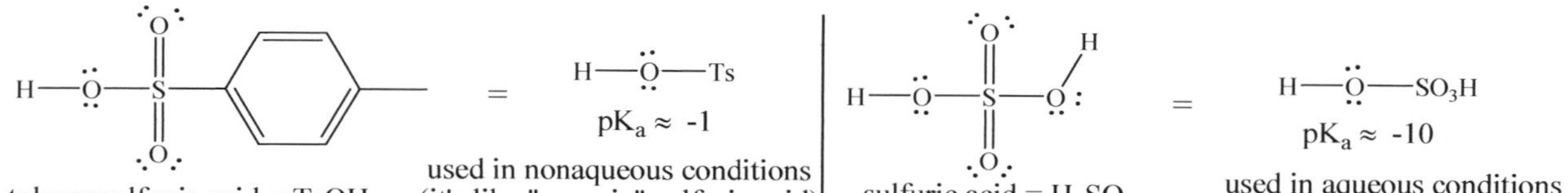

H—O—S(=O)(=O)—C_6H_4—CH_3 = H—O—Ts

$pK_a \approx -1$

toluenesulfonic acid = TsOH

used in nonaqueous conditions (it's like "organic" sulfuric acid)

H—O—S(=O)(=O)—O—H = H—O—SO_3H

$pK_a \approx -10$

sulfuric acid = H_2SO_4

used in aqueous conditions

Problem 15 (p 384) – Show the expected product and provide a mechanism for each of the following reactions.

a

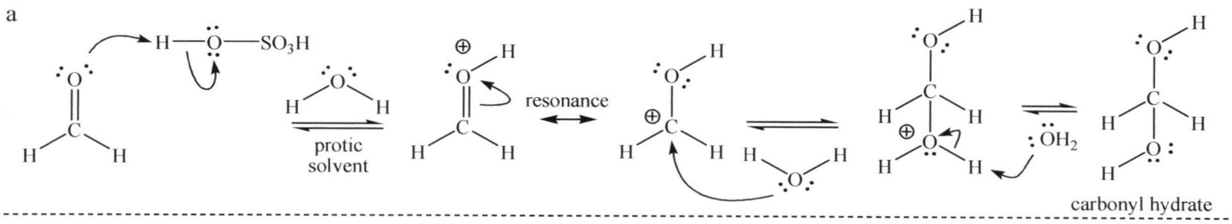

O
H C H
H—O—SO_3H
H O H
protic solvent
⊕ O H
H C H
resonance
O H
⊕ C
H H
H O H
O H
H C H
H ⊕ O H
H
: OH_2
O H
H C H
H O
carbonyl hydrate

b

O
H_3C C CH_3
H—O—SO_3H
H O H
protic solvent
⊕ O H
H_3C C CH_3
resonance
O H
⊕ C
H_3C CH_3
H O H
O H
H_3C C CH_3
H ⊕ O H
H
: OH_2
O H
H_3C C CH_3
H O
carbonyl hydrate

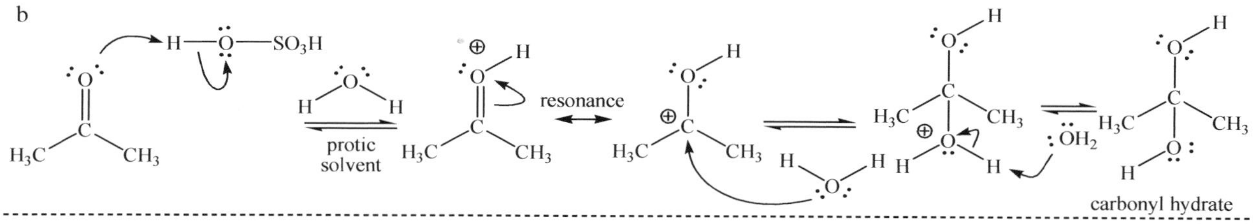

c

O

H_3C C H

H—O—SO_3H

H O CH_3

protic solvent

⊕ O H

H_3C C H

resonance

O H

⊕ C

H_3C H

H_3C O H

O H

H_3C C H

⊕ O

H_3C H

H O CH_3

HO_3S—O—H

O H

H_3C C H

O

H_3C

hemi-acetal

H_3C O

H_3C C H

O :

H_3C

acetal

H O CH_3

H_3C ⊕ O H

H_3C C H

O :

H_3C

H_3C O H

H_3C C H

⊕ O

H_3C

H O H

resonance

H_3C ⊕ C H

O :

H_3C

H ⊕ O H

H_3C C H

O :

H_3C

d

O

H_3C C CH_3

H—O—SO_3H

H O CH_3

protic solvent

⊕ O H

H_3C C CH_3

resonance

O H

⊕ C

H_3C CH_3

H_3C O H

O H

H_3C C CH_3

⊕ O

H_3C H

H O CH_3

HO_3S—O—H

O H

H_3C C CH_3

O :

H_3C

hemi-ketal

H_3C O

H_3C C CH_3

O :

H_3C

ketal

H O CH_3

H_3C ⊕ O H

H_3C C CH_3

O :

H_3C

H_3C O H

H_3C C CH_3

⊕ O

H_3C

H O H

resonance

H_3C ⊕ C CH_3

O :

H_3C

H ⊕ O H

H_3C C CH_3

O :

H_3C

Problem 16 (p 385) – Write the reverse mechanism for formation of the carbonyl compound and two alcohols from the acetal using aqueous acid solution.

ketal

resonance

hemi-ketal

resonance

ketone

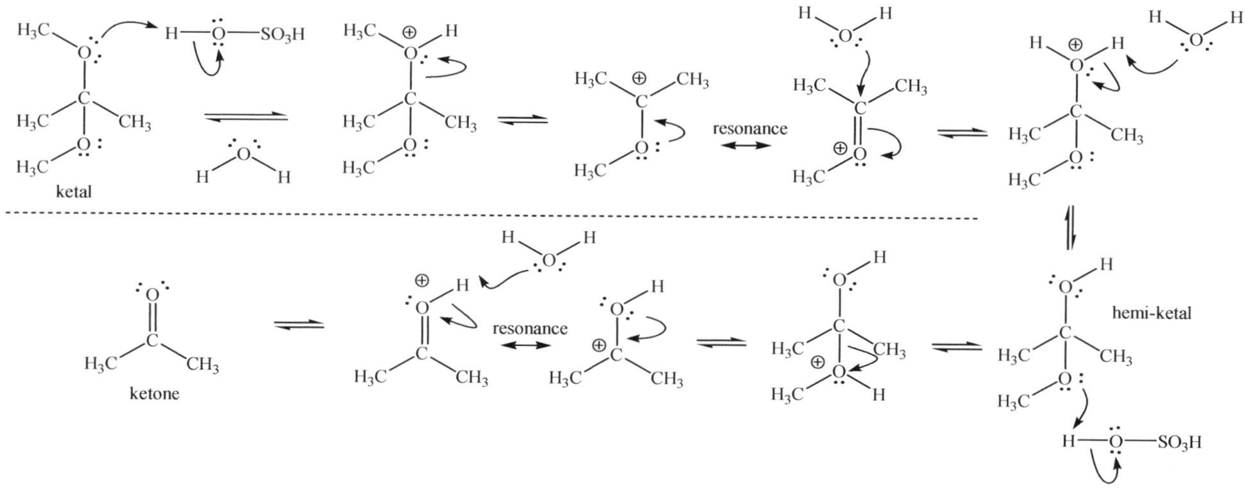

Problem 17 (p 385) – Write the reverse mechanism for formation of the carboxylic acid and alcohol from the ester using aqueous acid solution.

ester = **C**

proton transfer

resonance

resonance

Acid and lots of water drives the equilibrium

proton transfer

resonance

resonance

T.I. = **B**

carboxylic acid = **A**

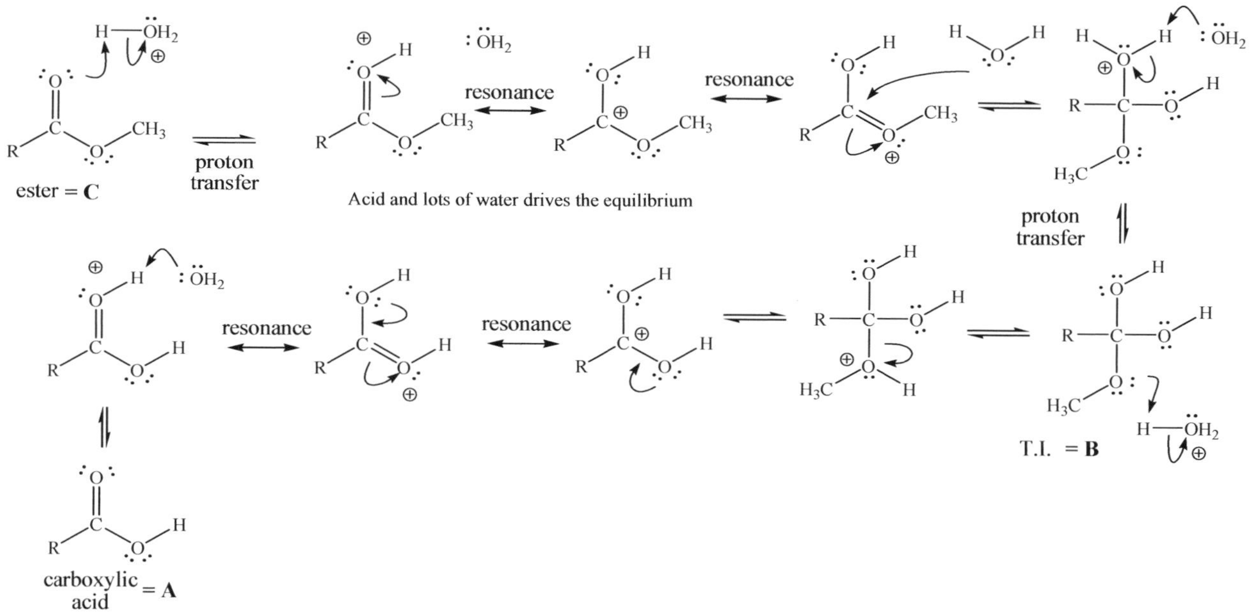

Problem 18 (p 386) – Show the expected product and provide a mechanism for each of the following reactions.

a

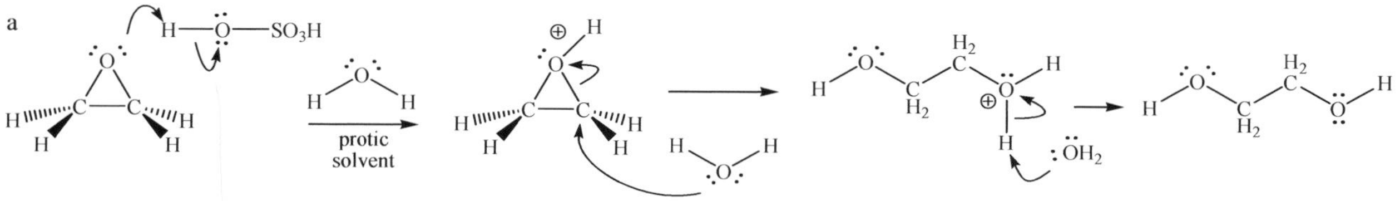

protic solvent

b

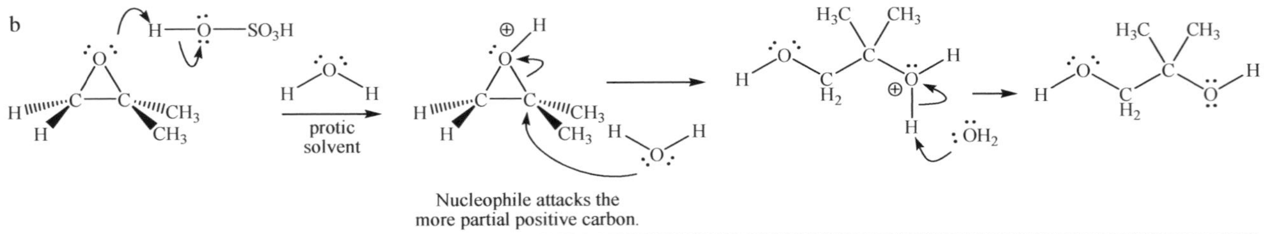

protic solvent

Nucleophile attacks the more partial positive carbon.

c

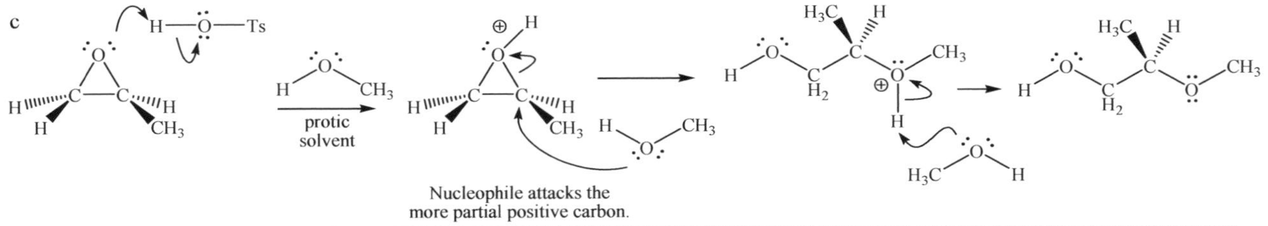

protic solvent

Nucleophile attacks the more partial positive carbon.

d

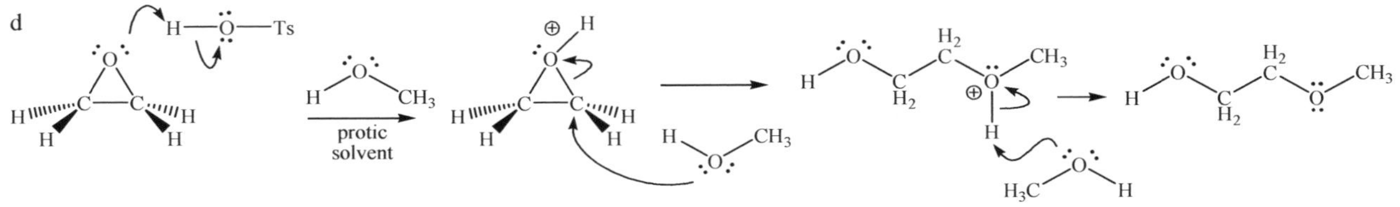

protic solvent

Problem 19 – Propose a reasonable synthetic approach for the following target molecules. Remember, not every double bond we see has to come from a Wittig reaction. Wittig reactions are most useful when a "less stable" double bond has to be prepared, or if other functionality is present in molecule that cannot survive harsh elimination conditions (more typical of real life situations). When the molecule is uncomplicated by other functionality and the most stable double bond is the target, straight forward E1 and/or E2 reactions might be acceptable alternatives. You can use our usual sources of carbon and typical organic reagents.

a

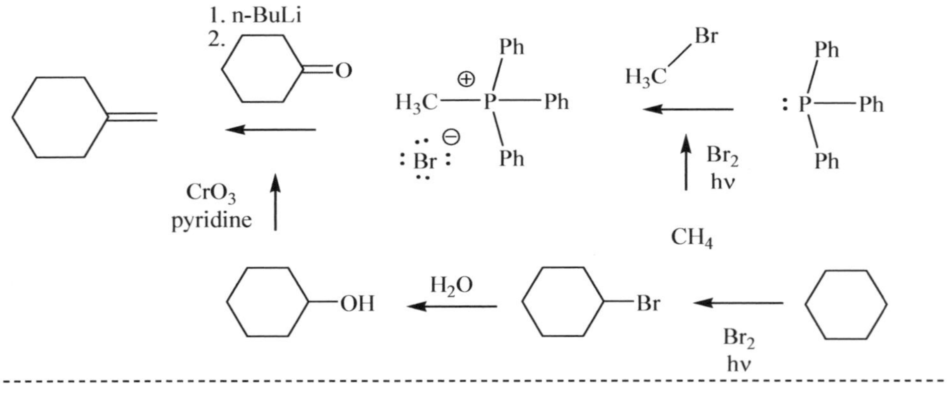

1. n-BuLi
2. (cyclohexanone)

CrO_3 pyridine

Br_2 hν

CH_4

H_2O

Br_2 hν

b.

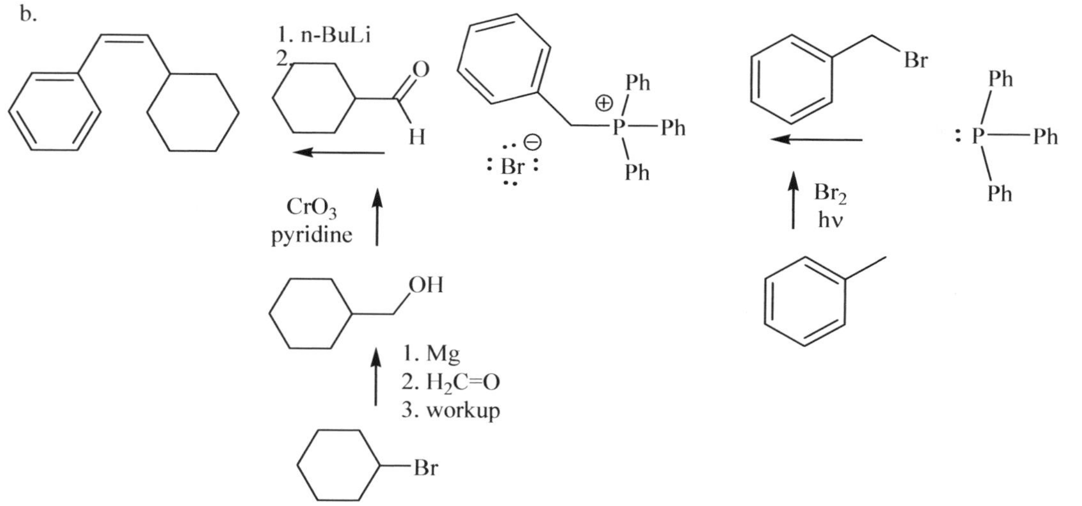

1. n-BuLi
2.

CrO_3
pyridine

OH

1. Mg
2. $H_2C{=}O$
3. workup

Br

Br

Ph
P—Ph
Ph
$:\ddot{Br}:^{\ominus}$

Br

Br_2
hν

: P—Ph
Ph
Ph

c. Just like "b" except while still cold, add 1 more equivalent of n-BuLi to remove proton and allow the more stable configuration to form, then add mild acid to lock in the trans configuration. Schlosser modification of the Wittig.

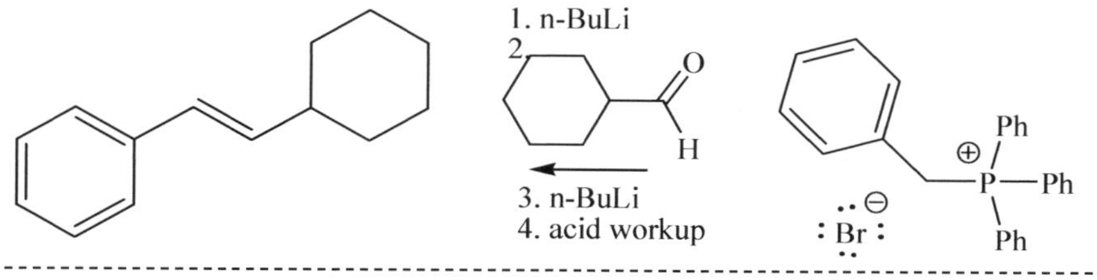

1. n-BuLi
2.
3. n-BuLi
4. acid workup

Ph
P—Ph
Ph
$:\ddot{Br}:^{\ominus}$

d

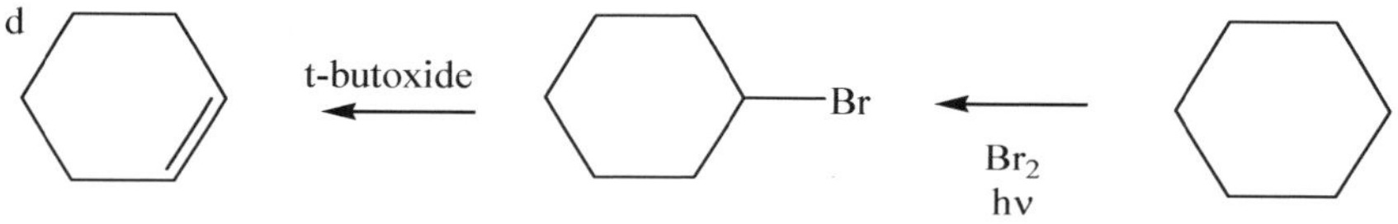

t-butoxide

Br

Br_2
hν

e

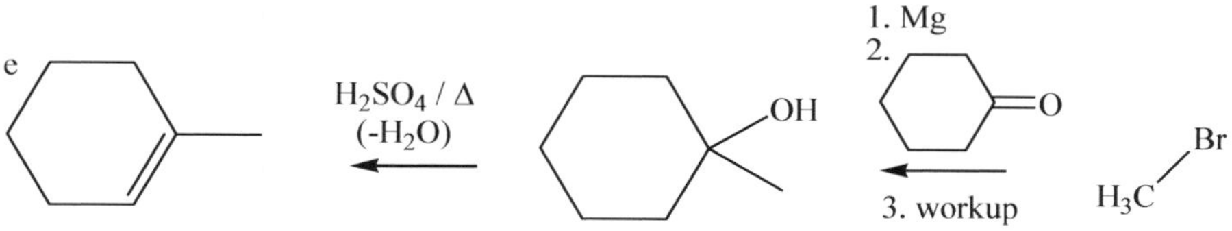

H_2SO_4 / Δ
$(-H_2O)$

OH

1. Mg
2.
3. workup

H_3C—Br

f

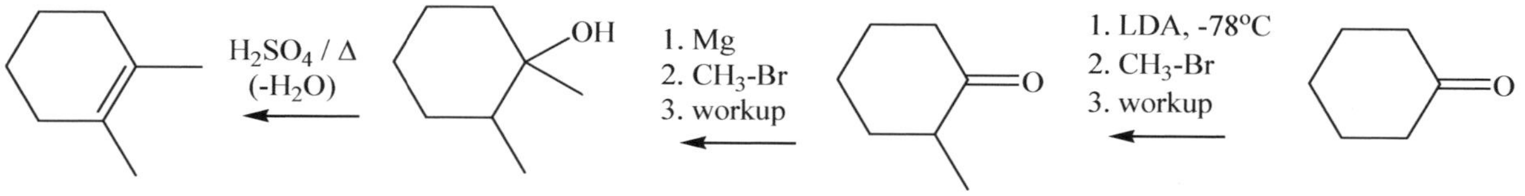

H_2SO_4 / Δ
$(-H_2O)$

OH

1. Mg
2. CH_3-Br
3. workup

O

1. LDA, -78°C
2. CH_3-Br
3. workup

O

g

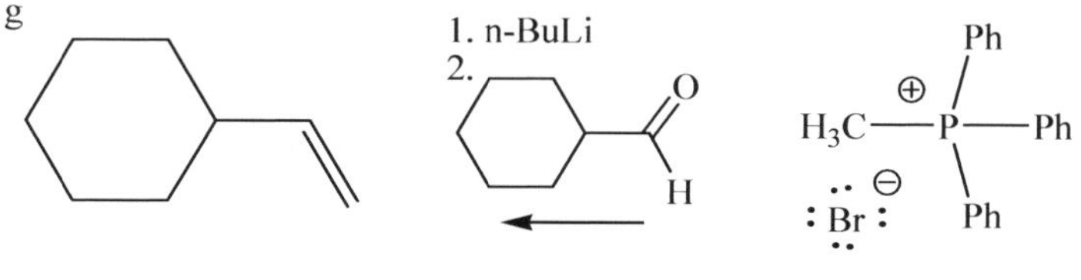

1. n-BuLi
2.

H_3C—P—Ph
Ph
Ph
$:\ddot{Br}:^{\ominus}$

h

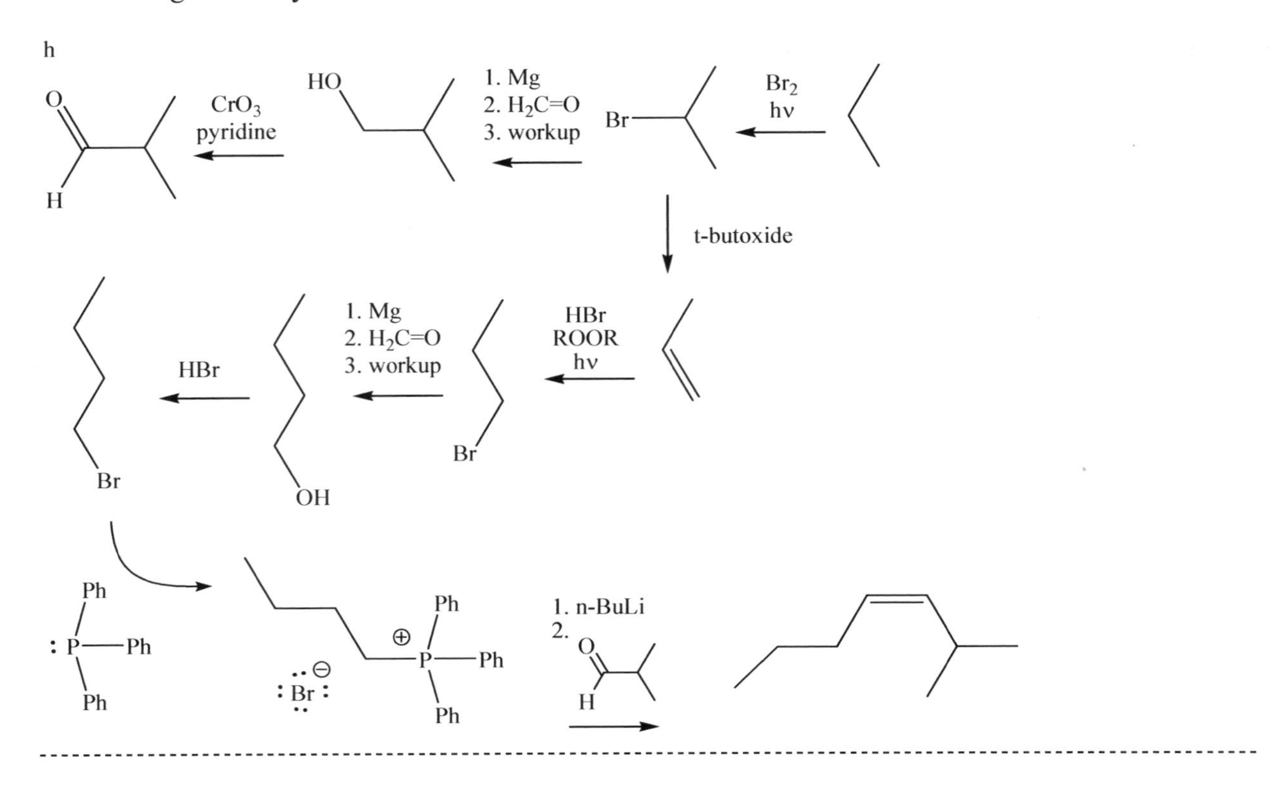

i. same as "h", except add 1 more equivalent of n-BuLi and then mild acid. Schlosser modification of Wittig gives E alkene

Problem 20 (p 394) – Supply the missing starting structure and reagents to prepare the indicated target molecules.

a. Br NaOH → OH

b. Br H_2O → OH

c. Br ⊖O, O → O, O

d. OH 1. NaH 2. CH_3Br → S_N2 O

S_N1 with CH_3OH and 2-bromobutane is also possible

e. Br NaSH → SH

f. SH 1. NaOH 2. Br → S

g. Br O⊖ →

h. Br NaOH →

i. Br 1. NaN_3 2. $LiAlH_4$ 3. workup → NH_2

j. Br → NaCN C≡N

k. Br ⊖ →

l. O 1. LDA 2. Br → O

m. OH HBr → Br

n. OH HBr → Br

o. OH CH_3 CrO_3 H_2O (Jones) → O H H

p. OH CrO_3 pyridine (PCC) → O H

q. OH CrO_3 H_2O (Jones) → O

r. OH CrO_3 H_2O (Jones) → O OH

s. O OH 1. NaOH 2. Br → O O

t. H_3C Br 1. Mg 2. O H 3. workup → OH

u. H_3C Br 1. Mg 2. O 3. workup → OH

v. O H 1. $LiAlH_4$ 2. workup → OH

w. O 1. $NaBH_4$ 2. workup → OH

x. Br 1. Mg 2. CO_2 3. workup → O OH

y. H_3C Br 1. Mg 2. PhCN 3. workup → O

z. C≡N 1. $LiAlH_4$ 2. workup → NH_2

aa. Br 1. $NaBD_4$ 2. workup → D

bb. O H 1. $NaBD_4$ 2. workup → OH D H

cc. O 1. $LiAlD_4$ 2. workup → HO D

dd. C≡N 1. $LiAlD_4$ 2. workup → D D NH_2

ee. OH HBr → Br

ff. OH CrO_3 pyridine (PCC) → O

gg. O H 1. $LiAlD_4$ 2. workup → OH D H

hh. OH CrO_3 H_2O (Jones) → O OH

ii. Br 1. Mg 2. C≡N 3. workup → O

jj. C≡N 1. $LiAlH_4$ 2. workup → NH_2

kk. Br 1. $NaBD_4$ 2. workup → D

ll. Br O⊖ →

mm. OH H_2SO_4 Δ →

Problem 21 (p 395) – Show the necessary synthetic steps to make each functional group below. Most of them are compounds missing from our chemical catalog. The first sequence (a) shows some generic transformations possible. Answers are only given for d in the key, since they are all pretty similar to the generic patterns and cover most of the possibilities.

d.

Start with C_3... Propose synthetic sequence to make...

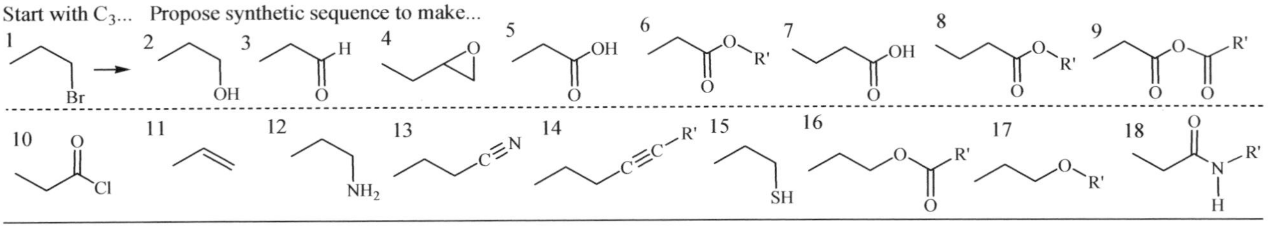

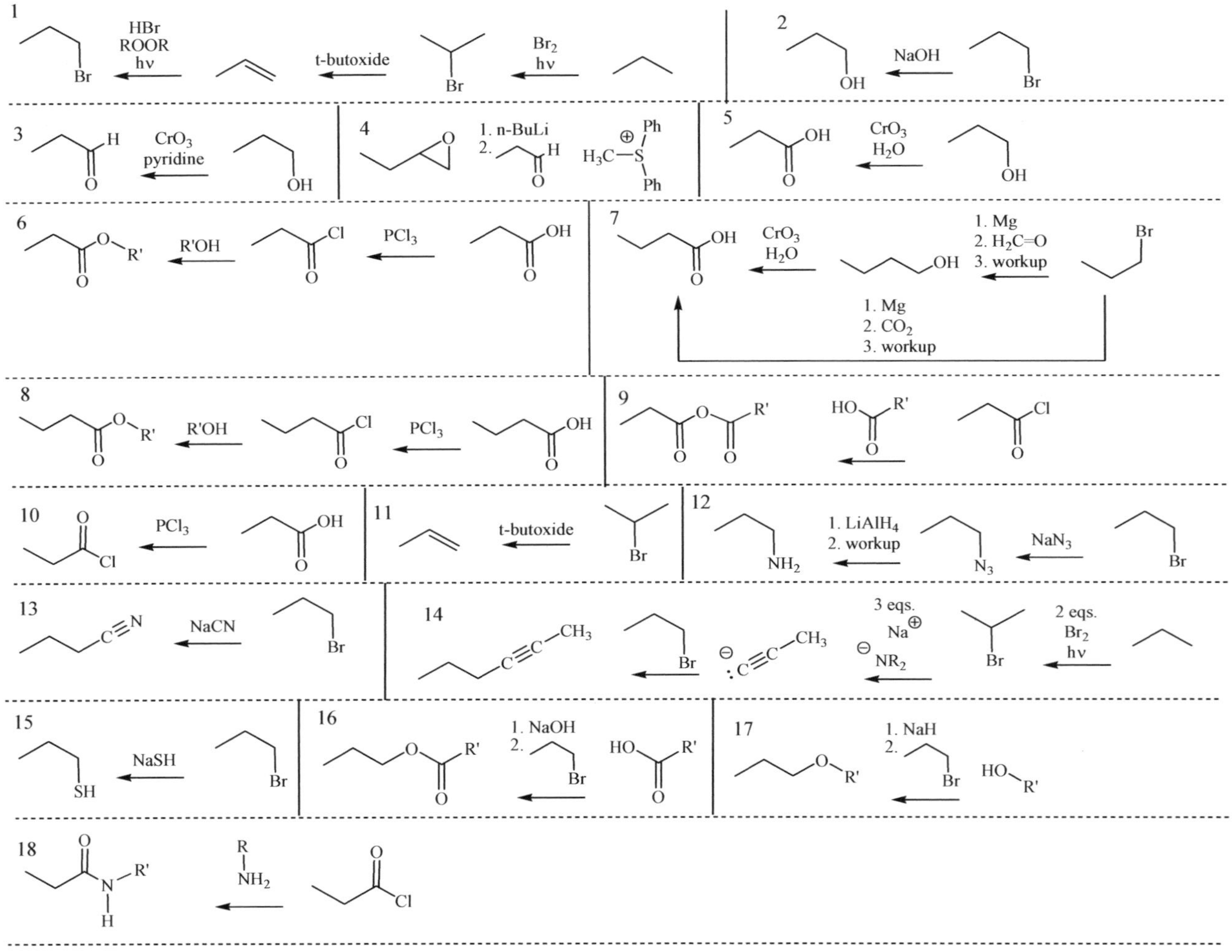

1 Br → 2 OH 3 H O 4 O 5 OH O 6 O R' O 7 OH O 8 O R' O 9 O R' O O

10 O Cl 11 12 NH_2 13 C≡N 14 C≡C R' 15 SH 16 O R' O 17 O R' 18 O N R' H

1 Br ← HBr ROOR hν ← t-butoxide ← Br ← Br_2 hν

2 OH ← NaOH ← Br

3 H O ← CrO_3 pyridine ← OH

4 O ← 1. n-BuLi 2. H O, $H_3C–S^{\oplus}(Ph)_2$

5 OH O ← CrO_3 H_2O ← OH

6 O R' O ← R'OH ← Cl O ← PCl_3 ← OH O

7 OH O ← CrO_3 H_2O ← OH ← 1. Mg 2. $H_2C{=}O$ 3. workup ← Br; Br → 1. Mg 2. CO_2 3. workup → OH O

8 O R' O ← R'OH ← Cl O ← PCl_3 ← OH O

9 O R' O O ← HO R' O + Cl O

10 O Cl ← PCl_3 ← OH O

11 ← t-butoxide ← Br

12 NH_2 ← 1. $LiAlH_4$ 2. workup ← N_3 ← NaN_3 ← Br

13 C≡N ← NaCN ← Br

14 C≡C–CH_3 ← Br + $^{\ominus}$:C≡C–CH_3 ← 3 eqs. $Na^{\oplus}$ $^{\ominus}NR_2$ ← Br ← 2 eqs. Br_2 hν

15 SH ← NaSH ← Br

16 O R' O ← 1. NaOH 2. Br ← HO R' O

17 O R' ← 1. NaH 2. Br ← HO R'

18 O N R' H ← R–NH_2 ← O Cl

List of $C_6H_{13}X$ targets (X = Br, OH, NH_2, SH, SR, CN, CO_2H, etc.)

1 X

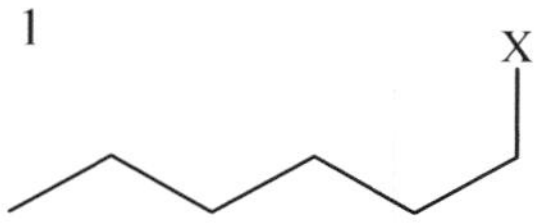

2 X

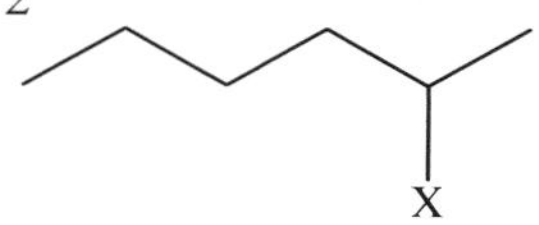

3 X

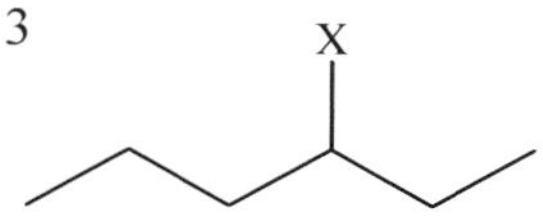

4 X

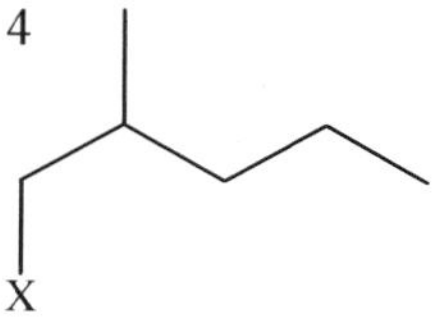

5 X

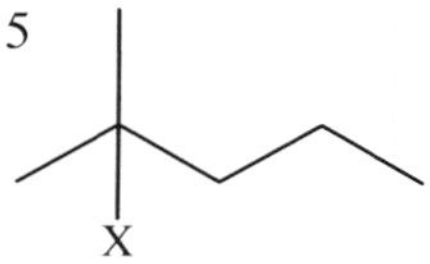

6 X

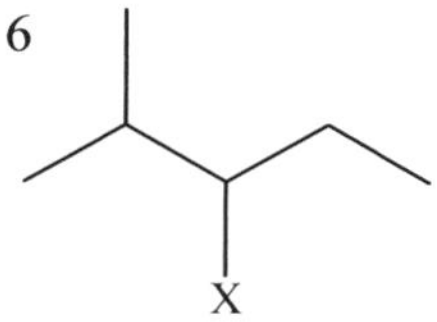

7 X

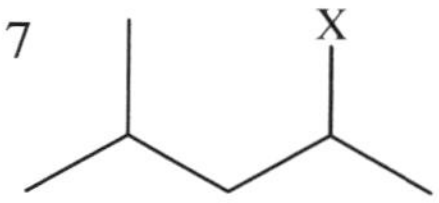

8 X

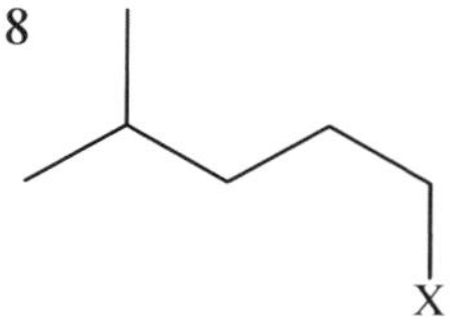

9 X

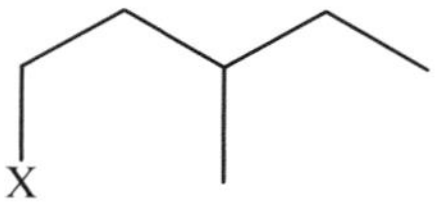

10 X

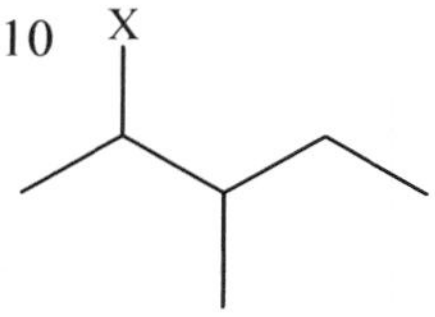

11 X

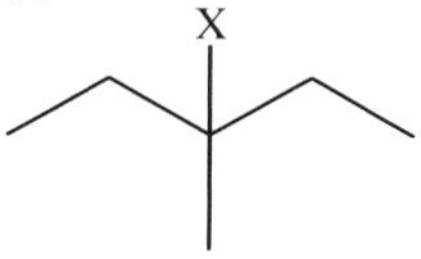

12 X

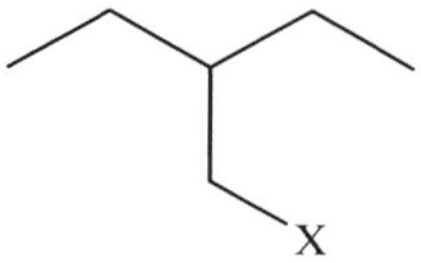

13 X

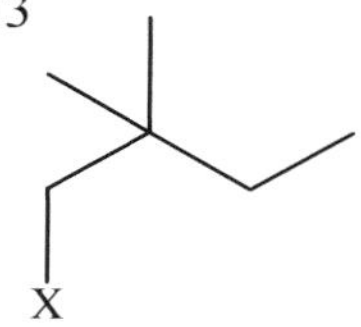

14 X

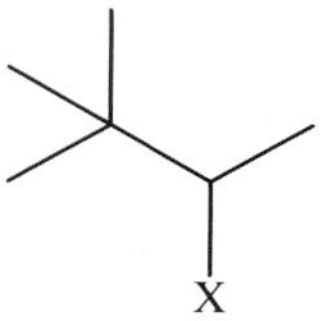

15 X

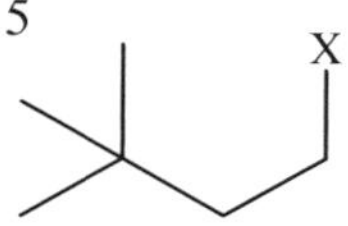

16 X

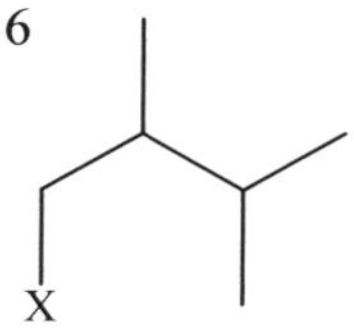

17 X

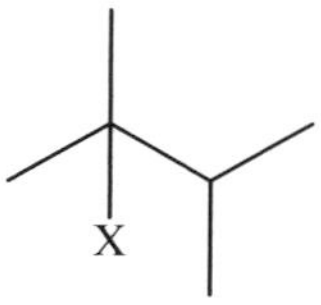

Chapter 13

Problem 1 (p 406) – Specify each C=C pi bond below as electron rich or electron poor or neither?

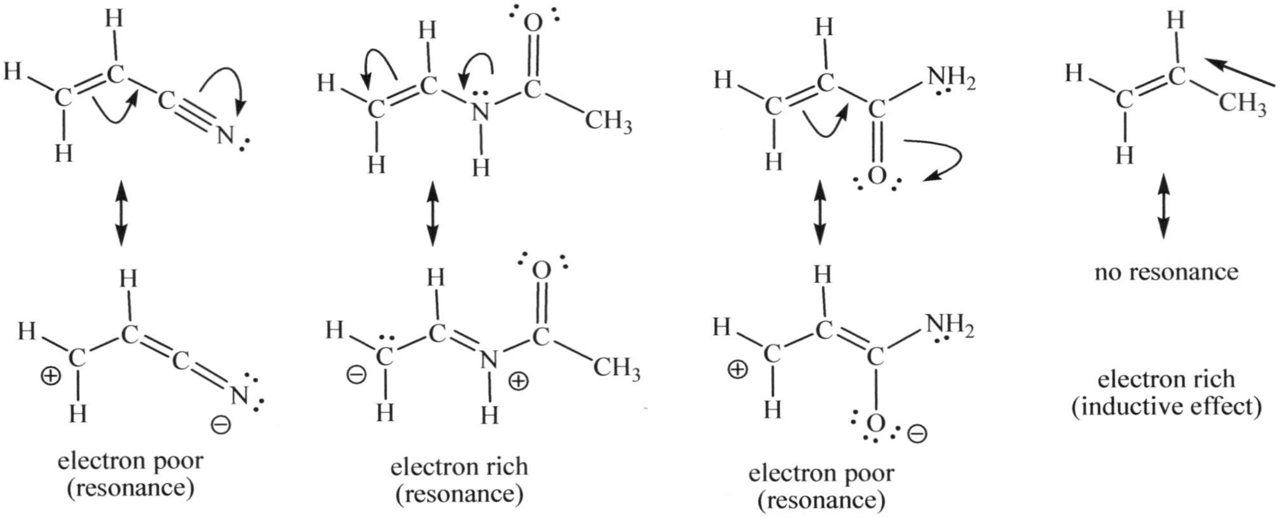

electron poor (resonance)

electron rich (resonance)

electron poor (resonance)

no resonance

electron rich (inductive effect)

Problem 2 (p 409) – What stereochemical relationships exists in the products of the following electrophilic addition reactions?

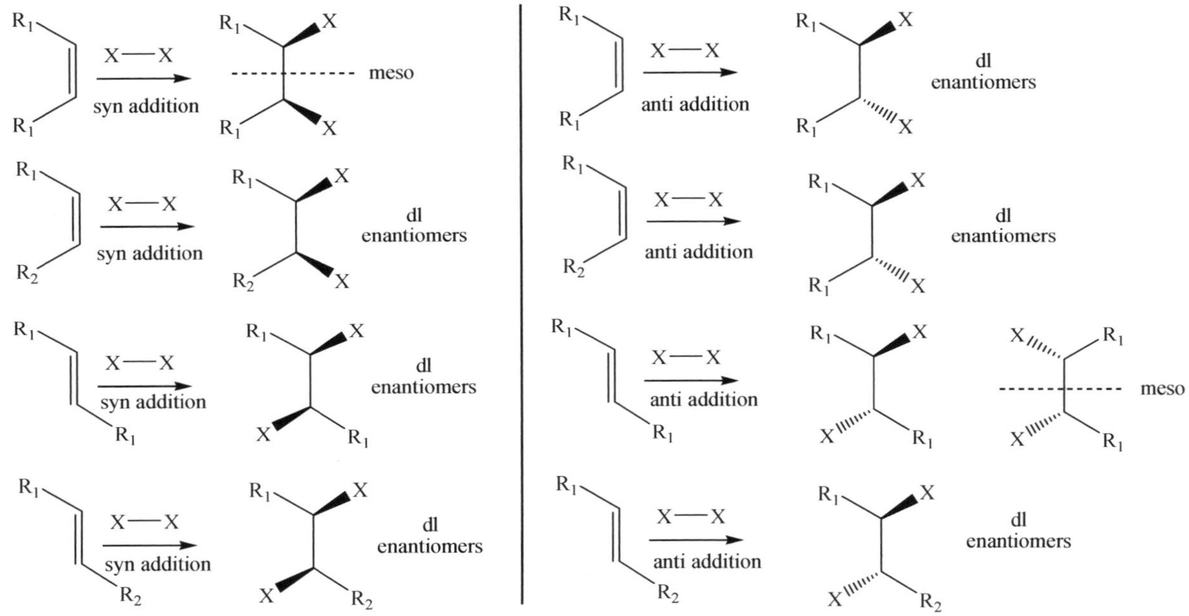

X—X syn addition — meso

X—X anti addition — dl enantiomers

X—X syn addition — dl enantiomers

X—X anti addition — dl enantiomers

X—X syn addition — dl enantiomers

X—X anti addition — meso

X—X syn addition — dl enantiomers

X—X anti addition — dl enantiomers

R_1 / R_1 (cis) → X—X, syn addition → dl enantiomers
R_1 / R_1 (cis) → X—X, anti addition → dl enantiomers

R_1 / R_2 (cis) → X—X, syn addition → dl enantiomers
R_1 / R_2 (cis) → X—X, anti addition → dl enantiomers

R_1 / R_1 (trans) → X—X, syn addition → dl enantiomers
R_1 / R_1 (trans) → X—X, anti addition → dl enantiomers

R_1 / R_2 (trans) → X—X, syn addition → dl enantiomers
R_1 / R_2 (trans) → X—X, anti addition → dl enantiomers

Problem 3 (p 410) – Can regioselectivity be observed in an electrophilic addition to an alkene or alkyne when X = Y?

R 3 2 1 nucleophilic alkenes → electrophilic X—X → Not regioselective?

R 2 1 → electrophilic X—X → Not regioselective?

Problem 4 (p 415)– Point out if any stereoisomers are formed in the following reactions. Indicate if they are enantiomers or diastereomers. If chiral centers are drawn, indicate if the absolute configuration is R or S in your drawings. Could the opposite configurations also form? Are the reactions regioselective and/or stereoselective? D (deuterium) is an isotope of hydrogen and reacts in a similar way, but is observably different.

a.

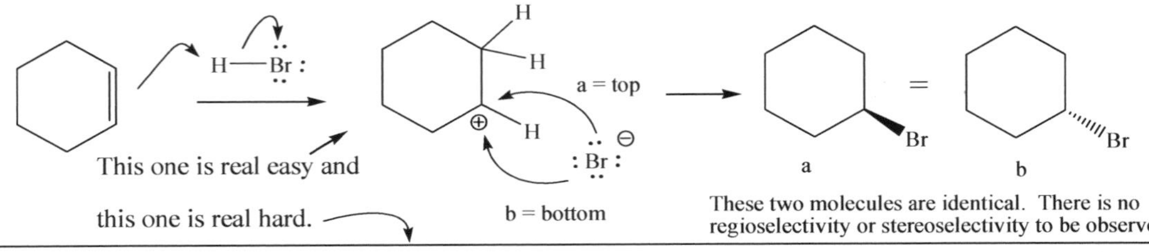

H—Br :

a = top

b = bottom

This one is real easy and

this one is real hard.

a b

These two molecules are identical. There is no regioselectivity or stereoselectivity to be observed.

b. "D" can add to the top carbon or the bottom carbon AND "D" can add to the top face or the bottom face.

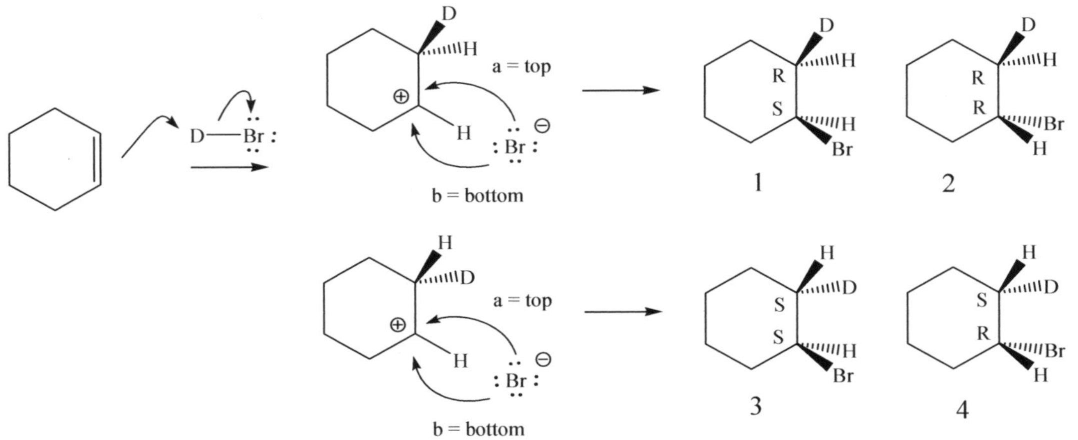

D—Br :

a = top

b = bottom

1 2 3 4

identical molecules = (3,5) (2,8) (1,7) (4,6)
enantiomers = (2,3) (2,5) (3,8) (5,8) (1,4) (1,6) (4,7) (6,7)
diastereomers = (1, 2) (1, 3) (1, 5) (1, 8) and a whole lot more...

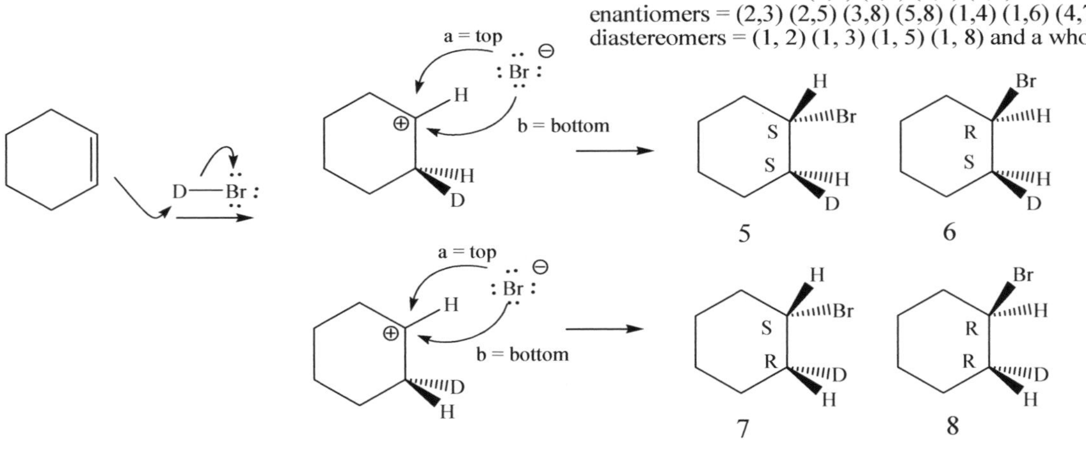

5 6 7 8

Problem 5 (p 417) – Provide a mechanistic explanation for the following reactions. Are there any stereocenters to consider? If so, point out R/S and/or E/Z possibilities.

more stable product because the alkene is more substituted and there is resonance with the pi systems

H—Br :

forms fast

possible intermediate

Br

Br

regioselective to make the most stable carbocation.

: Br :

: Br :

CH_2

C

C

H

CH_3

rearrangement

resonance

(lots of resonance with the aromatic ring and the alkene)

Problem 6 (p 417) – Explain the regioselectivity in the following reaction. Is this reasonable considering the triple bond is substituted on both ends? Was the reaction stereoselective, as shown? If so, was the addition syn, anti or random? (How did that chlorine sneak in here?)

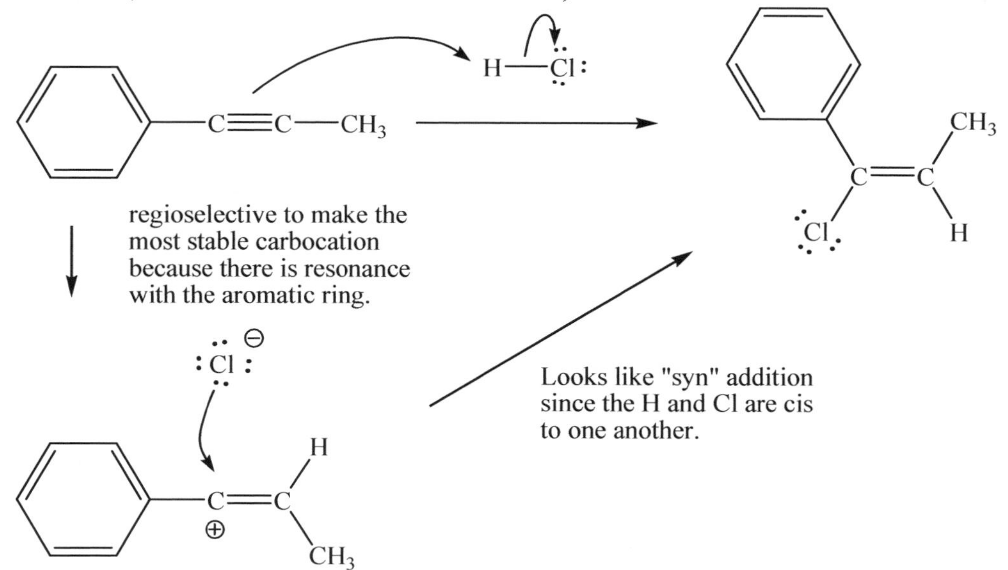
H—Cl :
C≡C—CH3
CH3
C=C
Cl
H
regioselective to make the most stable carbocation because there is resonance with the aromatic ring.
: Cl :
H
C=C
CH3
Looks like "syn" addition since the H and Cl are cis to one another.

(p 418-420)**10 Alkenes to practice on** (This isn't much different than adding H-Br.)

1. Ethane cannot show regioselectivity or stereoselectivity with any of our reagents. A complete mechanism is provided.

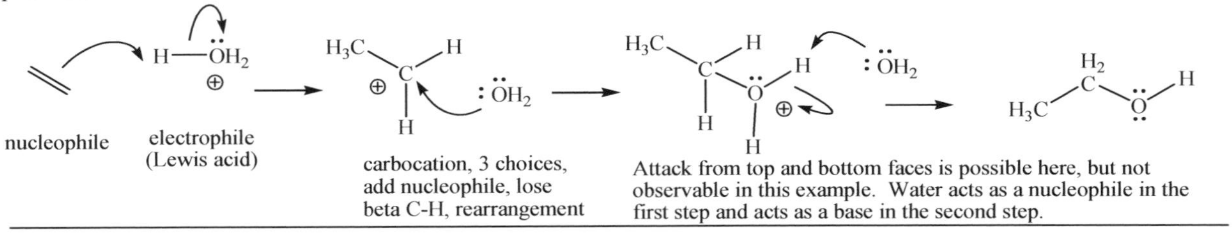

nucleophile | electrophile (Lewis acid)

carbocation, 3 choices, add nucleophile, lose beta C-H, rearrangement

Attack from top and bottom faces is possible here, but not observable in this example. Water acts as a nucleophile in the first step and acts as a base in the second step.

2. Propene can show regioselectivity, but cannot show stereoselectivity with any of our reagents. Add in any necessary mechanistic details.

nucleophile | electrophile (Lewis acid)

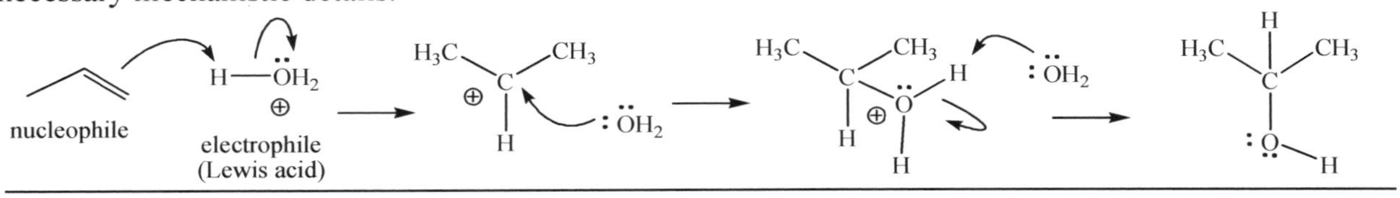

3. 2-methylprop-1-ene can show regioselectivity, but cannot show stereoselectivity with any of our reagents. Add in any missing mechanistic details. Add in any necessary mechanistic details.

nucleophile | electrophile (Lewis acid)

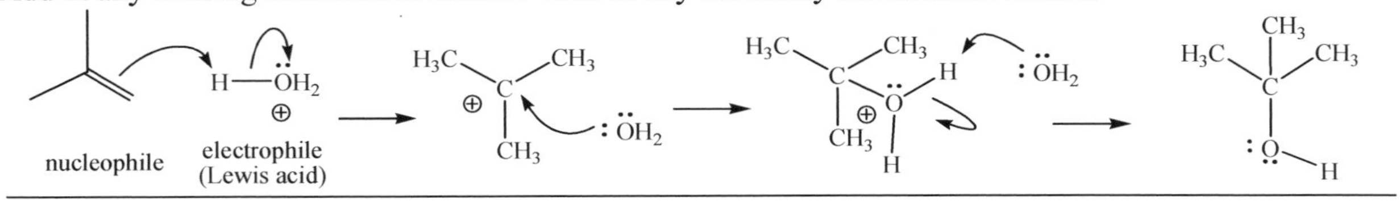

4. trans-but-2-ene cannot show regioselectivity, but can show stereoselectivity, but not with HBr. Write a complete mechanism.

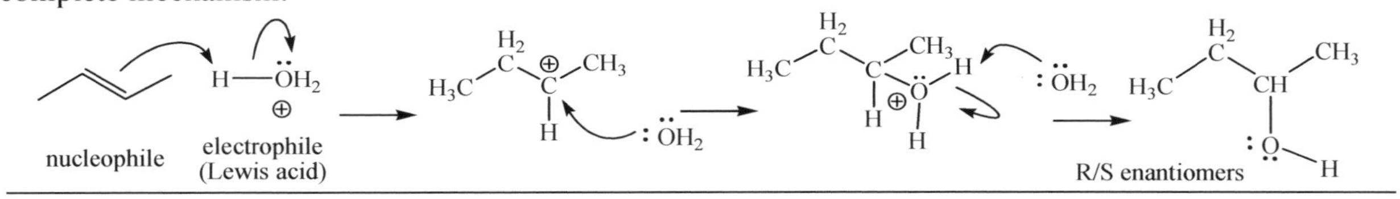

nucleophile | electrophile (Lewis acid)

R/S enantiomers

5. Cyclohexene cannot show regioselectivity, but can show stereoselectivity with any of our reagents. Write in a complete mechanism.

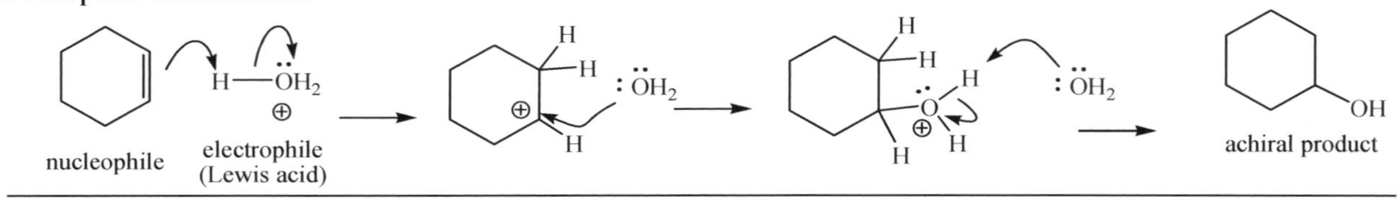

nucleophile | electrophile (Lewis acid)

achiral product

6. 3-methylbut-1-ene can show regioselectivity, but does cannot show stereoselectivity with any of our reagents. This example will also rearrange whenever carbocations form. Add in any missing mechanistic details.

nucleophile | electrophile (Lewis acid) | rearrangement | achiral product

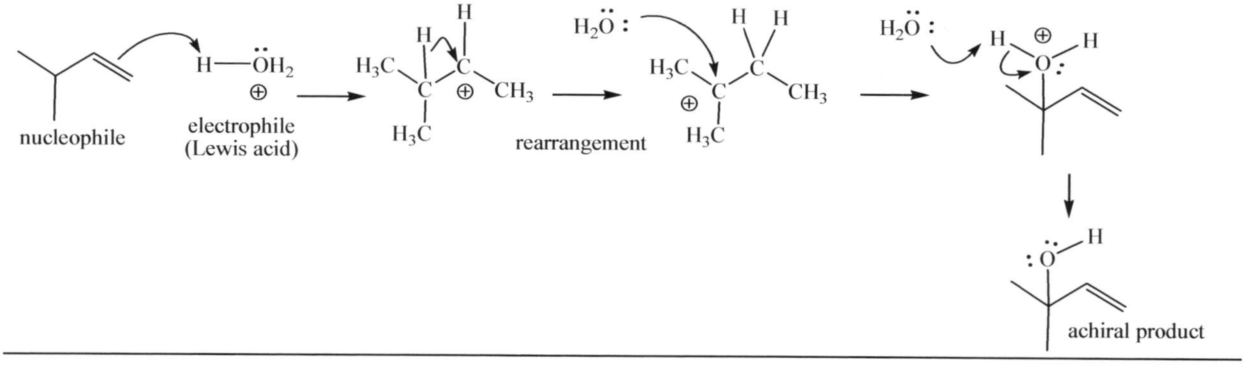

7. Styrene (vinylbenzene, ethenylbenzene) can show regioselectivity, but cannot show stereoselectivity with any of our reagents. Add in any necessary mechanistic details.

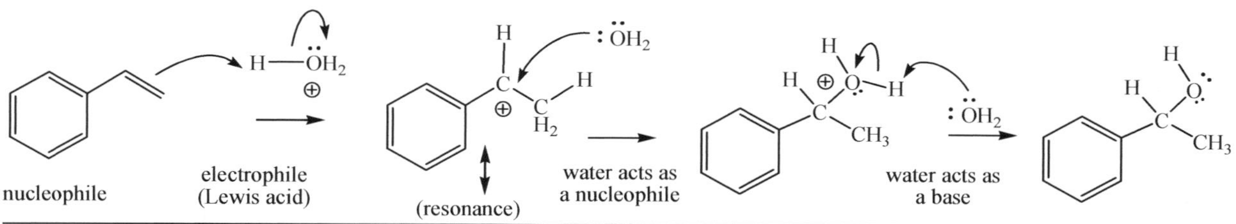

nucleophile | electrophile (Lewis acid) | (resonance) | water acts as a nucleophile | water acts as a base

8. 3-methylcyclohex-1-ene is good for showing when rearrangements are possible. Add in any missing mechanistic details.

nucleophile | electrophile (Lewis acid) | rearrangement

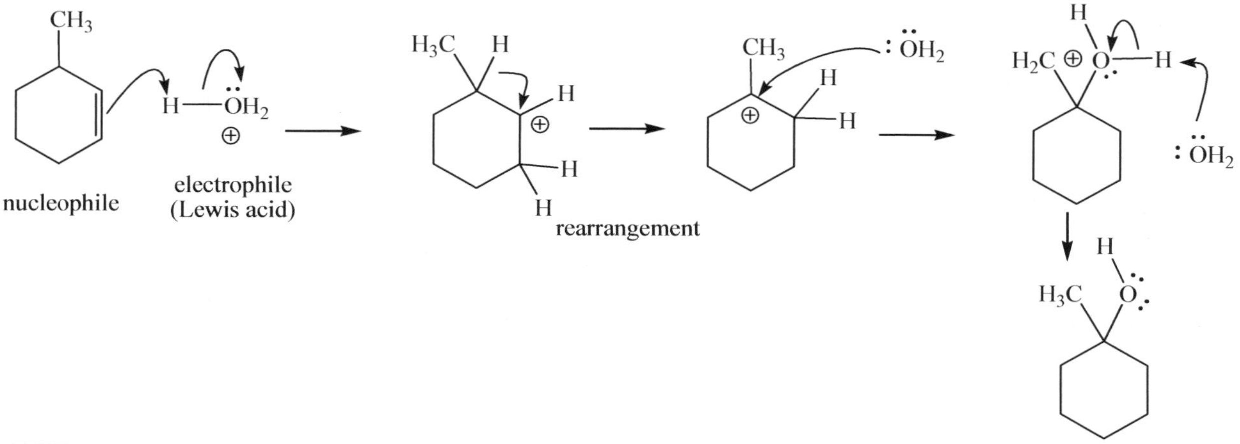

9. 1-methylcyclohex-1-ene can show regioselectivity and can show stereoselectivity depending on the reagents reacting with it. Write a complete mechanism.

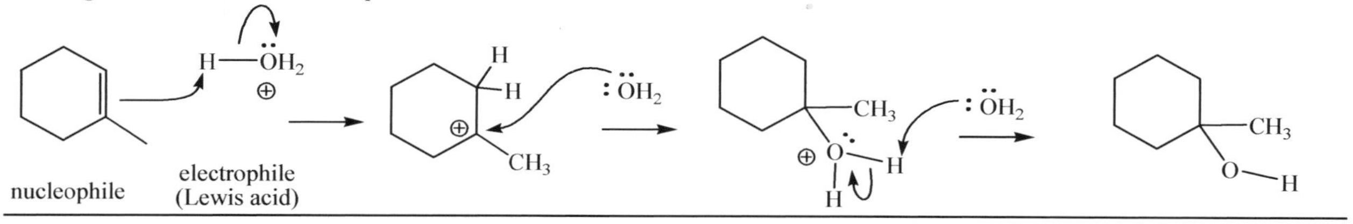

nucleophile | electrophile (Lewis acid)

10. 3-methyl-4-phenylpent-1-ene is good for showing complicated rearrangement possibilities. Add in any missing mechanistic details.

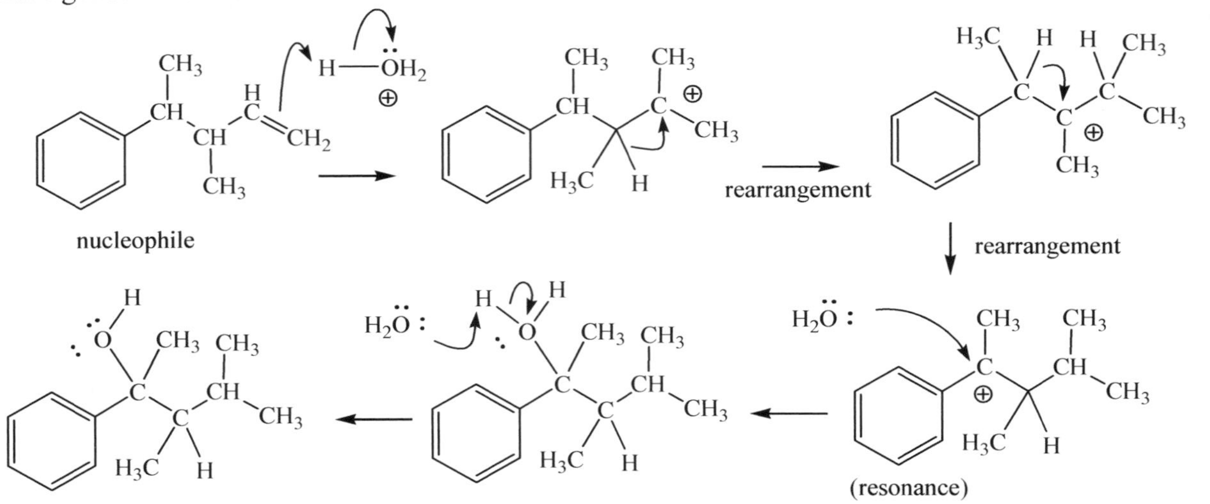

nucleophile

rearrangement

rearrangement

(resonance)

Problem 7 (p 421) – Explain the following observation. Include a complete mechanism.

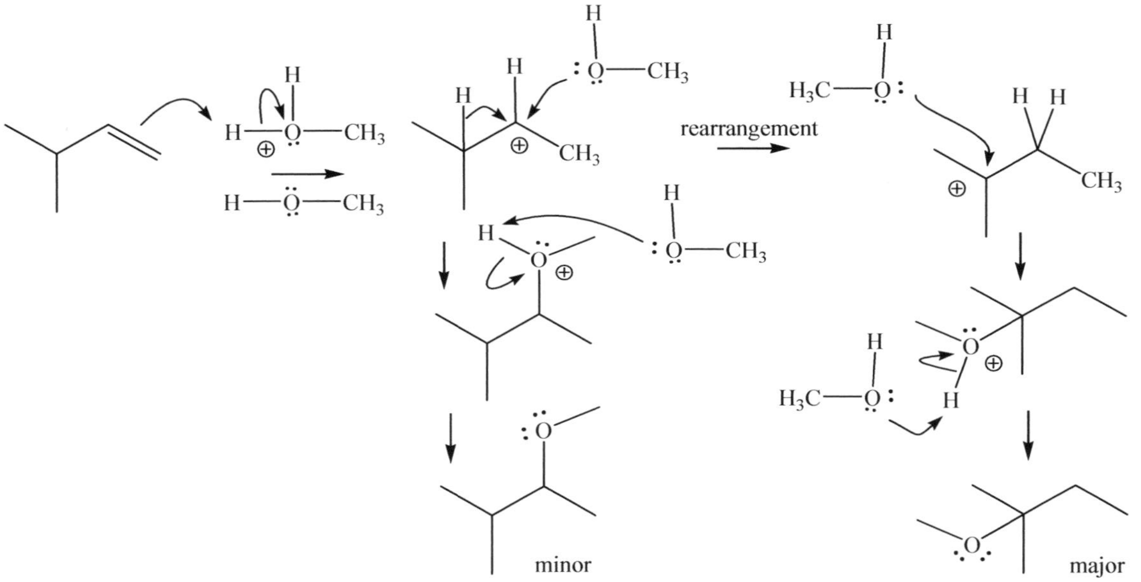

rearrangement

minor

major

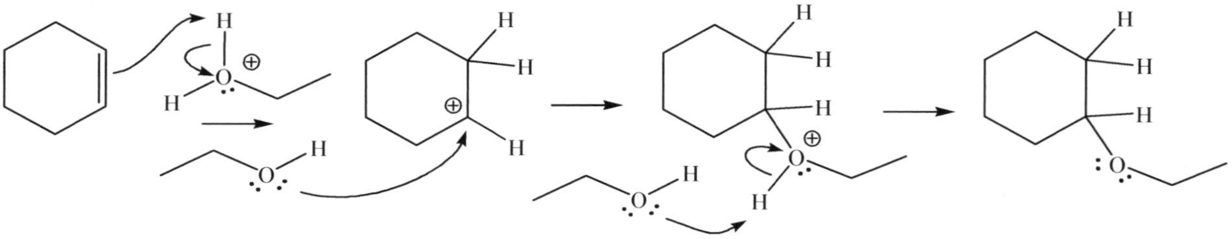

Problem 8 (p 422) – Point out if any stereoisomers are formed in the following reactions. Indicate if they are enantiomers or diastereomers or neither. If chiral centers are drawn, indicate if the absolute configuration is R or S in your drawings. Could the opposite configurations also form? Are the reactions regioselective and/or stereoselective?

a.

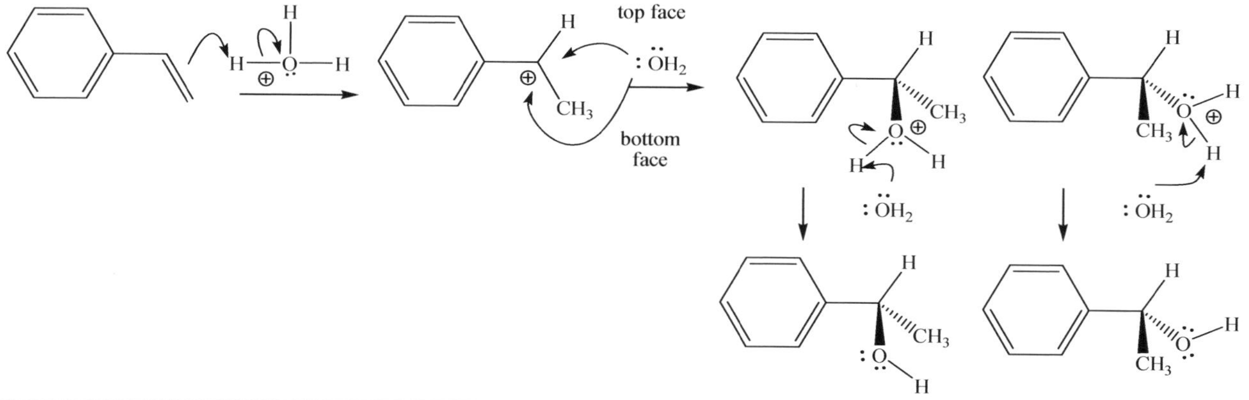

H H–O–H ⊕

H ⊕ CH_3

top face

: $\ddot{O}H_2$

bottom face

H CH_3 O ⊕ H H

: $\ddot{O}H_2$

H CH_3 O ⊕ H H

: $\ddot{O}H_2$

H CH_3 : O H

H CH_3 O H

b

D D–O–D ⊕

regioselectivity forms most stable carbocation due to resonance with aromatic ring

top face

: $\ddot{O}D_2$

bottom face

H ⊕ S C–CH_3 H D

1. add water
2. lose proton

top face

: $\ddot{O}D_2$

bottom face

H ⊕ R C–CH_3 D H

1. add water
2. lose proton

1 OD H S S C–CH_3 H D

3 OD H S R C–CH_3 D H

enantiomers

2 H OD R S C–CH_3 H D

4 H OD R R C–CH_3 D H

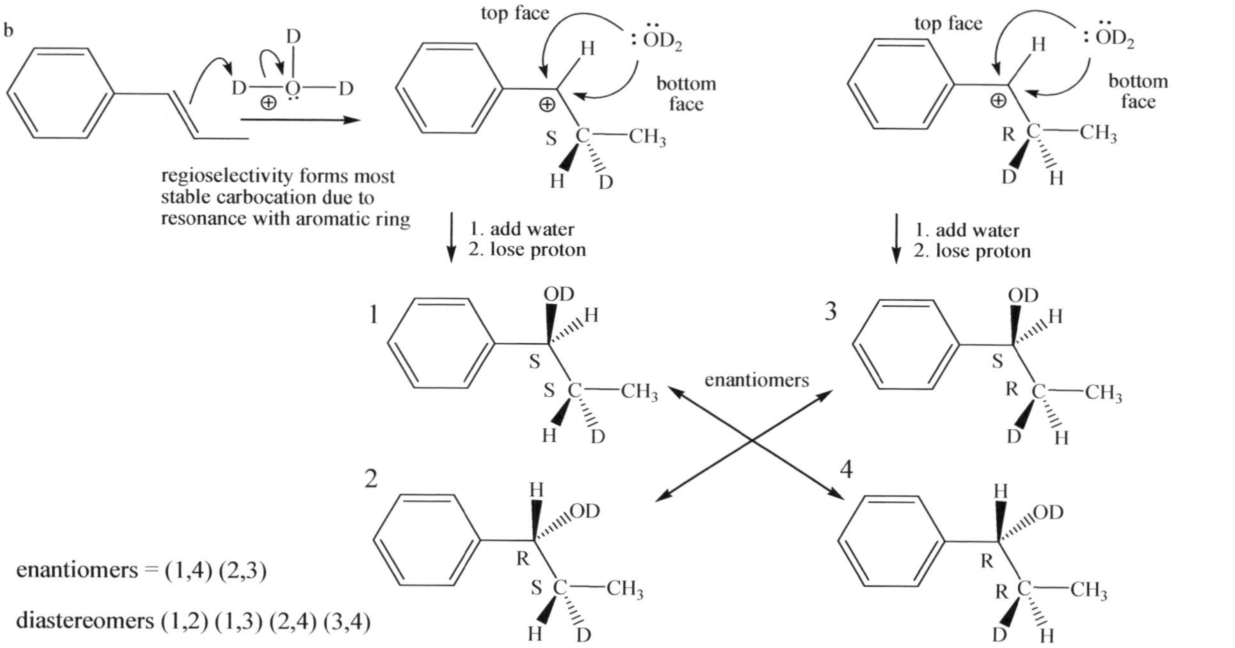

enantiomers = (1,4) (2,3)

diastereomers (1,2) (1,3) (2,4) (3,4)

Problem 9 (p 422) – Provide an arrow pushing mechanism for the following transformation. Hint: Find the extra proton to decide where the reaction began, then find the carbon-carbon bonds that formed, then where the nucleophile added. A more complicated version of this reaction occurs every time your body makes cholesterol (that's a 24/7 reaction).

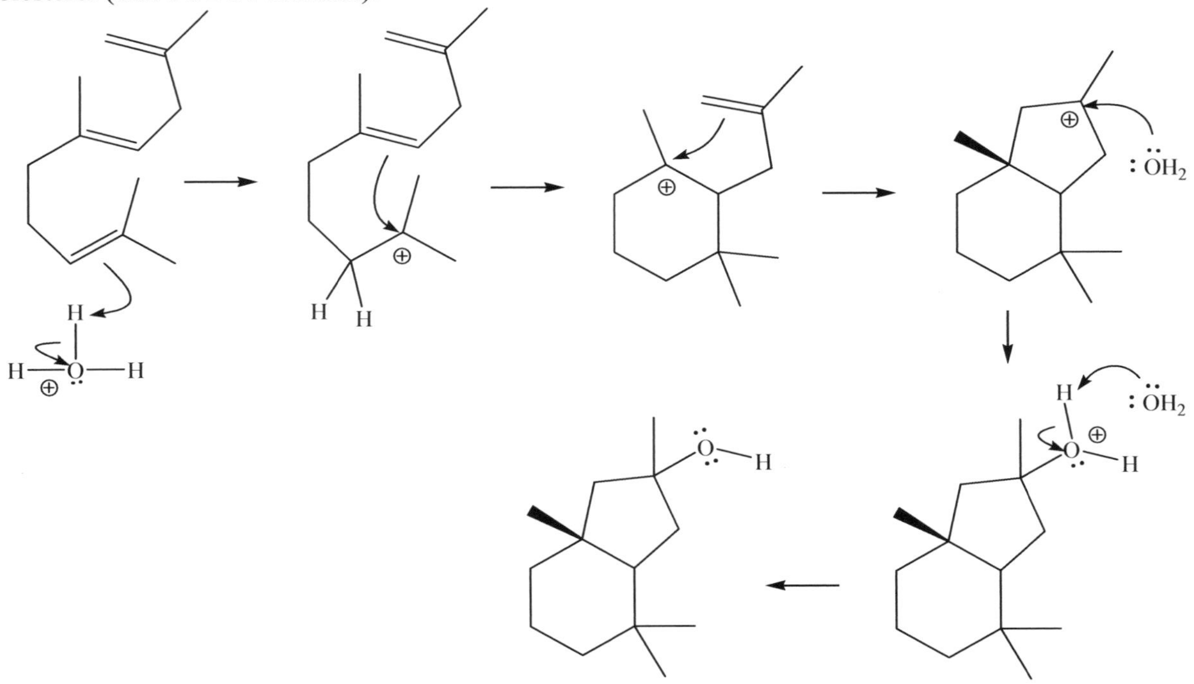

Problem 10 (p 422) – Dihydropyran (DHP) is a special alkene-ether (called an enol ether). It adds alcohols with high regioselectivity (as shown) and is, kinetically, a fast reaction. It is also used to protect alcohols (disguised as ethers), while a different reaction is run that normally would react with an alcohol group (called a protecting group). Explain the regioselectivity and speed of the reaction. Hint: An oxygen lone pair of the enol ether is important to your answer. Provide an arrow pushing mechanism.

dihydropyran

toluene sulfonic acid

alcohol

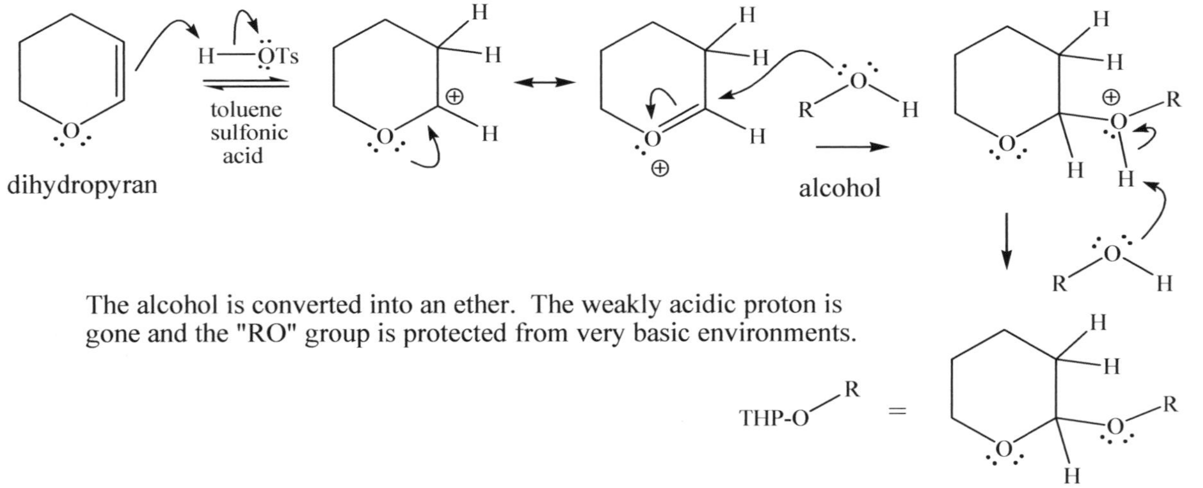

The alcohol is converted into an ether. The weakly acidic proton is gone and the "RO" group is protected from very basic environments.

THP-O–R =

Problem 11 (p 423) – The following are tautomer equilibria problems, without the mechanistic details first and with the mechanistic details next (so you can check your arrow pushing). Supply all of the mechanistic details in both directions, using H_3O^+/H_2O conditions or H_2O/HO^- conditions, as specified. There are four potential problems with each set of equilibrium arrows. (keto to enol in acid, keto to enol in base, enol to keto in acid and enol to keto in base). All of the reactions follow a similar pattern: 1. start off with proton transfer, 2. resonance intermediates, and 3. finish with proton transfer. If the first step puts on a proton (using strong acid, H_3O^+), the last step will take off a proton (using H_2O as the base). If the first step takes off a proton (using strong base, HO^-), the last step will put on a proton (always using H_2O as the acid). Water is always the weak partner (conjugate) to the stronger component (acid or base).

Mechanism steps with the details (curved arrows show proton transfers and resonance).

Tautomers in acid

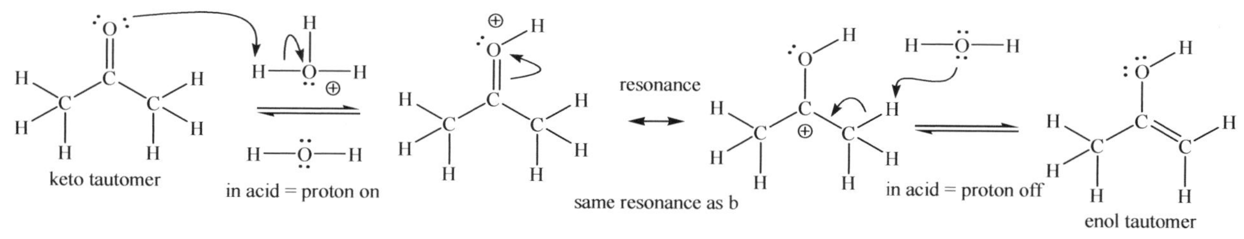

keto tautomer — in acid = proton on — resonance — same resonance as b — in acid = proton off — enol tautomer

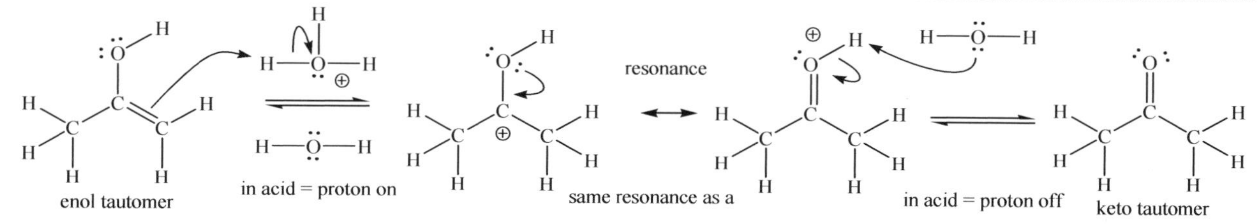

enol tautomer — in acid = proton on — resonance — same resonance as a — in acid = proton off — keto tautomer

Tautomers in base

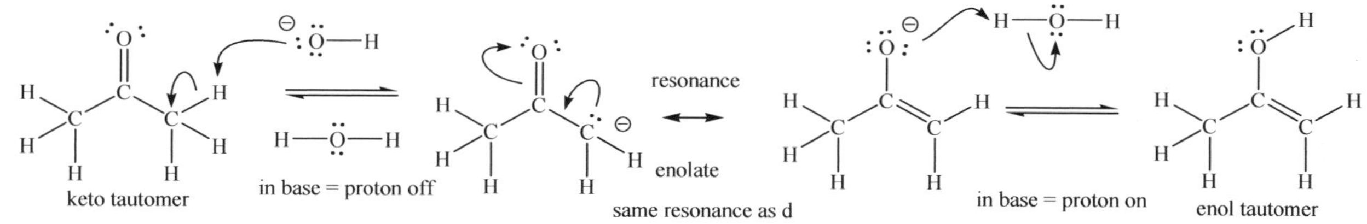

keto tautomer — in base = proton off — resonance — enolate — same resonance as d — in base = proton on — enol tautomer

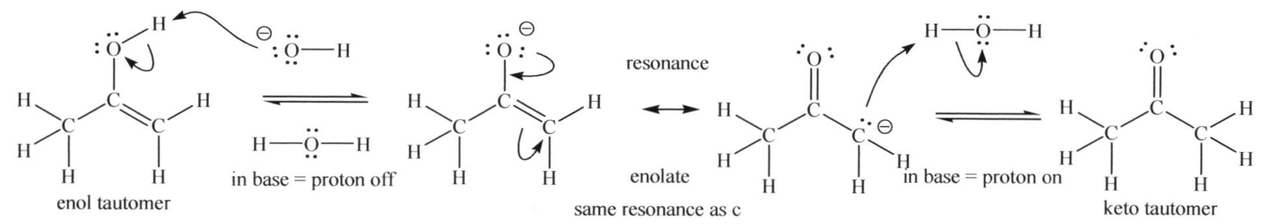

enol tautomer — in base = proton off — resonance — enolate — same resonance as c — in base = proton on — keto tautomer

Problem 12 (p 425) – Most students are familiar with glycolysis from biology classes. There are two steps in glycolysis that involve double tautomer reactions. The first is the transformation of glucose to fructose at the beginning and the second is an equilibrium situation between glyceraldehyde and dihydroxy acetone in the middle. Supply the necessary mechanistic details to show these transformations. Stereochemistry features have been omitted. Use B: as the base and B^+-H as the acid for proton transfers.

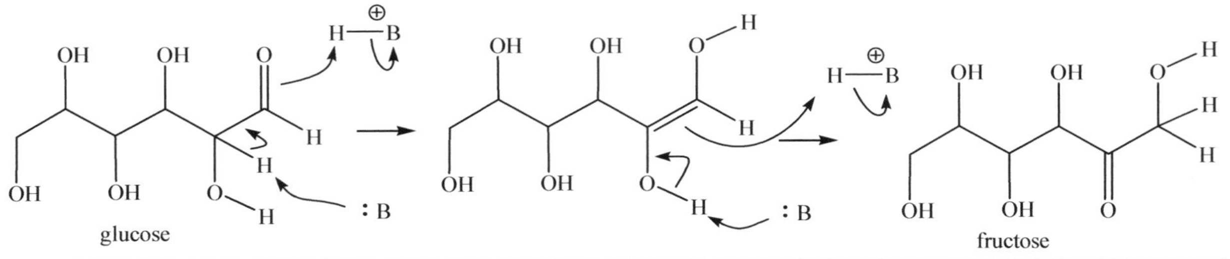

glucose → fructose

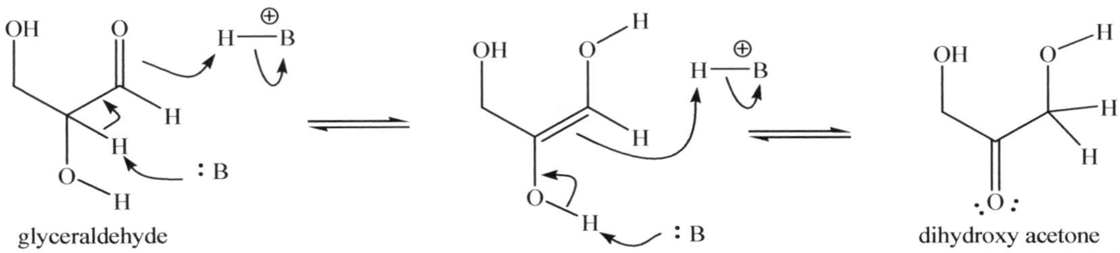

glyceraldehyde ⇌ dihydroxy acetone

Problem 13 (p 428-430) – Do the following pi systems allow one to observe anti addition (as opposed to syn addition or random addition)? Write the products, clearly showing stereochemical features. Indicate if products are enantiomers, diastereomers, meso compounds or achiral structures.

10 Alkenes to practice on

1. Ethane cannot show regioselectivity or stereoselectivity with any of our reagents.

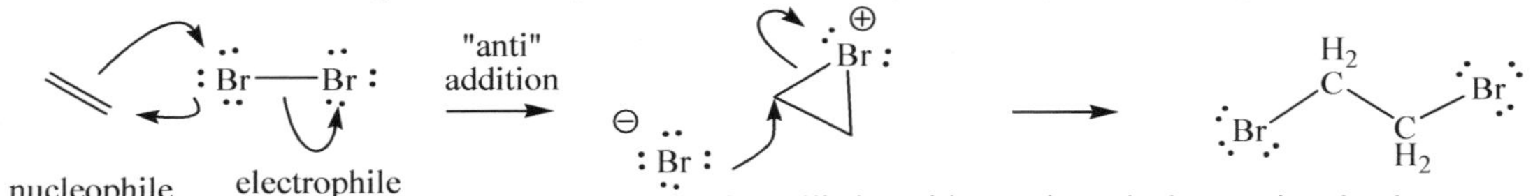

nucleophile — electrophile (Lewis acid) — "anti" addition

Nucleophilic bromide attack on the bromonium ion is anti. Attack at either carbon leads to the same product.

2. Propene cannot show regioselectivity or stereoselectivity with bromine addition.

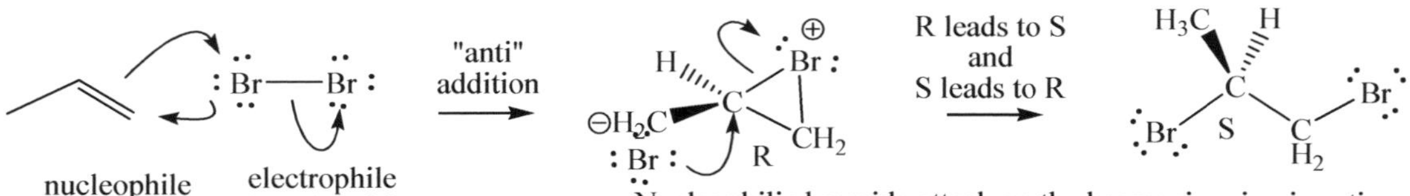

nucleophile — electrophile (Lewis acid) — "anti" addition — R leads to S and S leads to R

Nucleophilic bromide attack on the bromonium ion is anti. Attack at the more partial positive carbon can occur with either R or S absolute configuration since a chiral center forms in the bridging intermediate leading to enantiomers.

3. 2-methylprop-1-ene cannot show regioselectivity or stereoselectivity with bromine addition. Add in any missing mechanistic details.

nucleophile

electrophile (Lewis acid)

"anti" addition

Nucleophilic bromide attack on the bromonium ion is anti. Attack at the more partial positive carbon can only occur from the anti face. In this case the product is achiral because the two methyl groups are identical

4. trans-but-2-ene cannot show regioselectivity, but can show stereoselectivity with bromine addition. Add in any necessary mechanistic details.

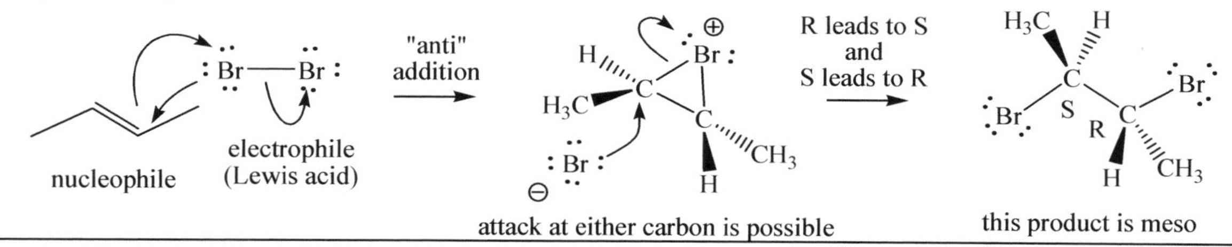

5. Cyclohexene cannot show regioselectivity, but can show stereoselectivity with any of our reagents. Write in a complete mechanism. Supply the necessary mechanistic details.

nucleophile

electrophile (Lewis acid)

from the bromonium bridge

An enantiomeric pair of stereoisomers is obtained (SS) and (RR).

6. 3-methylbut-1-ene cannot show regioselectivity or stereoselectivity with bromine addition. Normally we would expect rearrangement with this molecule, but the bromonium bridge prevents rearrangement. Write a complete mechanism.

nucleophile

electrophile (Lewis acid)

"anti" addition

R leads to S and S leads to R

S and R enantiomers

7. Styrene (vinylbenzene, ethenylbenzene) cannot show regioselectivity or stereoselectivity with bromine addition. Write a complete mechanism.

"anti" addition

nucleophile

electrophile (Lewis acid)

(resonance)

enantiomers are obtained

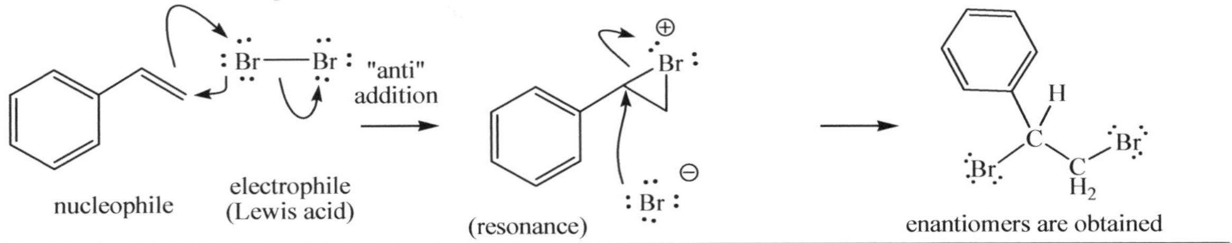

8. 3-methylcyclohex-1-ene cannot show regioselectivity or stereoselectivity with bromine addition. Normally we would expect rearrangement with this molecule, but the bromonium bridge prevents rearrangement.

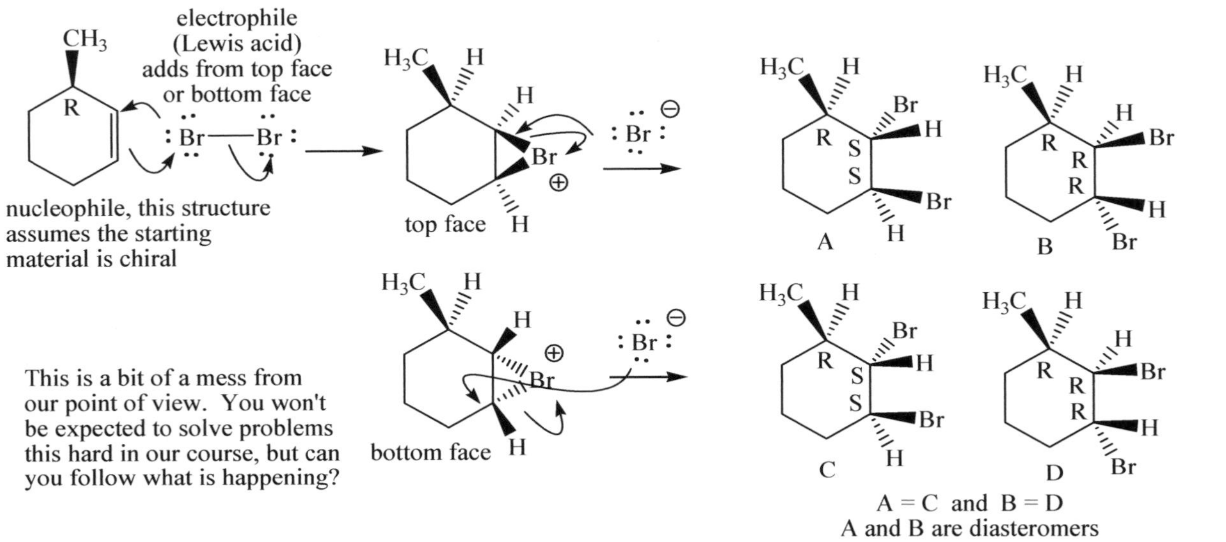

electrophile (Lewis acid) adds from top face or bottom face

nucleophile, this structure assumes the starting material is chiral

top face

bottom face

This is a bit of a mess from our point of view. You won't be expected to solve problems this hard in our course, but can you follow what is happening?

A = C and B = D

A and B are diasteromers

9. 1-methylcyclohex-1-ene can show regioselectivity and can show stereoselectivity with any of our reagents. Write a complete mechanism.

nucleophile

electrophile (Lewis acid)

from the bromonium bridge

An enantiomeric pair of stereoisomers is obtained (SS) and (RR).

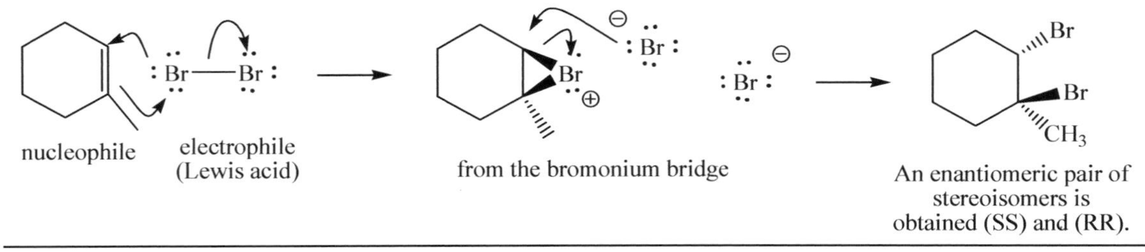

10. 3-methyl-4-phenylpent-1-ene cannot show regioselectivity or stereoselectivity with bromine addition. Normally we would expect rearrangement with this molecule, but the bromonium bridge prevents rearrangement. Add in any necessary mechanistic details.

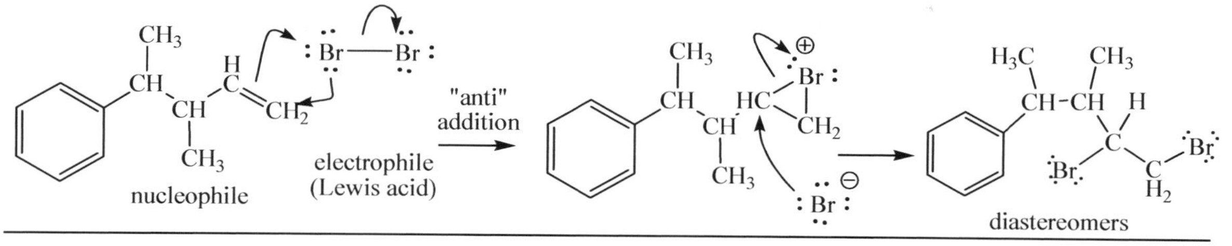

nucleophile — electrophile (Lewis acid) — "anti" addition — diastereomers

Problem 14 (p 431)– Fill in the expected products for each reaction of the indicated sequence.

a. Br, $K^{\oplus}$ tert-butoxide → propene; Br_2 → dibromide; 1. excess $Na^{\oplus}\ NR_2^{\ominus}$ 2. H_3O^+ workup → H-terminal alkyne

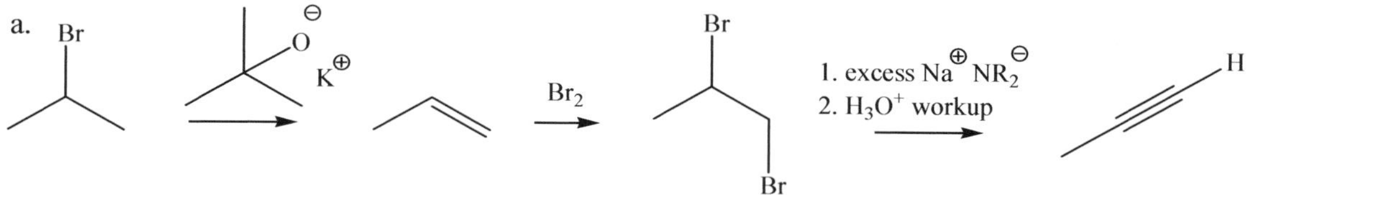

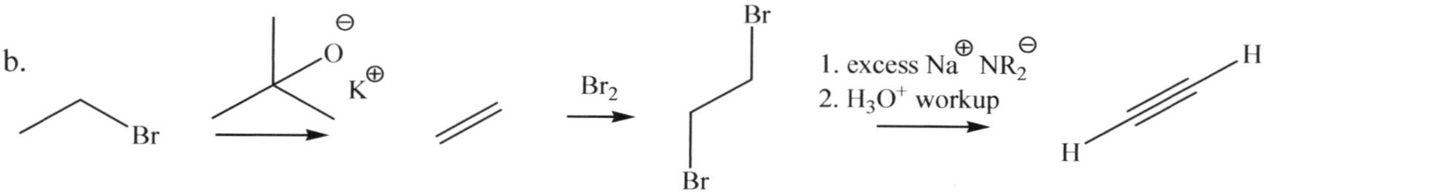

b. Br, $K^{\oplus}$ tert-butoxide → ethene; Br_2 → dibromide; 1. excess $Na^{\oplus}\ NR_2^{\ominus}$ 2. H_3O^+ workup → H–C≡C–H

c. Br, $K^{\oplus}$ tert-butoxide → styrene; Br_2 → dibromide; 1. excess $Na^{\oplus}\ NR_2^{\ominus}$ 2. H_3O^+ workup → phenylacetylene (H)

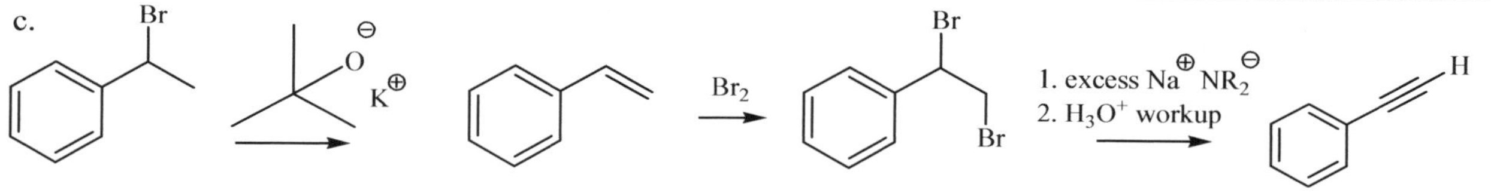

d. OH; H_2SO_4 / Δ → propene; Br_2 → dibromide; 1. excess $Na^{\oplus}\ NR_2^{\ominus}$ 2. epoxide (O) 3. H_3O^+ workup → alkyne-OH

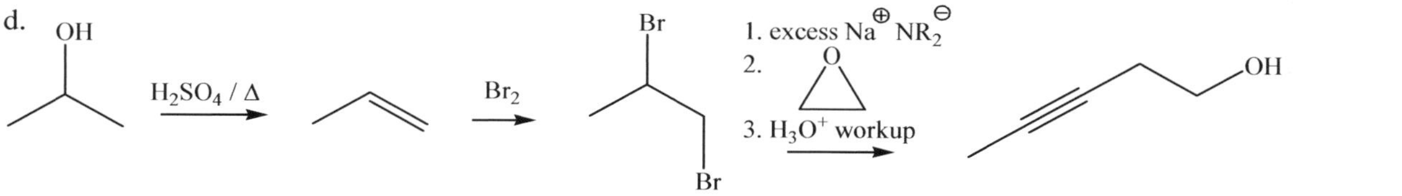

e. O–H; H_2SO_4/Δ → propene; Br_2 → dibromide; 1. excess $Na^{\oplus}\ NR_2^{\ominus}$ 2. $H_2C{=}O$ 3. H_3O^+ workup → alkyne-OH

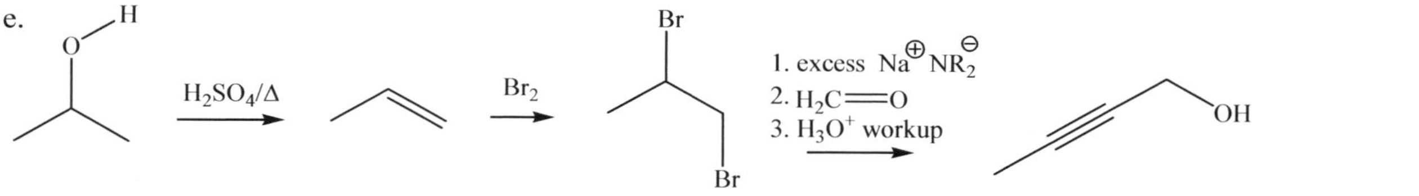

f. Don't forget free radical substitution using alkanes.

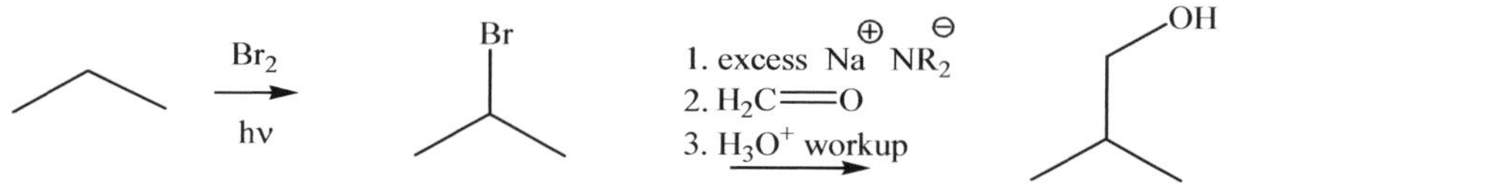

propane; Br_2 / hν → 2-bromopropane (Br); 1. excess $Na^{\oplus}\ NR_2^{\ominus}$ 2. $H_2C{=}O$ 3. H_3O^+ workup → OH

Problem 15 (p 432) – Bromine and water addition to methylcyclohexene produces a product having two favored chair conformations. Which chair conformation is "most" preferred and why?

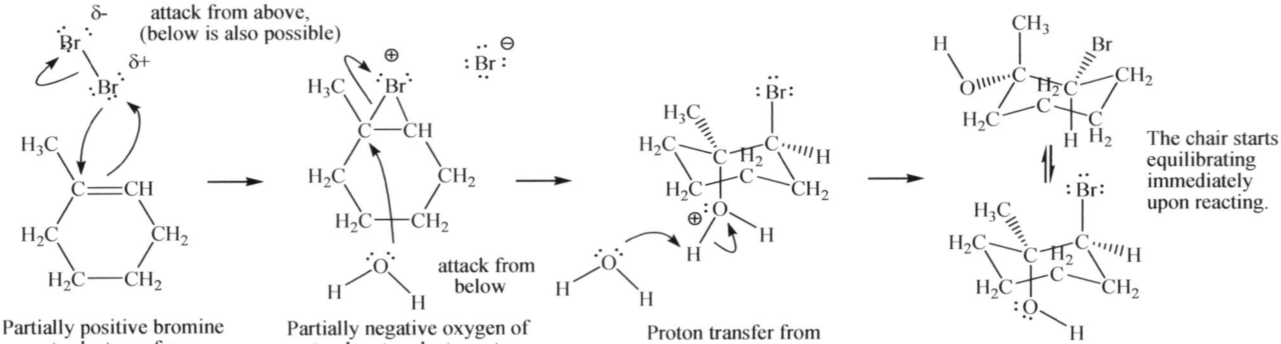

δ- attack from above, (below is also possible) δ+

Partially positive bromine accepts electrons from alkene pi bond and shares its lone pair to make a bridge.

attack from below

Partially negative oxygen of water donates electrons to more partially positive tertiary carbon over secondary carbon, and attacks from the opposite side of the bromine bridge.

Proton transfer from trivalent oxygen is required to the best base available, water.

The chair starts equilibrating immediately upon reacting.

This result shows both the regioselectivity, and the "anti" stereoselectivity (trans product).

Problem 16 (p 432-434) – Do the following pi systems allow one to observe anti addition (as opposed to syn addition or random addition)? Write the products, clearly showing stereochemical features. Indicate if products are enantiomers, diastereomers, meso compounds or achiral structures.

<u>Ten alkenes to practice on</u> (Br_2 / H_2O addition to alkenes)

1. Ethane cannot show regioselectivity or stereoselectivity with any of our reagents.

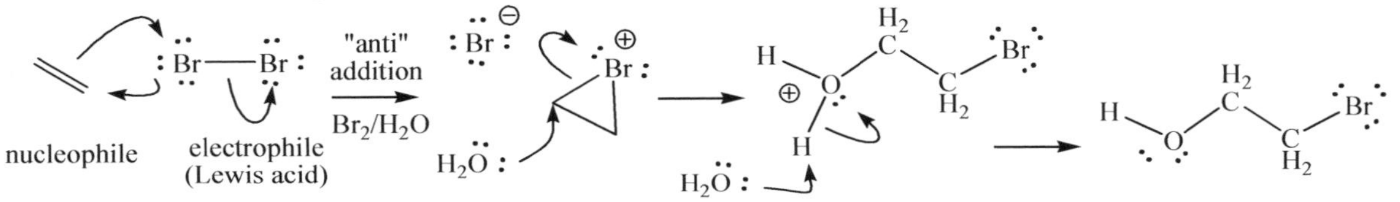

nucleophile — electrophile (Lewis acid) — "anti" addition — Br_2/H_2O

Nucleophilic bromide attack on the bromonium ion is anti. Attack at either carbon leads to the same product.

2. Propene cannot show regioselectivity or stereoselectivity with bromohydrin addition.

nucleophile — electrophile (Lewis acid) — "anti" addition — Br_2/H_2O

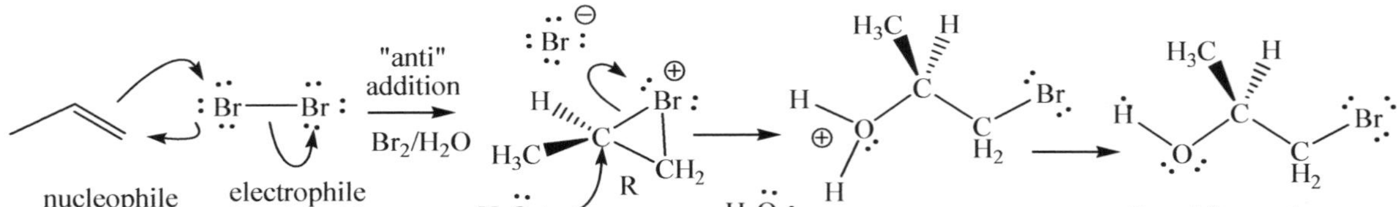

R and S enantiomers

3. 2-methylprop-1-ene cannot show regioselectivity or stereoselectivity with bromohydrin addition. Add in any missing mechanistic details.

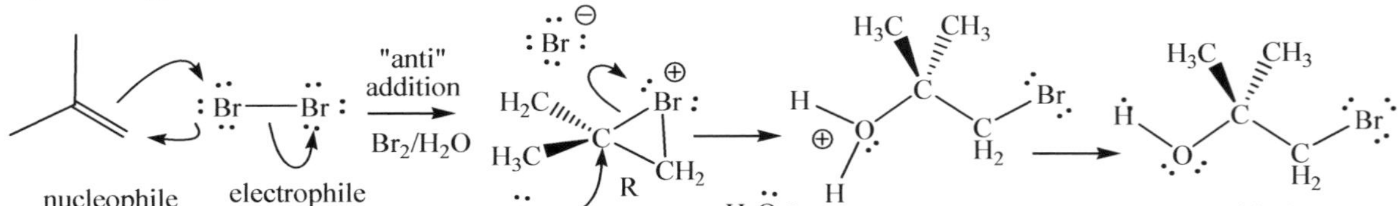

nucleophile — electrophile (Lewis acid) — "anti" addition — Br_2/H_2O

achiral

4. trans-but-2-ene cannot show regioselectivity, but can show stereoselectivity with bromine addition. Add in any necessary mechanistic details.

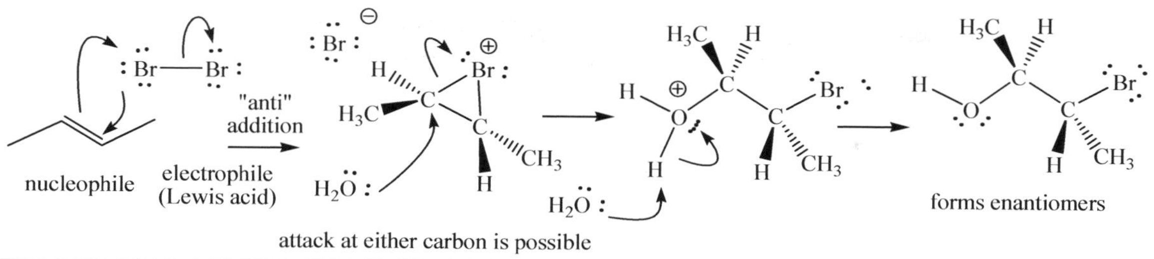

nucleophile | electrophile (Lewis acid) | "anti" addition | attack at either carbon is possible | forms enantiomers

5. Cyclohexene cannot show regioselectivity, but can show stereoselectivity with any of our reagents. Write in a complete mechanism. Supply the necessary mechanistic details.

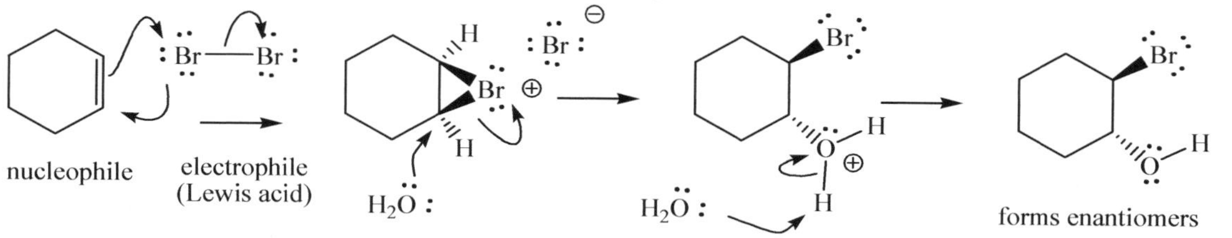

nucleophile | electrophile (Lewis acid) | forms enantiomers

6. 3-methylbut-1-ene cannot show regioselectivity or stereoselectivity with bromine addition. Normally we would expect rearrangement with this molecule, but the bromonium bridge prevents rearrangement. Write a complete mechanism.

nucleophile | electrophile (Lewis acid) | S and R enantiomers

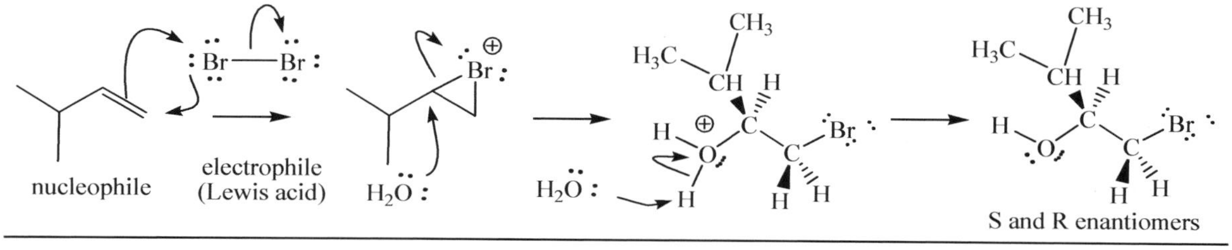

7. Styrene (vinylbenzene, ethenylbenzene) cannot show regioselectivity or stereoselectivity with bromine addition. Write a complete mechanism.

nucleophile | electrophile (Lewis acid) | S and R enantiomers

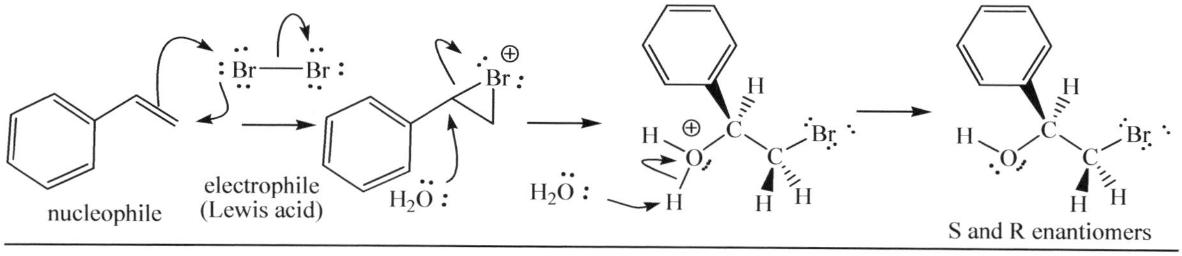

8. 3-methylcyclohex-1-ene cannot show regioselectivity or stereoselectivity with bromine addition. Normally we would expect rearrangement with this molecule, but the bromonium bridge prevents rearrangement. Add in any necessary mechanistic details.

electrophile (Lewis acid) adds from top face or bottom face

nucleophile, this structure assumes the starting material is chiral

top face — 2 steps — A, B

bottom face — 2 steps — C, D

This is a bit of a mess from our point of view. You won't be expected to solve problems this hard in our course, but can you follow what is happening?

A = C and B = D
A and B are diasteromers

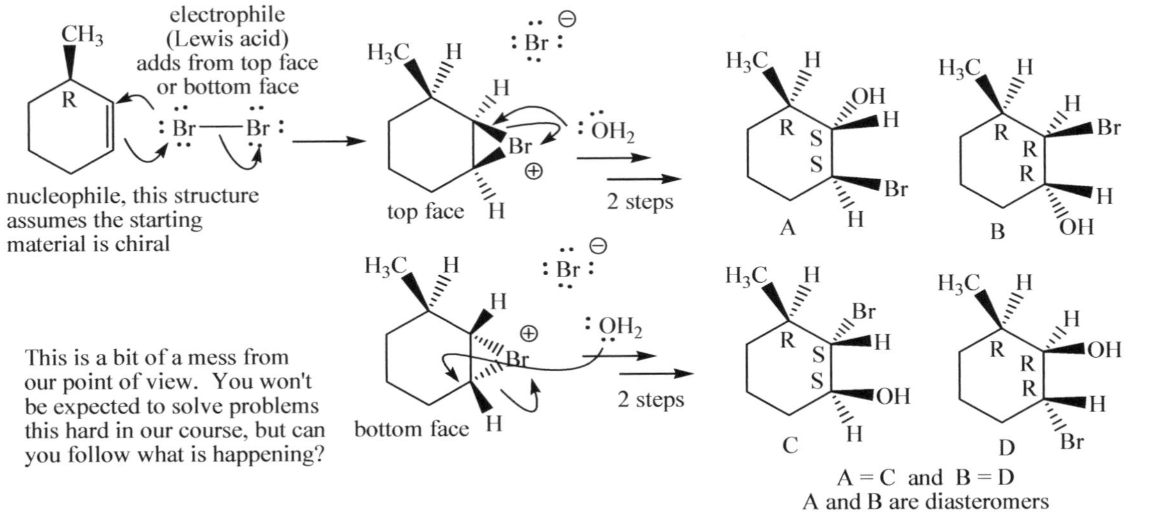

9. 1-methylcyclohex-1-ene can show regioselectivity and can show stereoselectivity with any of our reagents. Write a complete mechanism.

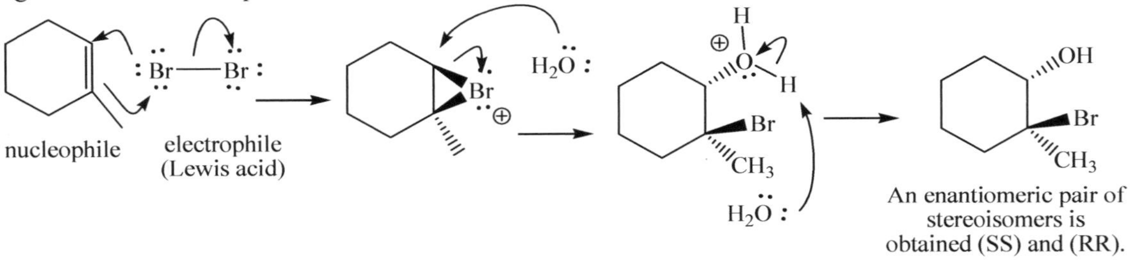

nucleophile — electrophile (Lewis acid)

An enantiomeric pair of stereoisomers is obtained (SS) and (RR).

10. 3-methyl-4-phenylpent-1-ene cannot show regioselectivity or stereoselectivity with bromine addition. Normally we would expect rearrangement with this molecule, but the bromonium bridge prevents rearrangement. Add in any necessary mechanistic details.

nucleophile — electrophile (Lewis acid) — 2 steps

diasstereomers

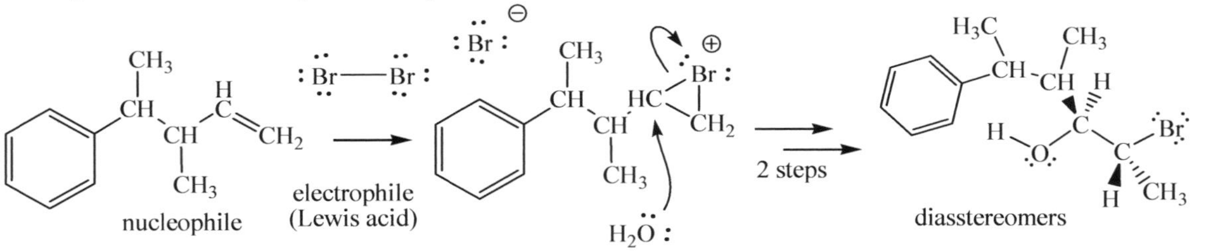

Problem 17 (p 435) – a. Develop an arrow pushing mechanism for epoxide formation in the following reaction. What is the nucleophilic atom, how is it formed and what approach does it need for a successful reaction? What conformation in a cyclohexane ring allows this approach? What conformation in an open chain allows for this approach (see part b)?

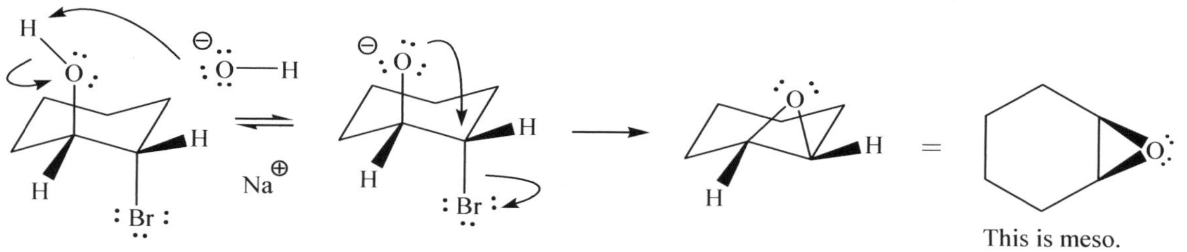

H O H H Br O H Na⊕ O H H Br O H H O

This is meso.

The Br has to be axial to be attacked from the backside. The oxygen has to be an anion to be a good nucleophile (exchanges protons with hydroxide).

b. Predict the product of the following reaction by showing a mechanism. Be very careful about the required conformation to react, which determines the stereochemistry of the product. You may have to rotate a bond to show the reaction.

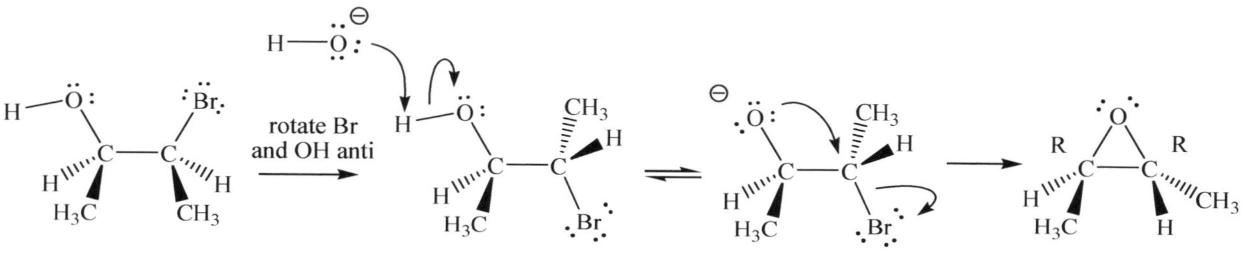

H–O C–C Br H H H_3C CH_3 → rotate Br and OH anti → H–O C–C CH_3 H H H_3C Br ⇌ O C–C CH_3 H H H_3C Br → R O R C–C H H_3C CH_3 H

enantiomers

Problem 18 (p 435) – Propose the necessary steps for the transformations indicated below. Part d requires that you make a carbon-carbon bond.

a. Br, E2, $K^{\oplus}$ $^{\ominus}$O–C(CH_3)$_3$ → cyclohexene → Br_2, H_2O → OH / Br (enantiomers) → NaOH → epoxide O (meso)

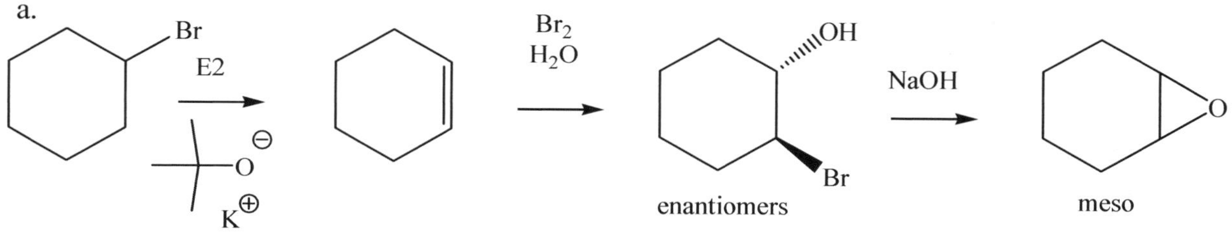

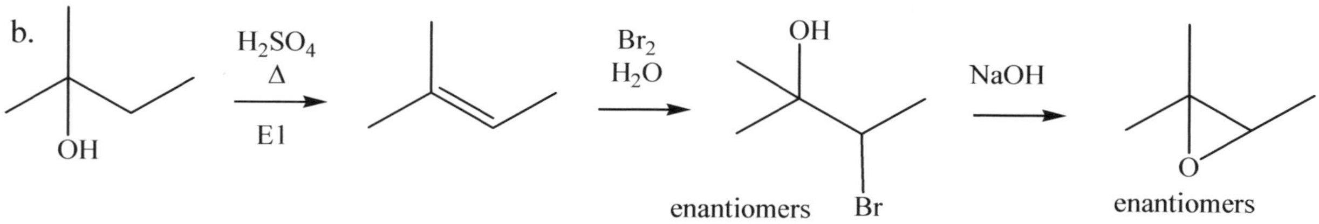

b. OH → H_2SO_4, Δ, E1 → alkene → Br_2, H_2O → OH / Br (enantiomers) → NaOH → epoxide O (enantiomers)

c. methylcyclohexane → Br_2, hν → Br → $K^{\oplus}$ $^{\ominus}$O–C(CH_3)$_3$, E2 → 1-methylcyclohexene → NaOH → epoxide O

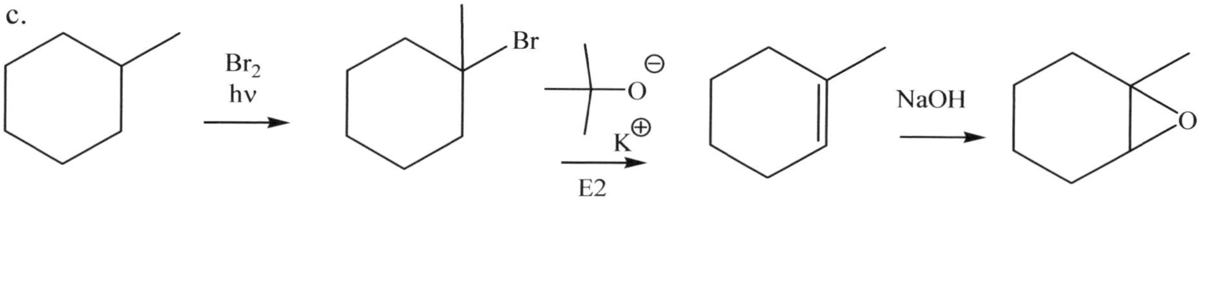

Problem 19 (p 436) – Show how each of the following transformations could be accomplished. Do you know the mechanistic steps for each reaction? Where possible, specify if the addition is Markovnikov or anti-Markovnikov. Regioselectivity is obvious in these problems. Is stereoselectivity also evident? Are stereocenters created? If so what sorts of stereoisomers are present, enantiomers or diastereomers or neither?

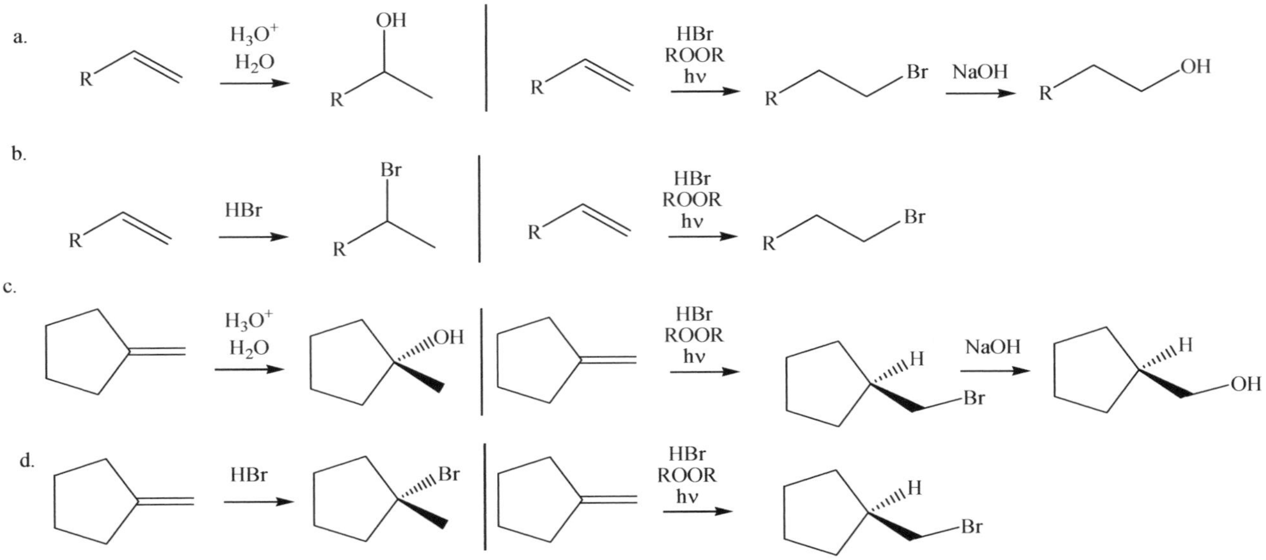

a. H_3O^+ / H_2O | HBr / ROOR / hν ; NaOH

b. HBr | HBr / ROOR / hν

c. H_3O^+ / H_2O | HBr / ROOR / hν ; NaOH

d. HBr | HBr / ROOR / hν

Problem 20 (p 437) – Write a mechanism for the following reaction in acidic alcohol and basic alcohol.

a. Open epoxides in alcoholic acid.

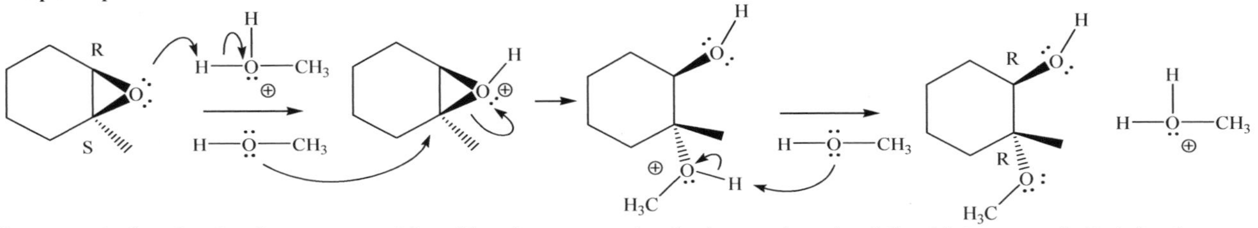

The more substituted carbon has greate partial positive charge and is attacked faster in S_N1 approach of weak nucleophile.

regioselective: weak nucleophile adds to more substituted carbon
stereoselective: weak nucleophile adds "anti" because of the bridge

b. Open epoxides in alcoholic base (supply any missing mechanistic details).

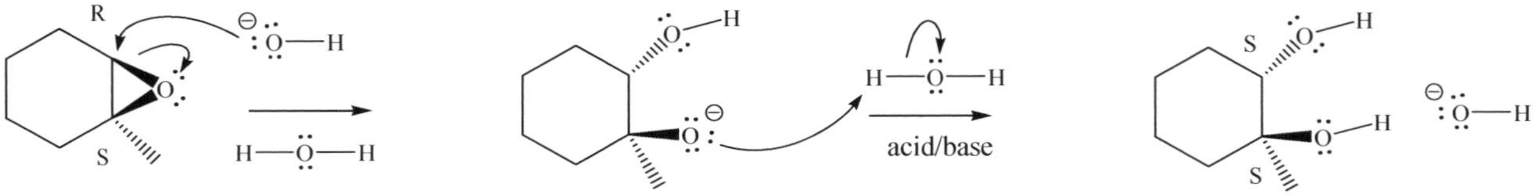

acid/base

Less substituted carbon has less steric hindrance and is attacked faster in S_N2 approach of strong nucleophile.

regioselective: strong nucleophile adds to less substituted carbon
stereoselective: strong nucleophile adds "anti" because of the bridge

Because of the high cost of osmium tetroxide and its toxicity, only a small catalytic amount is typically used. It is regenerated by an oxidizing agent (i.e. a peroxide or an amine oxide), which is much less expensive and reduces exposure to the chemist doing the reaction.

Problem 21 (p 438) – Propose a mechanism for re-oxidation of osmium trioxide to osmium tetraoxide using morpholine N-oxide.

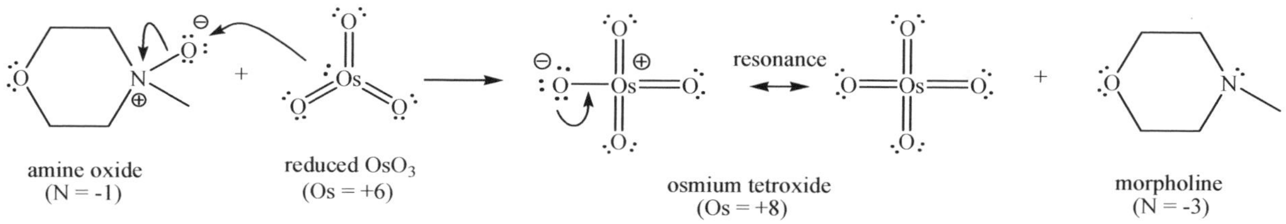

amine oxide (N = -1) + reduced OsO_3 (Os = +6) → osmium tetroxide (Os = +8) resonance + morpholine (N = -3)

Problem 22 (p 439-441) – Do the following pi systems allow one to observe anti addition (as opposed to syn addition or random addition)? Write the products, clearly showing stereochemical features. Indicate if products are enantiomers, diastereomers, meso compounds or achiral structures. Because the mechanism is so long we will only show one full mechanism (above), only fill in the missing mechanistic details in one and leave the rest for you to be able to write out a mechanism for yourself.

<u>10 Alkenes to practice on</u> (This isn't much different than adding H-Br.)

1. Ethane cannot show regioselectivity or stereoselectivity with any of our reagents.

nucleophile; electrophile (Lewis acid); proton transfers

Attack from top and bottom faces is possible here, but not observable in this example.

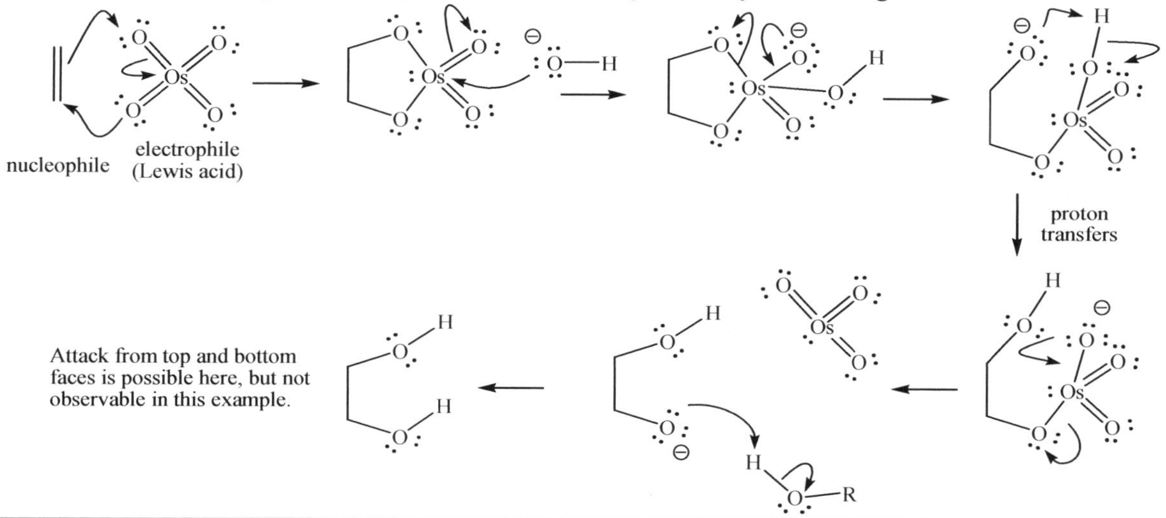

2. Propene can show regioselectivity, but cannot show stereoselectivity with any of our reagents.

nucleophile; electrophile (Lewis acid); proton transfers

Attack from top and bottom faces is possible here, but not observable in this example.

rracemic mixture of enantiomers

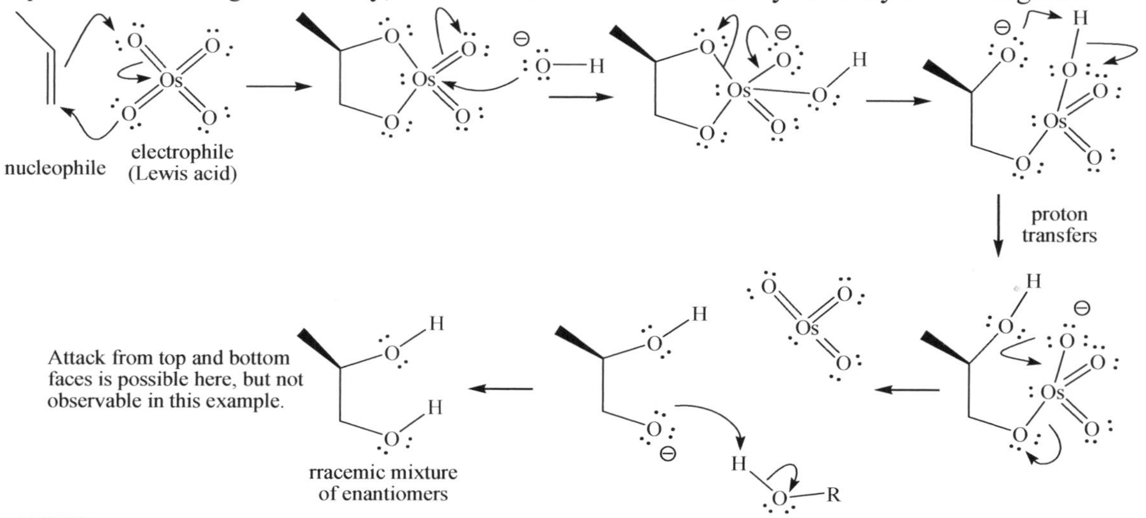

3. 2-methylprop-1-ene can show regioselectivity, but cannot show stereoselectivity with any of our reagents. Add in any missing mechanistic details.

nucleophile

electrophile (Lewis acid)

proton transfers

Attack from top and bottom faces is possible here, but not observable in this example.

rracemic mixture of enantiomers

4. trans-but-2-ene cannot show regioselectivity, but can show stereoselectivity, but not with HBr. Write a complete mechanism.

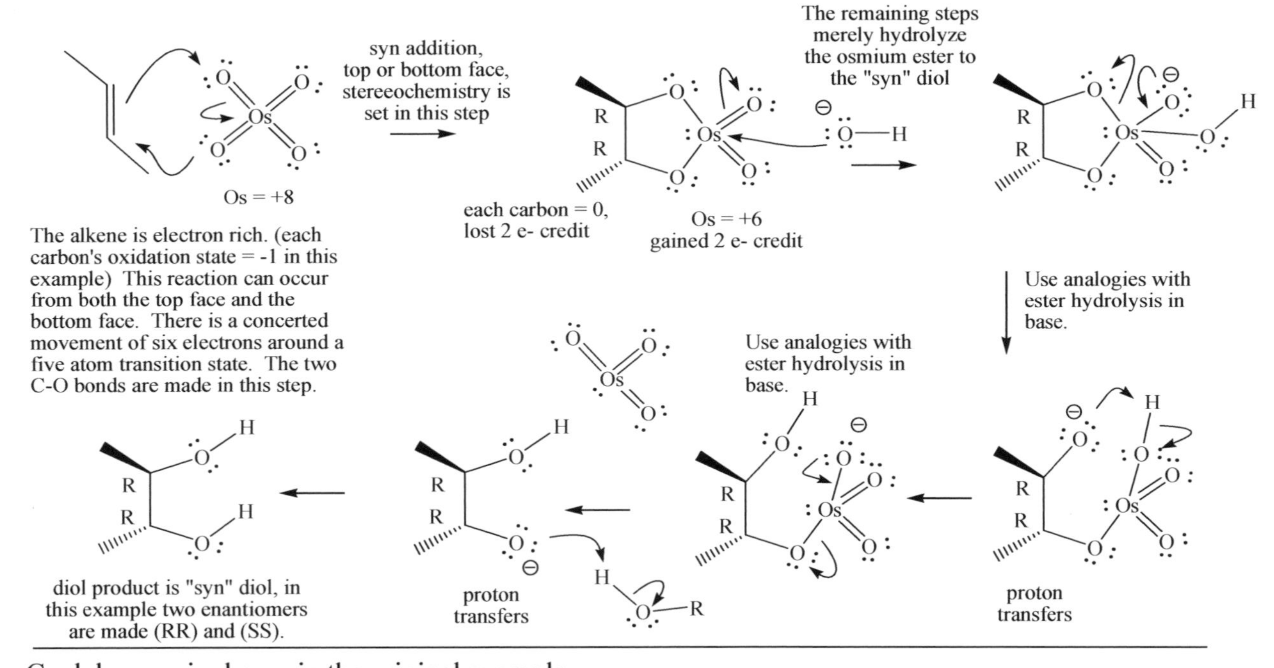

5. Cyclohexene is shown in the original example.

nucleophile

electrophile (Lewis acid)

similar to above mechanisms

OH

OH

Attack from top and bottom faces is possible here, and leads to a "meso" compound.

6. 3-methylbut-1-ene can show regioselectivity, but does cannot show stereoselectivity with any of our reagents. This example will also rearrange whenever carbocations form. Add in any missing mechanistic details.

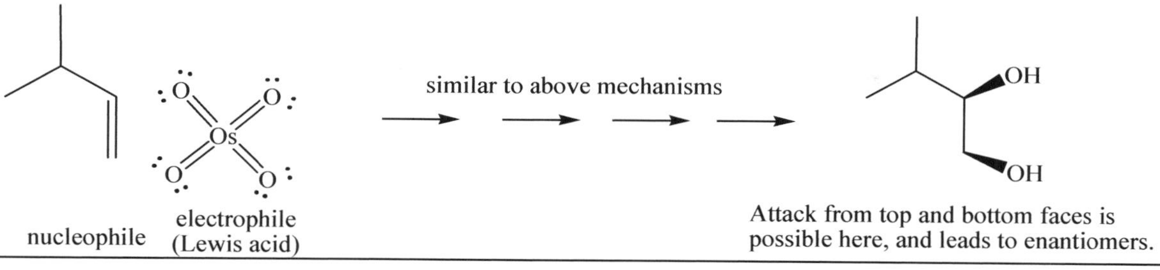
similar to above mechanisms
OH
OH
nucleophile
electrophile (Lewis acid)
Attack from top and bottom faces is possible here, and leads to enantiomers.

7. Styrene (vinylbenzene, ethenylbenzene) can show regioselectivity, but does cannot show stereoselectivity with any of our reagents. Write a complete mechanism.

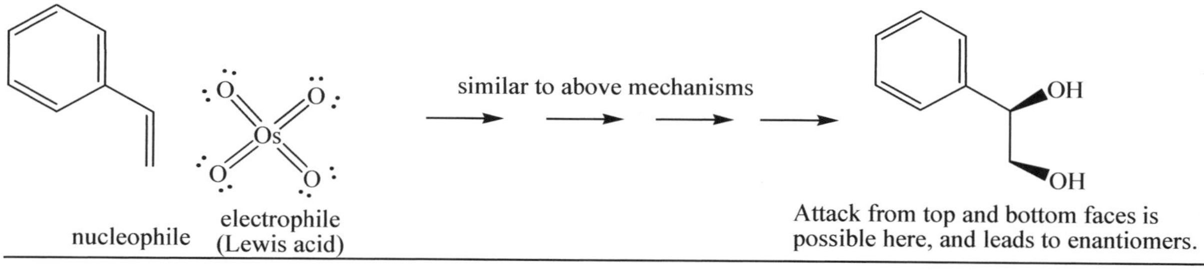
similar to above mechanisms
OH
OH
nucleophile
electrophile (Lewis acid)
Attack from top and bottom faces is possible here, and leads to enantiomers.

8. 3-methylcyclohex-1-ene is good for showing when rearrangements are possible. Add in any missing mechanistic details.

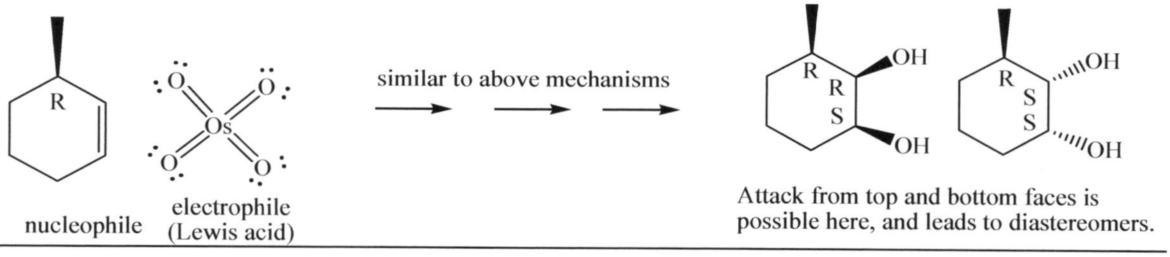
R
similar to above mechanisms
R R S OH OH
R S S OH OH
nucleophile
electrophile (Lewis acid)
Attack from top and bottom faces is possible here, and leads to diastereomers.

9. 1-methylcyclohex-1-ene can show regioselectivity and can show stereoselectivity with any of our reagents. Write a complete mechanism.

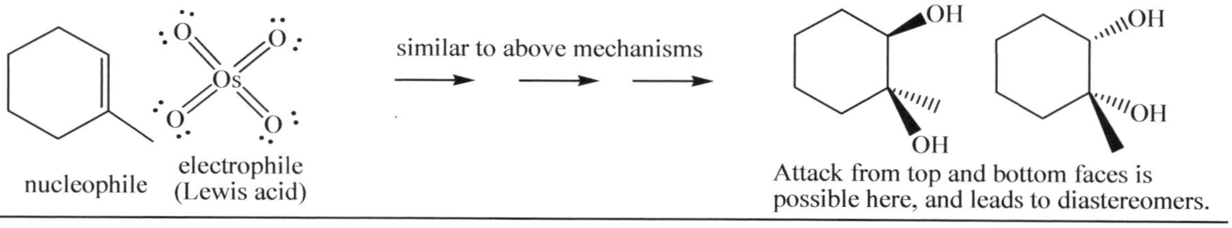
similar to above mechanisms
OH
OH
OH
OH
nucleophile
electrophile (Lewis acid)
Attack from top and bottom faces is possible here, and leads to diastereomers.

10. 3-methyl-4-phenylpent-1-ene is good for showing complicated rearrangement possibilities. Add in any missing mechanistic details.

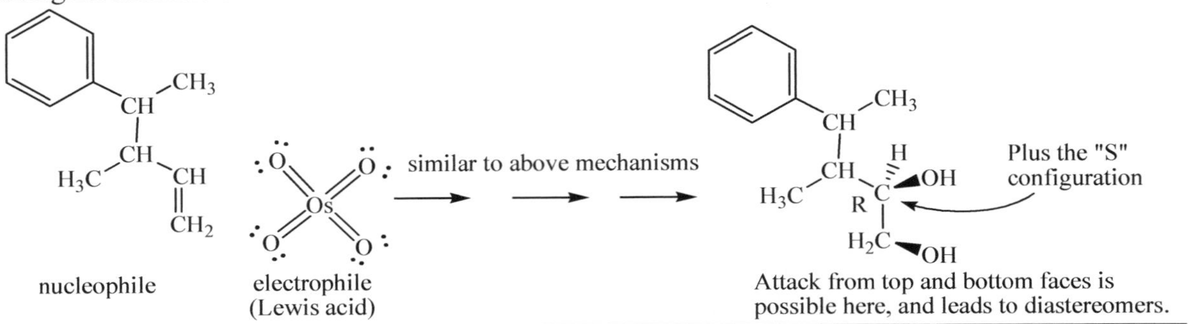

nucleophile — electrophile (Lewis acid) (OsO_4) — similar to above mechanisms → → →

Plus the "S" configuration

Attack from top and bottom faces is possible here, and leads to diastereomers.

Problem 23 (p 442) – What are the expected products and stereochemistries in the following reactions? How do the two approaches compare?

a.

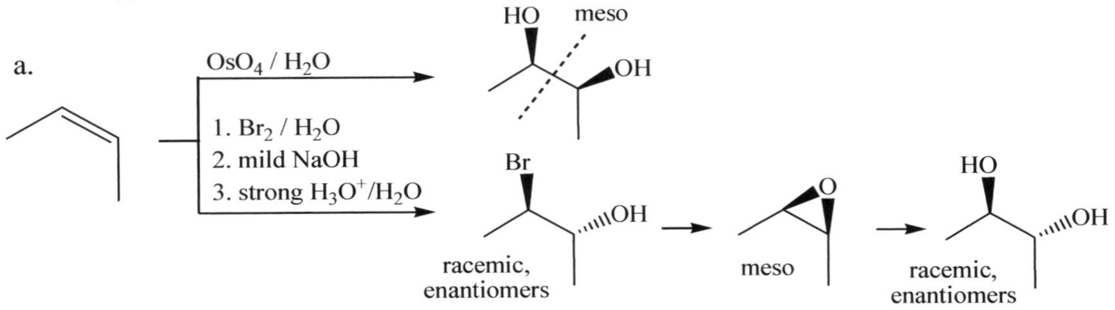

OsO_4 / H_2O → HO, OH — meso

1. Br_2 / H_2O
2. mild NaOH
3. strong H_3O^+/H_2O

Br, OH — racemic, enantiomers → O — meso → HO, OH — racemic, enantiomers

---

b.

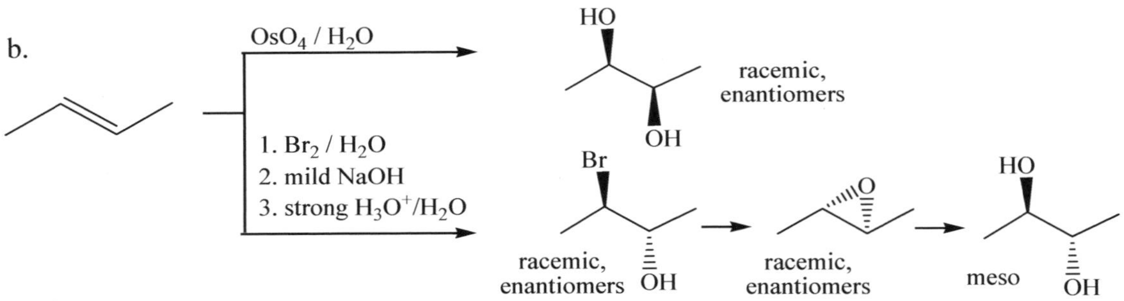

OsO_4 / H_2O → HO, OH — racemic, enantiomers

1. Br_2 / H_2O
2. mild NaOH
3. strong H_3O^+/H_2O

Br, OH — racemic, enantiomers → O — racemic, enantiomers → HO, OH — meso

---

c.

OsO_4 / H_2O → OH, OH — racemic, enantiomers

1. Br_2 / H_2O
2. mild NaOH
3. strong $HO^{\ominus}/H_2O$

OH, Br — racemic, enantiomers → O — racemic, enantiomers → OH, OH — racemic, enantiomers

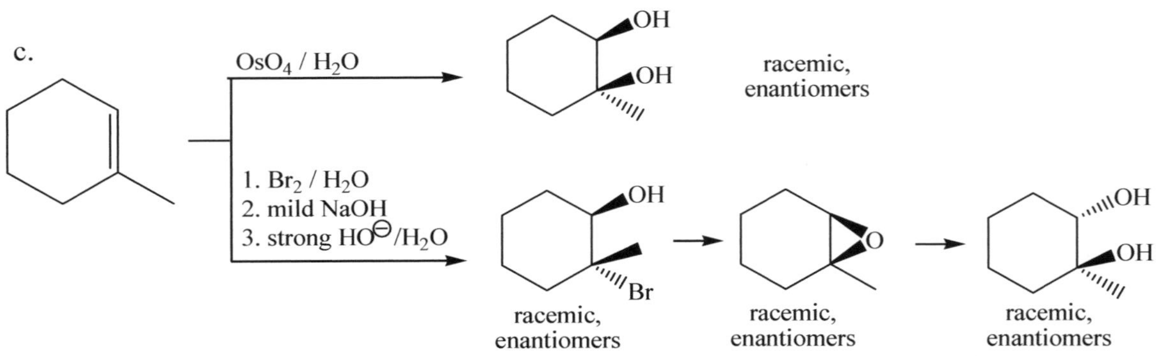

d.

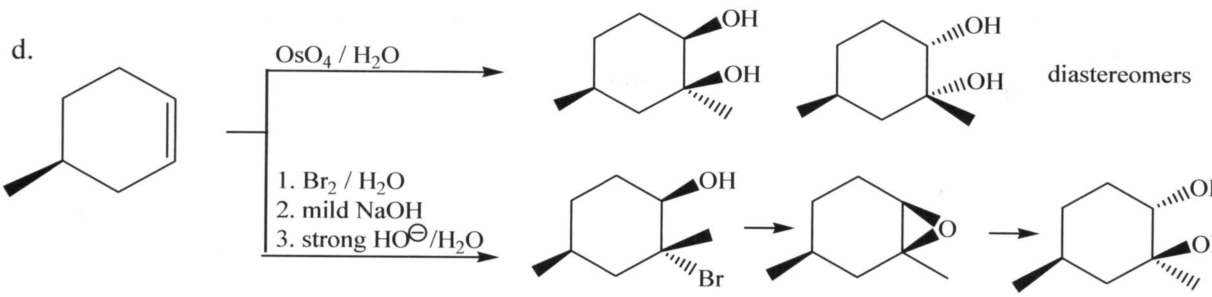

OsO_4 / H_2O

OH OH OH OH diastereomers

1. Br_2 / H_2O
2. mild NaOH
3. strong $HO^{\ominus}/H_2O$

OH Br O OH OH

all diastereomers, only one is shown

Problem 24 (p 444-446) – Do the following pi systems allow one to observe anti addition (as opposed to syn addition or random addition)? Write the products, clearly showing stereochemical features. Indicate if products are enantiomers, diastereomers, meso compounds or achiral structures. Because the mechanism is so long we will only show one full mechanism (above), only fill in the missing mechanistic details in one and leave the rest for you to be able to write out a mechanism for yourself.

10 Alkenes to practice on

1. Ethane cannot show regioselectivity or stereoselectivity with any of our reagents.

H H : Pd → D Pd D H H C C H H → D Pd D H H C C H H → D Pd H H C C H H D → D C H H C H H D achiral

Pd :

palladium repeats cycle

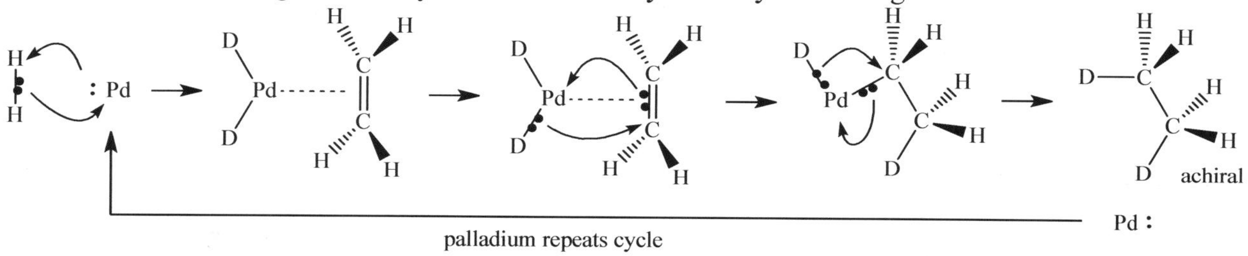

2. Propene can show regioselectivity, but cannot show stereoselectivity with any of our reagents.

H H : Pd → D Pd D H CH_3 C C H H → D Pd D H CH_3 C C H H → D Pd H CH_3 C C H H D → D C * H CH_3 C H H D

D_2 is delivered to both faces. Usually the less hindered face reacts faster.

* = chiral center
enantiomers

Pd :

palladium repeats cycle

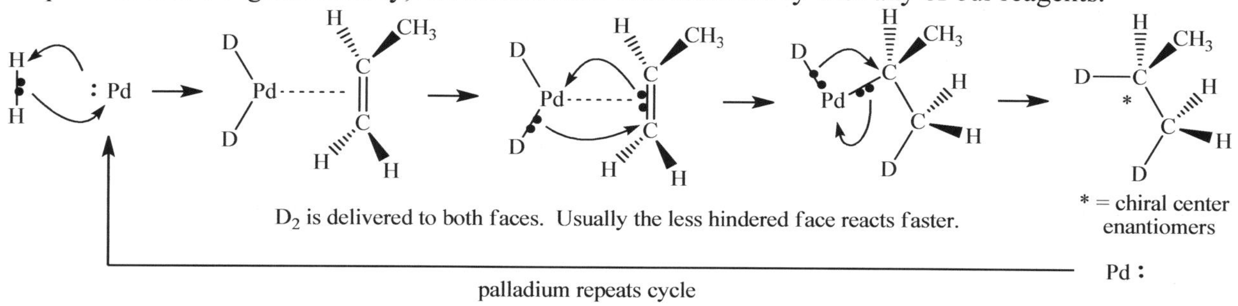

3. 2-methylprop-1-ene can show regioselectivity, but cannot show stereoselectivity with any of our reagents. Add in any missing mechanistic details.

H H : Pd → D Pd D H_3C CH_3 C C H H → D Pd D H_3C CH_3 C C H H → D Pd CH_3 CH_3 C C H H D → D C CH_3 CH_3 C H H D

D_2 is delivered to both faces. Usually the less hindered face reacts faster.

achiral

Pd :

palladium repeats cycle

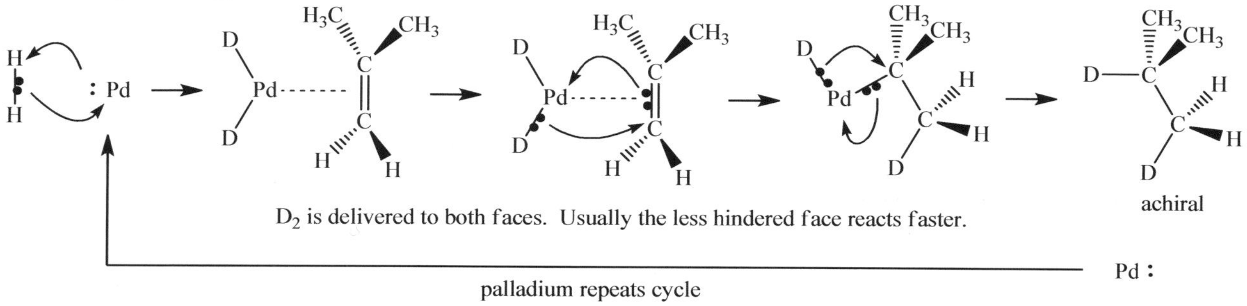

4. trans-but-2-ene cannot show regioselectivity, but can show stereoselectivity, but not with HBr. Write a complete mechanism.

H H : Pd → D D Pd ------ H_3C H C C H CH_3 → D D Pd ------ H_3C H C C H CH_3 → D Pd CH_3 H C D H C CH_3 → D * C CH_3 H * C H CH_3 D

D_2 is delivered to both faces. Usually the less hindered face reacts faster.

* = chiral centers
enantiomers

Pd :

palladium repeats cycle

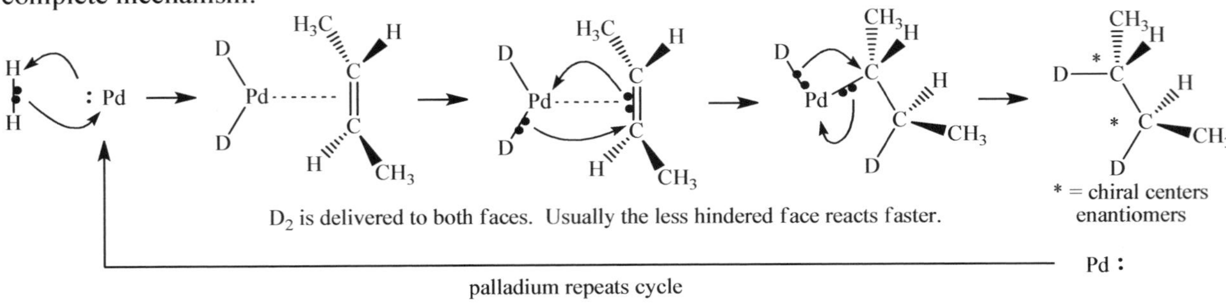

5. Cyclohexene cannot show regioselectivity, but can show stereoselectivity with any of our reagents. Write in a complete mechanism.

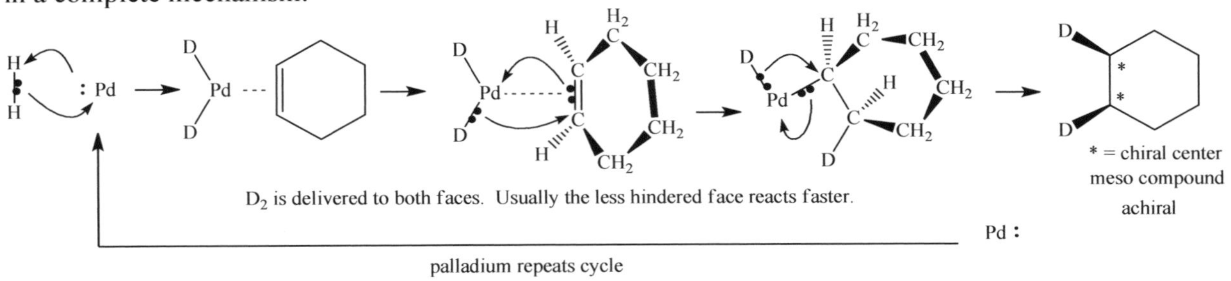

H H : Pd → D D Pd --- → D D Pd ------ H H_2 C C CH_2 C H CH_2 CH_2 → D Pd H H_2 C CH_2 C H C CH_2 CH_2 D → D * * D

D_2 is delivered to both faces. Usually the less hindered face reacts faster.

* = chiral center
meso compound
achiral

Pd :

palladium repeats cycle

6. 3-methylbut-1-ene can show regioselectivity, but does cannot show stereoselectivity with any of our reagents. This example will also rearrange whenever carbocations form. Add in any missing mechanistic details.

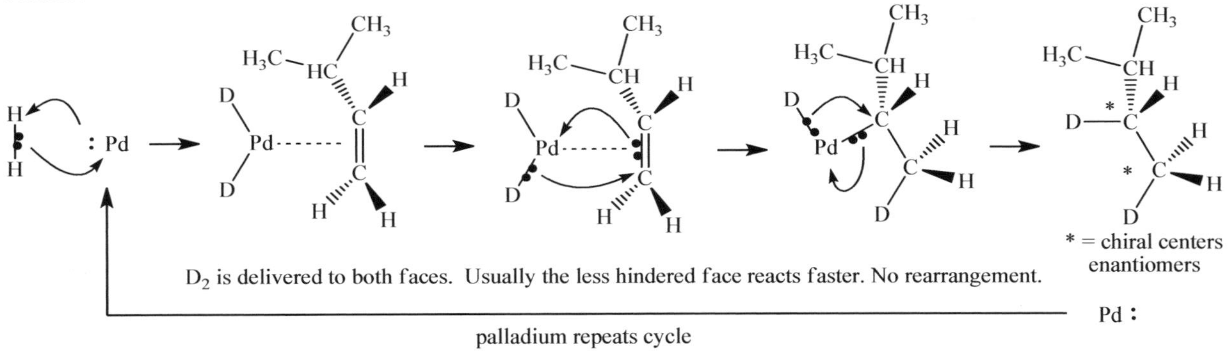

H H : Pd → D D Pd ------ CH_3 H_3C HC H C C H H → D D Pd ------ CH_3 H_3C CH H C C H H → D Pd CH_3 H_3C CH H C H C H D → CH_3 H_3C CH H D * C H * C H D

D_2 is delivered to both faces. Usually the less hindered face reacts faster. No rearrangement.

* = chiral centers
enantiomers

Pd :

palladium repeats cycle

7. Styrene (vinylbenzene, ethenylbenzene) can show regioselectivity, but does cannot show stereoselectivity with any of our reagents. Write a complete mechanism.

H H : Pd → D D Pd ------ H C C H H → D D Pd ------ H C C H H → D Pd H C H C H D → H D * C H * C H D

D_2 is delivered to both faces. Usually the less hindered face reacts faster. No rearrangement.

* = chiral centers
enantiomers

Pd :

palladium repeats cycle

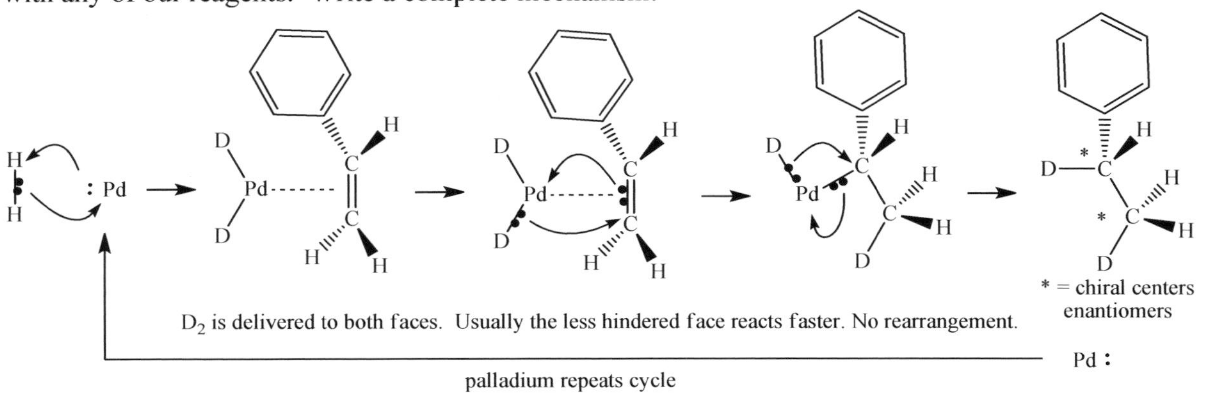

8. 3-methylcyclohex-1-ene is good for showing when rearrangements are possible. Add in any missing mechanistic details.

H H : Pd → D Pd D --- → D Pd D H C CH₃ H C CH₂ CH₂ H CH₂ → D Pd H H C CH₃ CH₂ H C CH₂ CH₂ D → D D CH₃ S R R ; D D CH₃ R R R

D_2 is delivered to both faces. Usually the less hindered face reacts faster.

Pd :

palladium repeats cycle

diasteromers

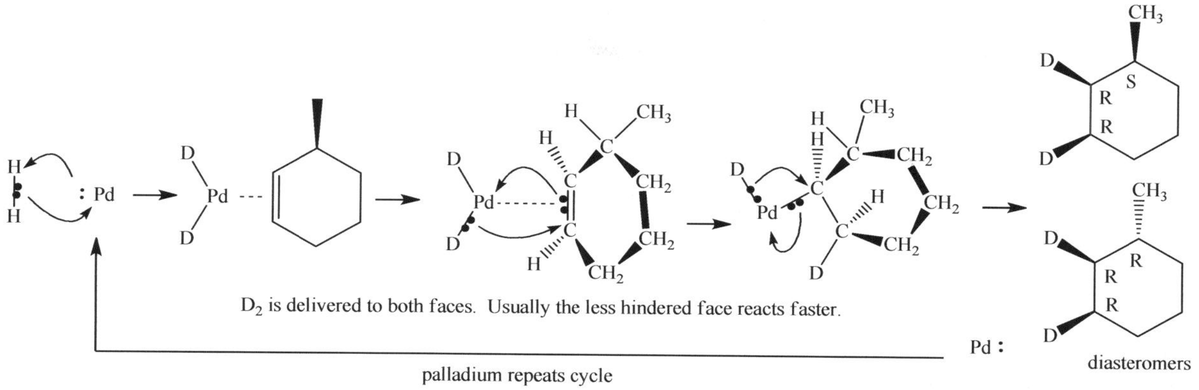

9. 1-methylcyclohex-1-ene can show regioselectivity and can show stereoselectivity with any of our reagents. Write a complete mechanism.

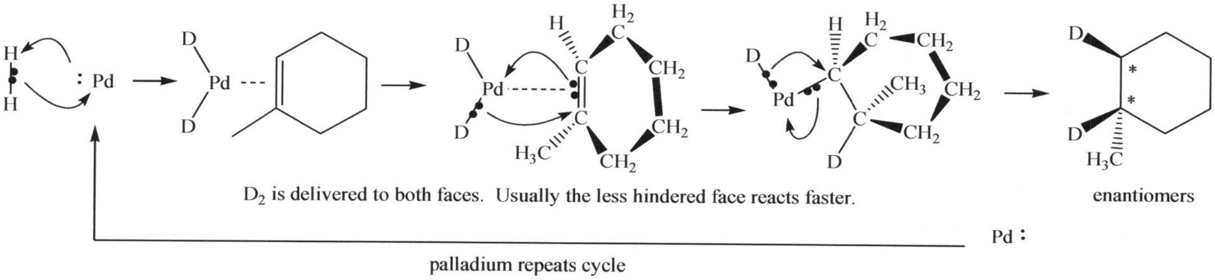

H H : Pd → D Pd D --- → D Pd D H H_2C C CH₂ H_3C C CH₂ CH₂ → D Pd H H_2C CH₂ C CH₃ C CH₂ CH₂ D → D D * * H_3C

D_2 is delivered to both faces. Usually the less hindered face reacts faster.

enantiomers

Pd :

palladium repeats cycle

10. 3-methyl-4-phenylpent-1-ene is good for showing complicated rearrangement possibilities. Add in any missing mechanistic details.

CH CH₃ CH H_3C CH CH₂ : Pd

nucleophile

similar to mechanisms above

* = chiral center

CH* CH₃ CH* H_3C H C D R * H_2C D

Also get S configuration here.

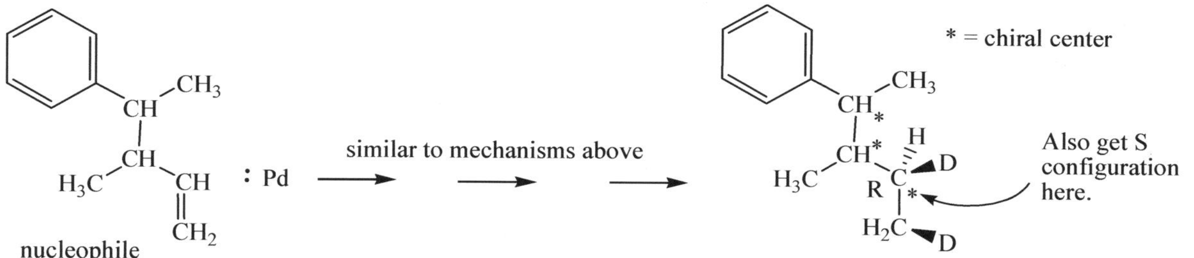

Attack from top and bottom faces is possible here, and leads to diastereomers.

Problem 25 (p 449) – Propose reasonable synthetic steps to accomplish the indicated transformations. Use the starting material and our much smaller chemical catalog (on the next page) to make the TM-1 and continue on from there. No mechanisms are needed, but write out the reagents, workup and products for each synthetic step. If you have already made something in a previous step, you can start there and just refer back to the number where you made it.

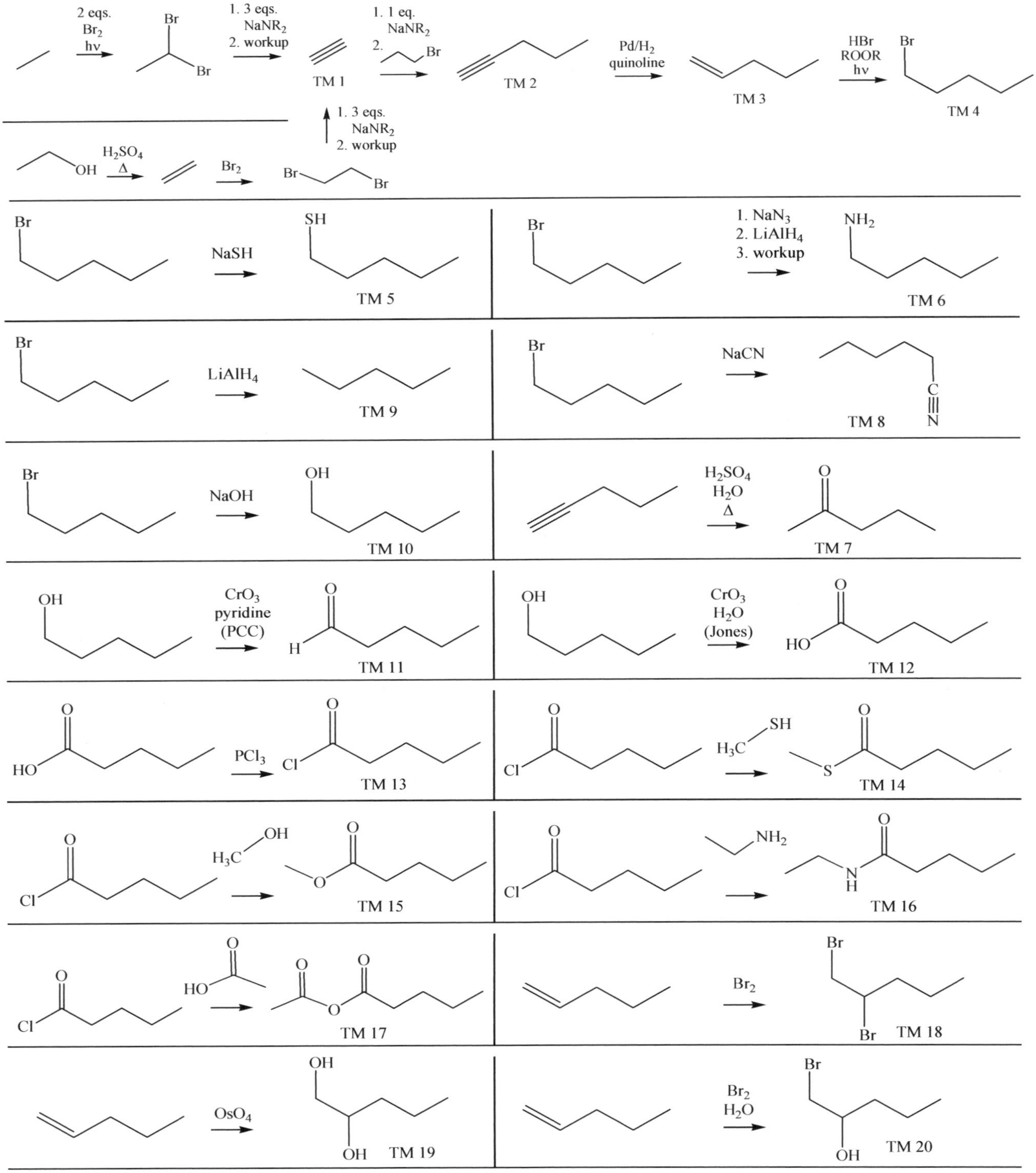

2 eqs. Br_2 hv; Br; Br; 1. 3 eqs. $NaNR_2$ 2. workup; TM 1; 1. 1 eq. $NaNR_2$ 2. Br; TM 2; Pd/H_2 quinoline; TM 3; HBr ROOR hv; Br; TM 4

1. 3 eqs. $NaNR_2$ 2. workup

OH; H_2SO_4 Δ; Br_2; Br; Br

Br; NaSH; SH; TM 5 | Br; 1. NaN_3 2. $LiAlH_4$ 3. workup; NH_2; TM 6

Br; $LiAlH_4$; TM 9 | Br; NaCN; C≡N; TM 8

Br; NaOH; OH; TM 10 | H_2SO_4 H_2O Δ; O; TM 7

OH; CrO_3 pyridine (PCC); O; H; TM 11 | OH; CrO_3 H_2O (Jones); O; HO; TM 12

O; HO; PCl_3; O; Cl; TM 13 | O; Cl; H_3C–SH; O; S; TM 14

O; Cl; H_3C–OH; O; O; TM 15 | O; Cl; NH_2; O; N H; TM 16

O; Cl; O; HO; O; O; O; TM 17 | Br_2; Br; Br; TM 18

OsO_4; OH; OH; TM 19 | Br_2 H_2O; Br; OH; TM 20

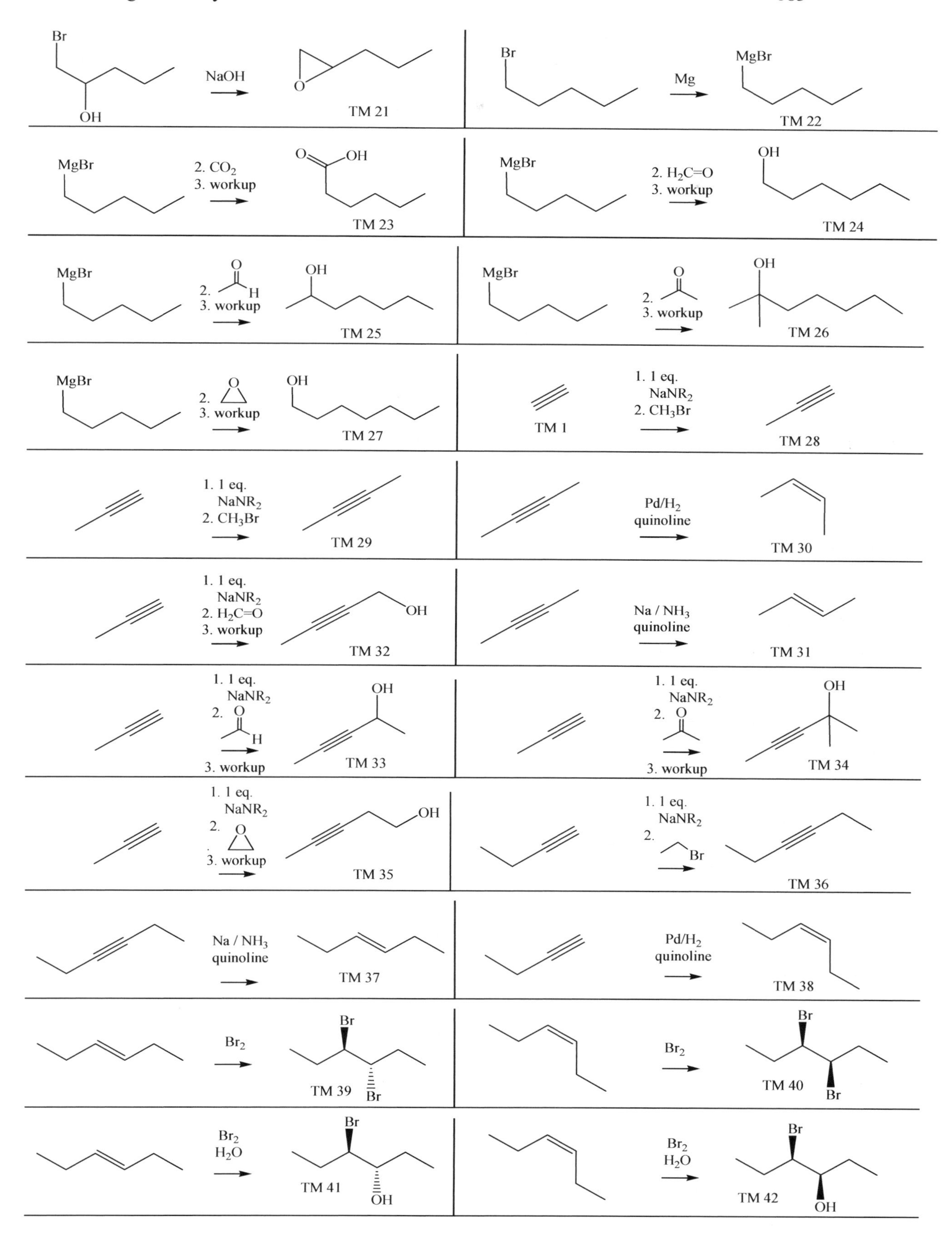

Br
OH
NaOH
O
TM 21
Br
Mg
MgBr
TM 22
MgBr
2. CO_2
3. workup
O
OH
TM 23
MgBr
2. $H_2C=O$
3. workup
OH
TM 24
MgBr
2.
H
3. workup
OH
TM 25
MgBr
2.
3. workup
OH
TM 26
MgBr
2.
3. workup
OH
TM 27
TM 1
1. 1 eq. $NaNR_2$
2. CH_3Br
TM 28
1. 1 eq. $NaNR_2$
2. CH_3Br
TM 29
Pd/H_2 quinoline
TM 30
1. 1 eq. $NaNR_2$
2. $H_2C=O$
3. workup
OH
TM 32
Na / NH_3 quinoline
TM 31
1. 1 eq. $NaNR_2$
2.
H
3. workup
OH
TM 33
1. 1 eq. $NaNR_2$
2.
3. workup
OH
TM 34
1. 1 eq. $NaNR_2$
2.
3. workup
OH
TM 35
1. 1 eq. $NaNR_2$
2.
Br
TM 36
Na / NH_3 quinoline
TM 37
Pd/H_2 quinoline
TM 38
Br_2
Br
TM 39
Br
Br_2
Br
TM 40
Br
Br_2
H_2O
Br
TM 41
OH
Br_2
H_2O
Br
TM 42
OH

OH OsO_4 → TM 43 OH | OsO_4 → OH TM 44 OH

OH NaOH → O TM 45 Br | OH NaOH → O TM 46 Br

Chapter 14

Problem 1 (p 453) – Provide an acceptable name for the following aromatic compounds. Look for a parent name in the above examples (=1) and use the lowest possible number to the next substituent. If there are only two substituents, you can use ortho, meta or para prefixes.

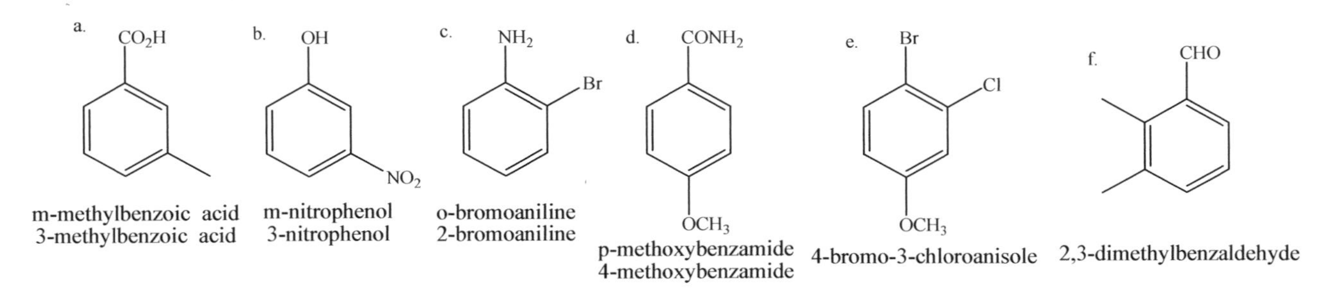

Problem 2 (p 455) – Multiple benzene rings can be connected together and many variation are known. Collectively, these are called *polycyclic aromatic hydrocarbons* (PAH). Three famous examples are shown below. Draw the other resonance structures. Use the resonance structures to predict which bonds might be the shortest in the structures below? Hint 1 – Draw the pi bonds all the way around the perimeter of the PAH ring system, then shift all of them over one time like we do with benzene: Next find any "benzene" looking rings and draw the "benzene" resonance. If you do that for every ring, you should have drawn all of the resonance structures. Hint 2 - Some experimental x-ray data is provided just below these structures from a 1951 journal article.

Benzene has 2 equivalent resonance structures.

benzene
(aromaticity = 36)

all bonds =140 pm

naphthalene - 3 possible resonance structures

A B C D

naphathelene
(aromaticity = 61)

A = 142 pm = 2 single, 1 double
B = 136 pm = 1 single, 2 double
C = 140 pm = 2 single, 1 double
D = 140 pm = 2 single, 1 double

anthracene - 4 possible resonance structures

A B C D E

anthracene
(aromaticity = 83)

A = 140 pm = 2 single, 2 double
B = 142 pm = 3 single, 1 double
C = 139 pm = 1 single, 3 double
D = 139 pm = 3 single, 1 double
E = 144 pm = 3 single, 1 double

---

phenanthrene - 5 possible resonance structures

phenanthrene
(aromaticity = 91)

F E G D H C B A

A = 134 pm = 1 single, 4 double
B = 142 pm = 4 single, 1 double
C = 142 pm = 3 single, 2 double
D = 134 pm = 2 single, 3 double
E = 137 pm = 3 single, 2 double
F = 138 pm = 2 single, 3 double
G = 138 pm = 3 single, 2 double
H = 146 pm = 4 single, 1 double

Problem 3 (p 460) - Propose a mechanism for the nitrosylation of benzene using nitrous acid and hydrochloric acid. Since the NO^{+} electrophile is less reactive, it requires a more reactive aromatic compound. The methyl side chain activates the ring towards electron donation. Can you think of a reason why electron donation is easier? The relative reactivity of aromatic compounds is discussed later in this chapter.

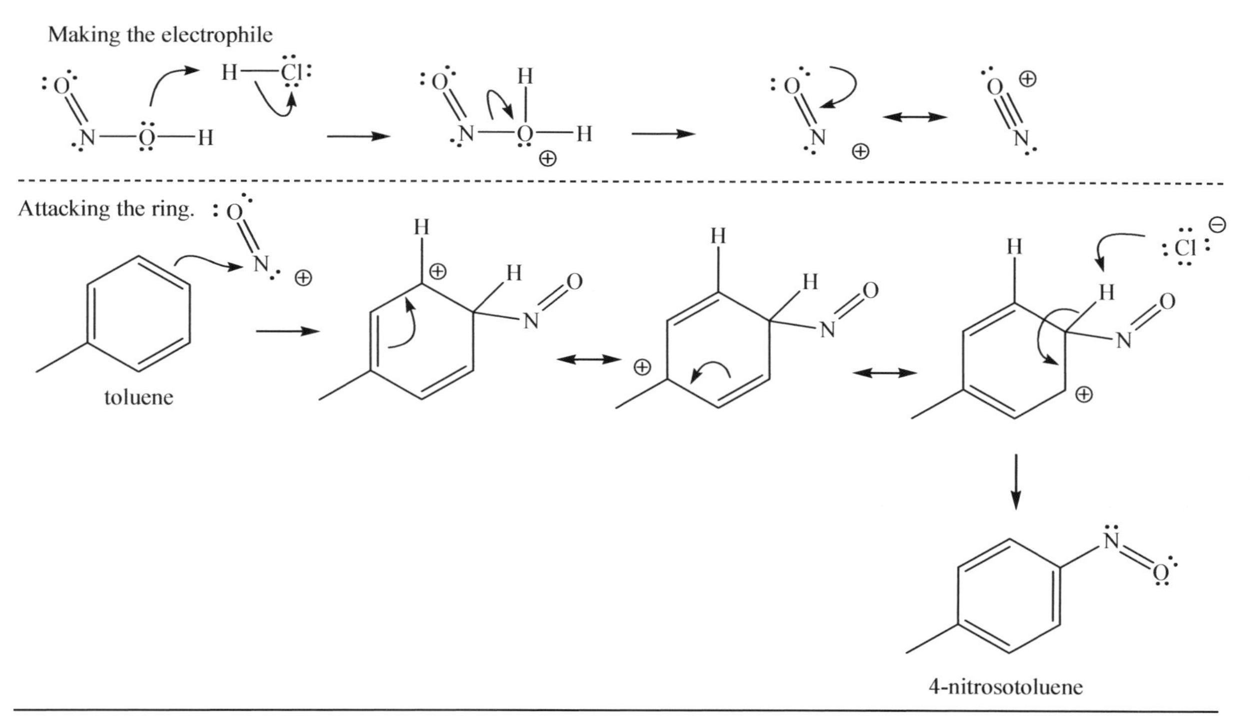

Problem 4 (460)– The amino group is often biologically important, and it has much synthetic potential, including diazonium chemistry, which we do not have time to cover. The following sequence is a very speculative mechanism for the reduction of a nitro group to an amine group, with a total of 6 electrons from 3 iron atoms. We show 2 electron transfers from the iron in the first two reduction steps and two single electron reduction steps in the last iron reduction step (to show a possible free radical variation). You don't need to know this, but speculation is fun. All of the mechanistic details are left out in a second scheme in case you want to try and fill them in yourself.

Problem 5 (p 462) – Fill in missing mechanistic details in the proposed mechanism for reduction of an aromatic nitro group to an aromatic amine.

nitrobenzene

: Fe

: Fe

Perhaps as a di cation?

: Fe

protonate "O" twice.

:Fe

resonance

Aniline would be protonated in the acidic medium. This could be neutralized in a workup step.

Aromatic amines are important in themselves, and are the starting point for versatile diazonium chemistry, discussed later.

Problem 6 (p 464) – Propose a reasonable mechanism for each of the following two reactions.

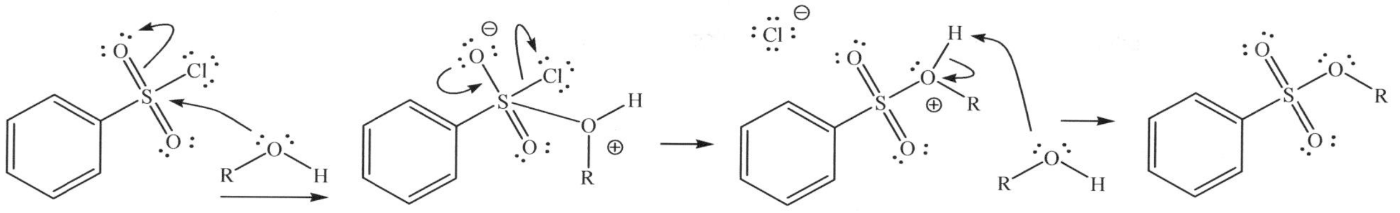

benzenesulfonyl chloride

alkyl benzenesulfonates

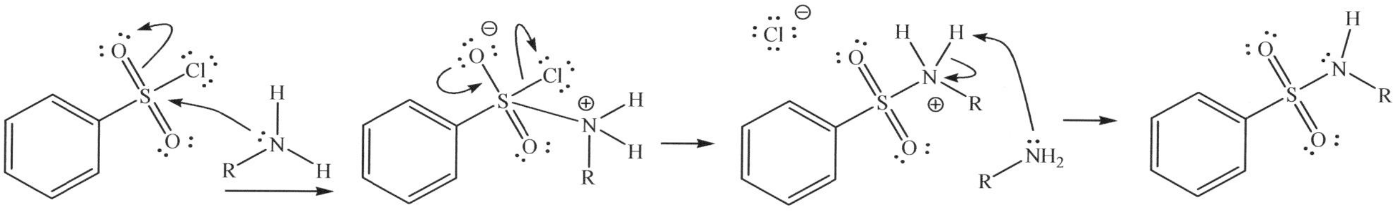

benzenesulfonyl chloride

alkyl benzenesulfonamides

Problem 7 (p 466) – Remember, aromatic R-X compounds do not undergo S_N1 or S_N2 reactions, but they can be used with the Mg organometallic reactions studied previously. Fill in the necessary reagents to accomplish the indicated transformations. Start with benzene.

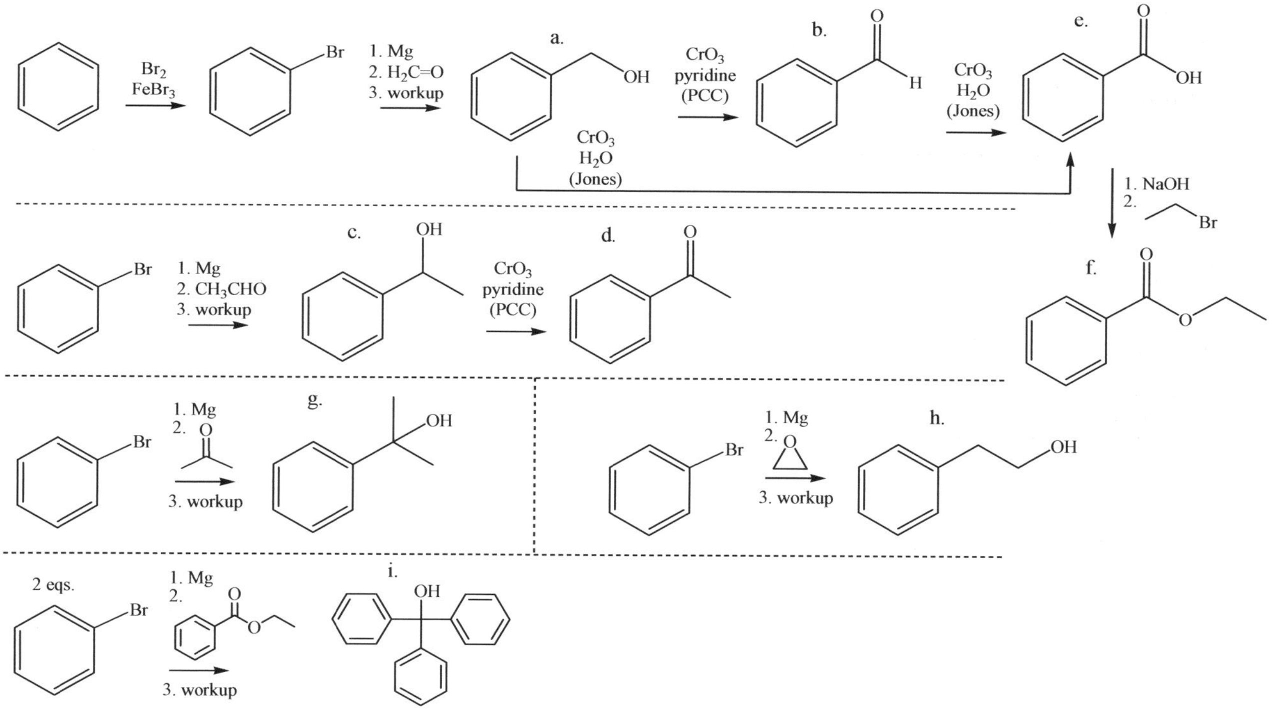

Br_2, $FeBr_3$

a. 1. Mg 2. $H_2C{=}O$ 3. workup

b. CrO_3 pyridine (PCC)

e. CrO_3 H_2O (Jones)

CrO_3 H_2O (Jones)

f. 1. NaOH 2. (bromoethane)

c. 1. Mg 2. CH_3CHO 3. workup

d. CrO_3 pyridine (PCC)

g. 1. Mg 2. (acetone) 3. workup

h. 1. Mg 2. (ethylene oxide) 3. workup

i. 2 eqs. 1. Mg 2. (ethyl benzoate) 3. workup

Problem 8 (p 466) – Warning – "hard problem." Provide a mechanism for the following reaction (this is just a minor product of the desired reaction, bromination of t-butylbenzene). What is the electrophile (what is newly present in the product)? Which aromatic carbon must be attacked? What is the leaving group and why is it relatively stable? What happens to the leaving group in this reaction? When an aromatic ring carbon with a substituent is attacked, it is called "ipso" attack.

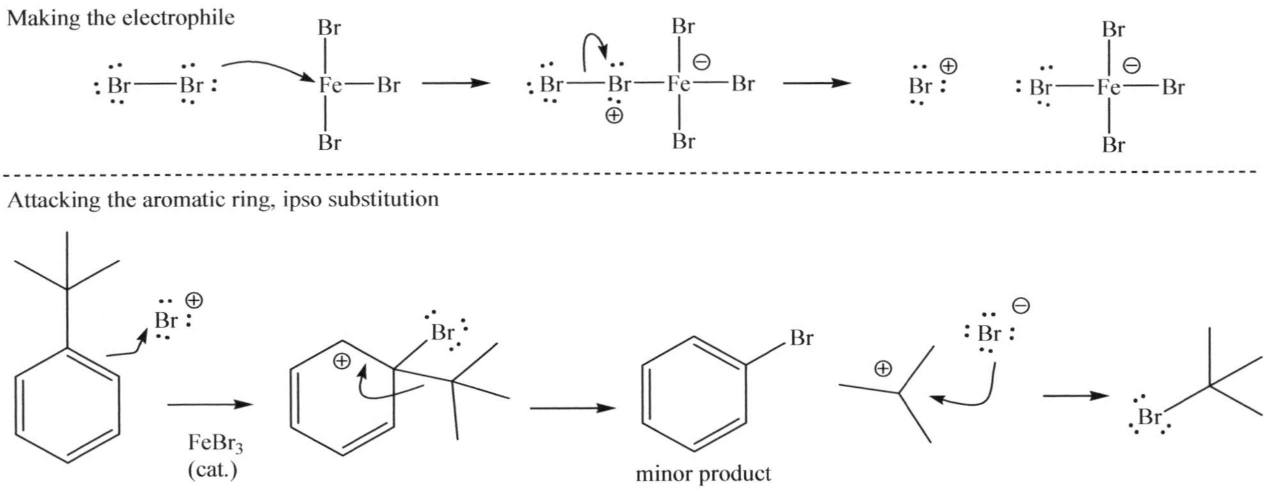

Making the electrophile

$Br-Br + FeBr_3 \rightarrow Br-Br^{+}-Fe^{-}Br_3 \rightarrow Br^{+} + Br-Fe^{-}Br_3$

Attacking the aromatic ring, ipso substitution

$FeBr_3$ (cat.)

minor product

Problem 9 (p 471) – Explain the data in the table below (percent mononitration products of alkylbenzenes).

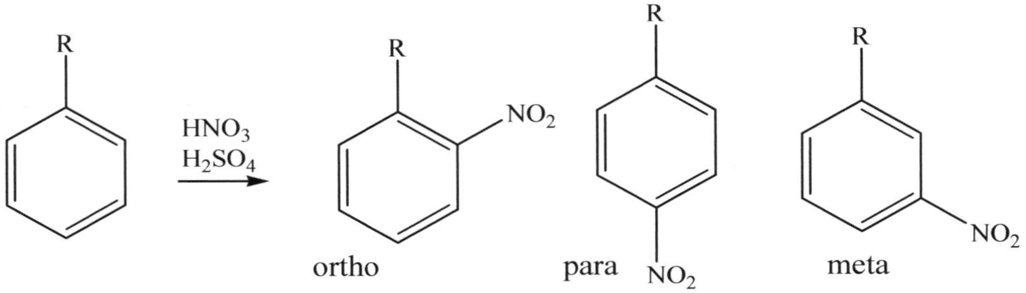

HNO_3 / H_2SO_4

ortho, para, meta

R =	% ortho	% para	% meta
$-CH_3$	62.1	34.6	3.3
$-CH_2CH_3$	50.3	46.6	3.6
$-CH(CH_3)_2$	44.2	53.5	2.3
$-C(CH_3)_3$	12.8	82.9	4.3

There are 2 ortho positions for every 1 para position. Based on simple numbers, there should be twice as much ortho attack as para attack. However, ortho attack is more crowded because the incoming electrophile and the adjacent "R" group are right next to one another. The bigger the "R" group the more severe the crowding and the slower ortho attack should be. When R= methyl (the smallest R group) there is 62% ortho and 35% para. When R = t-butyl (the largest R group), ortho attack drops to 13% and para rises to 83%.

Problem 10 (p 472) – Write out mechanisms for the following reactions, explaining the preferred regioselectivity.

Ortho and para attack are similar and place the positive charge in the intermediate next to the substituent. When the substituent helps stabilize positive charge ortho and para attack are faster, and electrophilic aromatic substitution will be faster than in benzene (the substituent is "activating"). Meta attack never places the positive charge on the carbon with the substituent. If the substituent is polarized positive or is fully positive this will be the better position to attack than ortho or para, though attack at any position will be slower than in benzene because the ring is "deactivated".

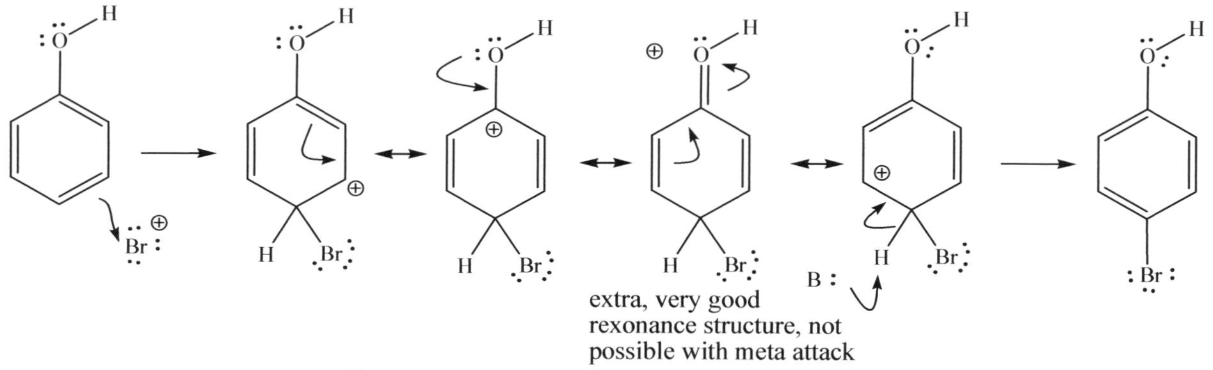

extra, very good rexonance structure, not possible with meta attack

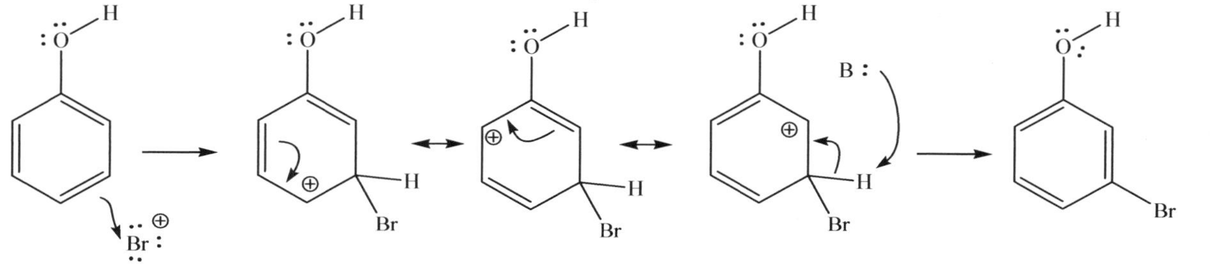

The ortho/para directing substituent never gets a chance to help stabilize the positive charge with meta attack. A tertiary carbocation is better than a secondary carbocation because of the extra inductive effect. This only possible with ortho and para attack.

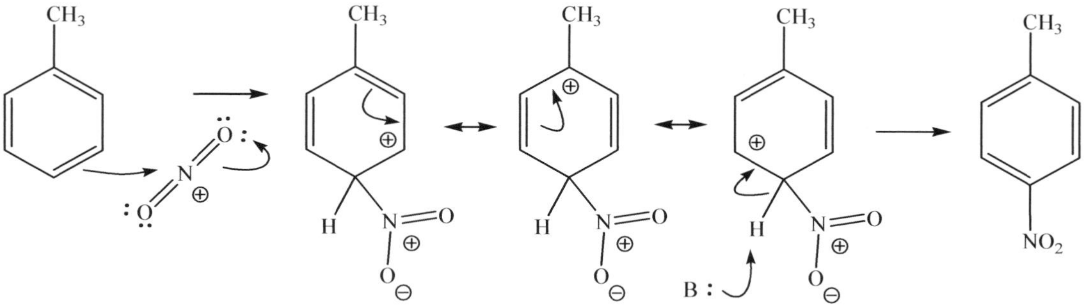

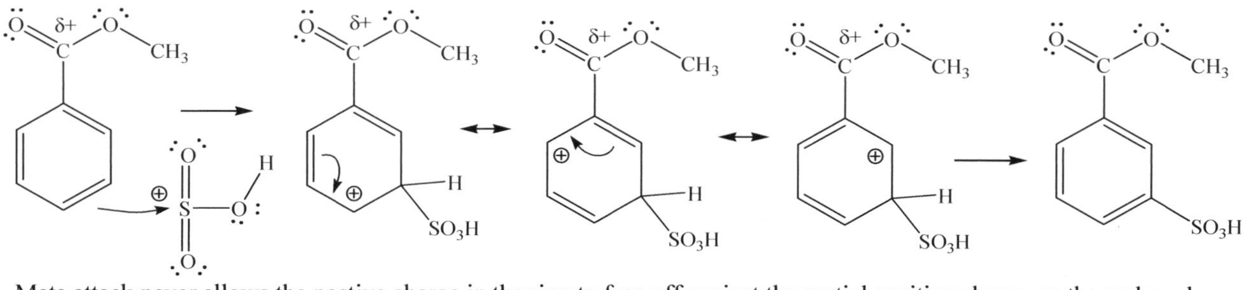

Meta attack never allows the postive charge in the ring to face off against the partial positive charge on the carbonyl carbon. The aromatic ring is more electron poor than benzene so overall attack is slower than benzene. However, meta attack is the best option and reacts faster than ortho and para.

Problem 11 (p 473) – Predict what type of directing effect each substituent has below (o/p or m). Predict the major expected product(s) in reactions with a. HNO_3 / H_2SO_4 b. Br_2 / $FeBr_3$ c. H_2SO_4 / Δ.

electropohile → HNO_3 H_2SO_4 | H_2SO_4 / Δ | Br_2 / $FeBr_3$

a

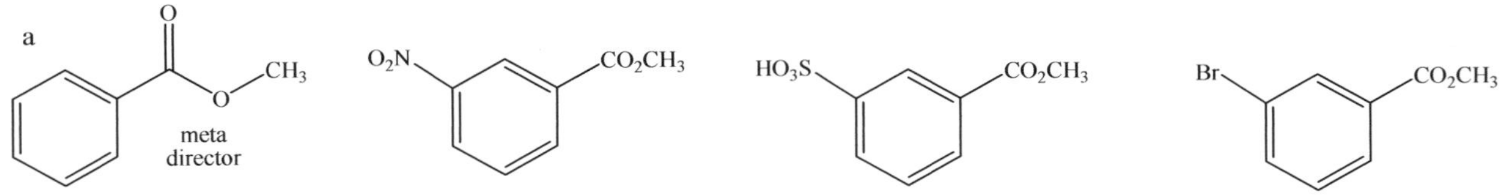
O
CH3
O
meta
director
O2N
CO2CH3
HO3S
CO2CH3
Br
CO2CH3

b

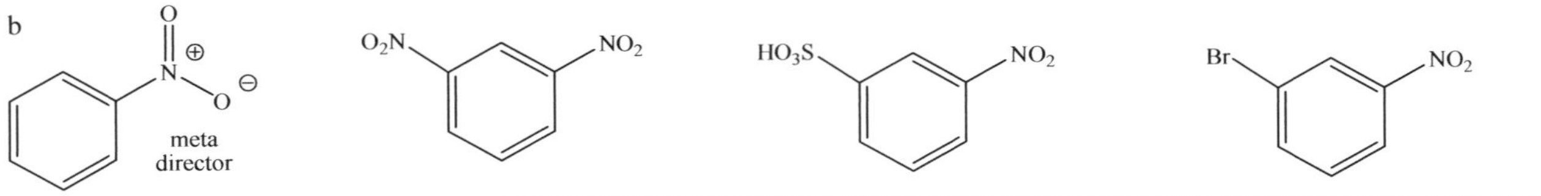
O
⊕
N
O
⊖
meta
director
O2N
NO2
HO3S
NO2
Br
NO2

c

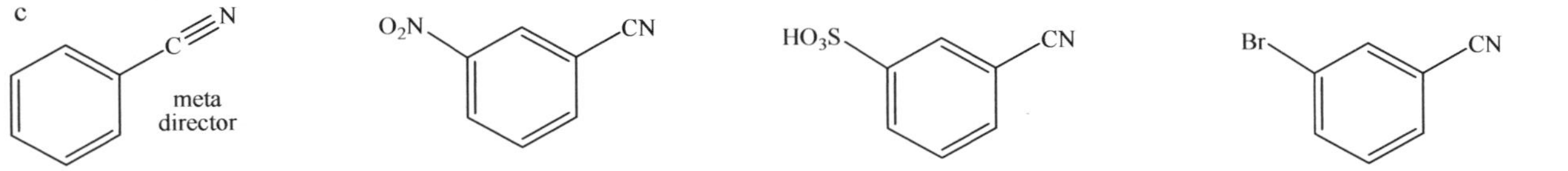
C
N
meta
director
O2N
CN
HO3S
CN
Br
CN

d

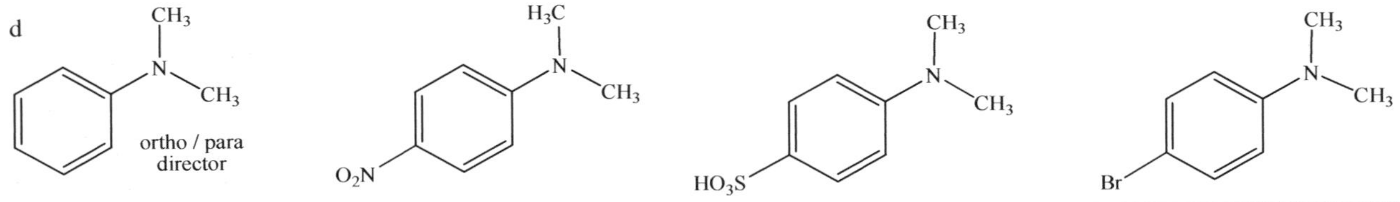
CH3
N
CH3
ortho / para
director
H3C
N
CH3
O2N
CH3
N
CH3
HO3S
CH3
N
CH3
Br

electropohile → HNO_3 H_2SO_4 | H_2SO_4 / Δ | Br_2 / $FeBr_3$

e

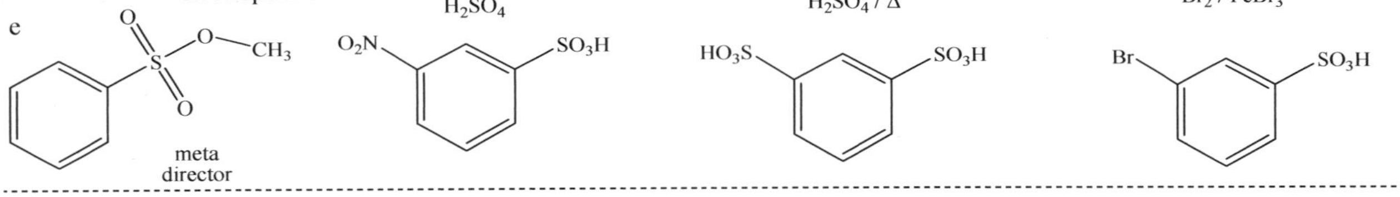
O
O
CH3
S
O
meta
director
O2N
SO3H
HO3S
SO3H
Br
SO3H

f

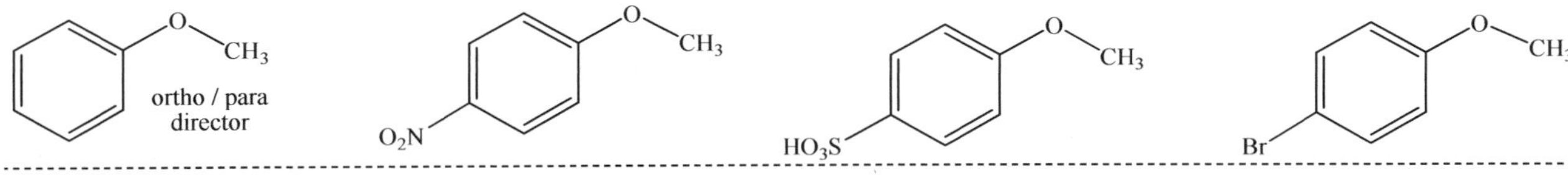
O
CH3
ortho / para
director
O
CH3
O2N
O
CH3
HO3S
O
CH3
Br

g

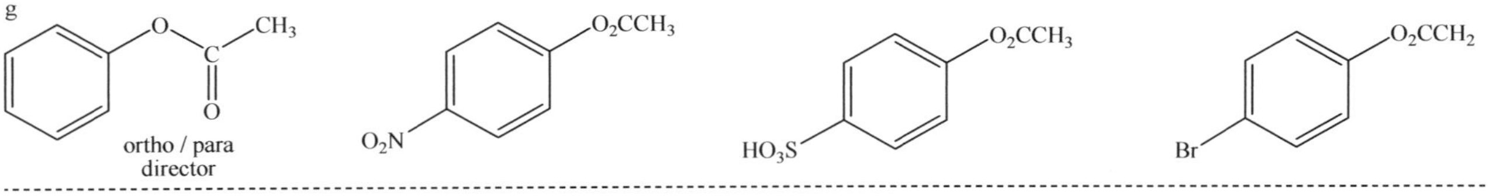
O
C
CH3
O
ortho / para
director
O2CCH3
O2N
O2CCH3
HO3S
O2CCH2
Br

h

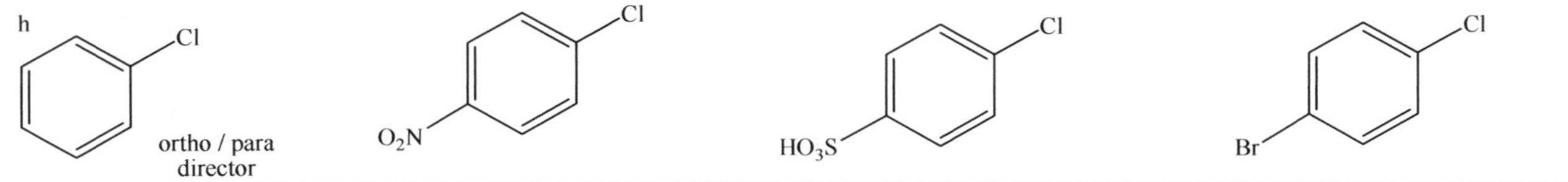
Cl
ortho / para
director
Cl
O2N
Cl
HO3S
Cl
Br

i

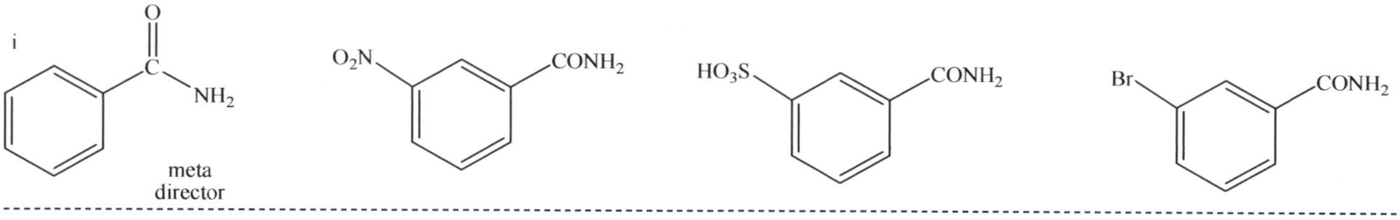

$C_6H_5CONH_2$ — meta director; products: O_2N–C_6H_4–$CONH_2$, HO_3S–C_6H_4–$CONH_2$, Br–C_6H_4–$CONH_2$

j

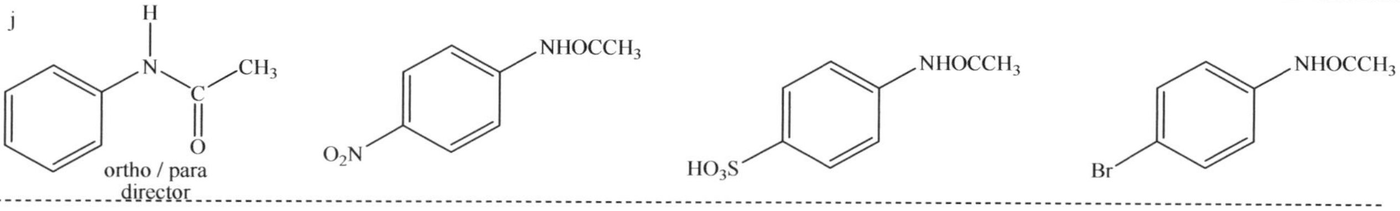

$C_6H_5NHCOCH_3$ — ortho / para director; products: O_2N–C_6H_4–$NHOCCH_3$, HO_3S–C_6H_4–$NHOCCH_3$, Br–C_6H_4–$NHOCCH_3$

k

$C_6H_5C(CH_3)_3$ — ortho / para director; products: O_2N–C_6H_4–$C(CH_3)_3$, HO_3S–C_6H_4–$C(CH_3)_3$, Br–C_6H_4–$C(CH_3)_3$

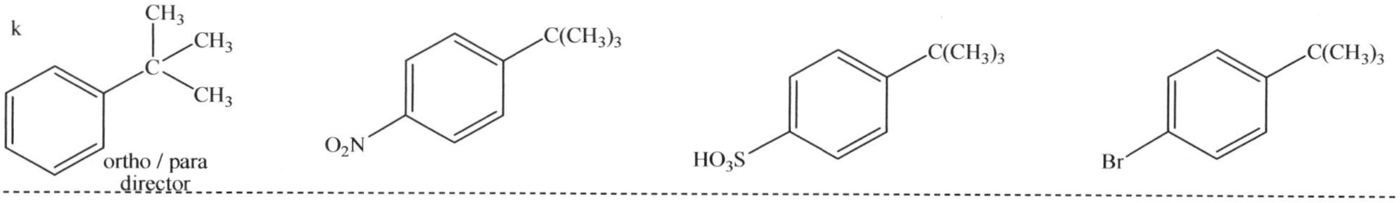

Problem 12 (p 473) – Predict what the product will be and where the substitution will occur in each of the following reactions.

a. 4-phenylanisol

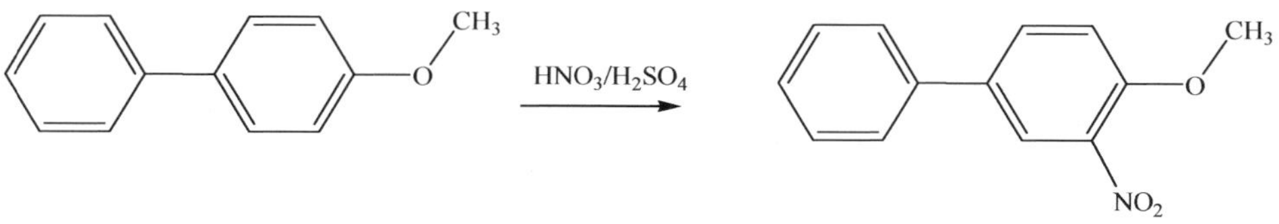

HNO_3/H_2SO_4 → product with NO_2 ortho to O–CH_3

Methoxy group is an activating group, so faster reaction will occur in this ring, and especially so at the ortho position (para is already occupied by the other phenyl ring). These are nitration conditions.

b. 4-phenylbenzoate

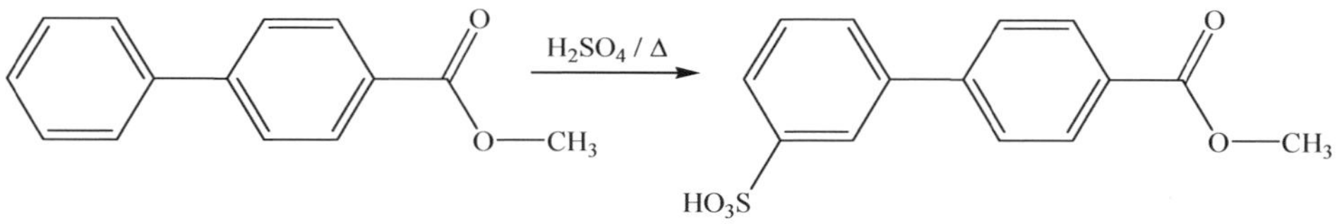

H_2SO_4 / Δ → product with HO_3S on the other ring (meta); O–CH_3

Ester group is an deactivating group, so faster reaction will occur in the other ring, and especially so at the meta position. These are sulfonation conditions.

c. 4-(4-nitrophenyl)phenol

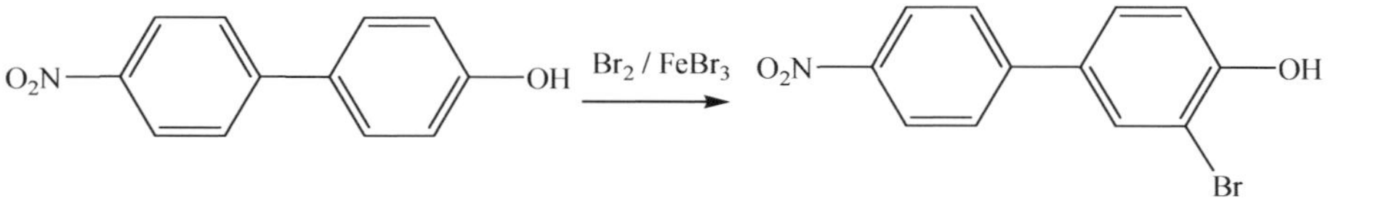

O_2N–…–OH, Br_2 / $FeBr_3$ → O_2N–…–OH with Br ortho to OH

Hydroxy group is an activating group, while nitro group is a deactivating group, so faster reaction will occur in phenol ring, and especially so at the ortho position (para is already occupied by the other phenyl ring). These are bromination conditions.

Made in the USA
San Bernardino, CA
21 January 2019